Development and Disaster Management

Amita Singh · Milap Punia
Nivedita P. Haran
Thiyam Bharat Singh
Editors

Development and Disaster Management

A Study of the Northeastern States of India

Editors
Amita Singh
Centre for the Study of Law
 and Governance
Jawaharlal Nehru University
New Delhi, Delhi, India

Milap Punia
Centre for the Study of Regional
 Development
Jawaharlal Nehru University
New Delhi, Delhi, India

Nivedita P. Haran
Disaster Research Programme (DRP)
Jawaharlal Nehru University
New Delhi, Delhi, India

Thiyam Bharat Singh
Centre for Study of Social Exclusion
 and Inclusive Policy
Manipur University
Imphal, Manipur, India

ISBN 978-981-10-8484-3   ISBN 978-981-10-8485-0   (eBook)
https://doi.org/10.1007/978-981-10-8485-0

Library of Congress Control Number: 2018934670

© The Editor(s) (if applicable) and The Author(s) 2018
This work is subject to copyright. All rights are solely and exclusively licensed by the Publisher, whether the whole or part of the material is concerned, specifically the rights of translation, reprinting, reuse of illustrations, recitation, broadcasting, reproduction on microfilms or in any other physical way, and transmission or information storage and retrieval, electronic adaptation, computer software, or by similar or dissimilar methodology now known or hereafter developed.
The use of general descriptive names, registered names, trademarks, service marks, etc. in this publication does not imply, even in the absence of a specific statement, that such names are exempt from the relevant protective laws and regulations and therefore free for general use.
The publisher, the authors and the editors are safe to assume that the advice and information in this book are believed to be true and accurate at the date of publication. Neither the publisher nor the authors or the editors give a warranty, express or implied, with respect to the material contained herein or for any errors or omissions that may have been made. The publisher remains neutral with regard to jurisdictional claims in published maps and institutional affiliations.

Printed on acid-free paper

This Palgrave Macmillan imprint is published by the registered company Springer Nature Singapore Pte Ltd. part of Springer Nature
The registered company address is: 152 Beach Road, #21-01/04 Gateway East, Singapore 189721, Singapore

Foreword

Disasters do not kill, but lack of preparedness, non-compliance to laws and absence of community resilience kill. Making the whole world a safer and better prepared to deal with any disaster is an onerous responsibility. We have to recognize this challenge and the importance of including 'disaster' and 'climate risk management' as an integral part of developmental planning to realize the global goals of sustainable development and poverty eradication.

The region of the Northeast India is highly vulnerable. The geospatial and geomorphological mapping reveals that it falls in the high earthquake-prone zone V. Earthquakes on the hills of the Northeast India come with land sliding, mud-flooding and prolonged tremors following a high-intensity quake. During a short span of 53 years between 1897 and 1950, four great earthquakes in the Himalayan seismic belt (Shillong, 1897; Kangra, 1905; Bihar-Nepal, 1934; and Assam, 1950) exceeding magnitude of 8 on the Richter scale occurred in the region with vast devastation. Studies indicate that enough strains have accumulated to generate earthquakes with magnitude of 8 or more in the Himalayan region. However, when and where such an earthquake would occur is not known.

Disaster management in India has gone through a paradigm shift—from a reactive, relief-based approach to one of proactive disaster risk reduction, meticulous preparedness, effective response and well thought out build-back better programme. The impact of disasters can be reduced to a considerable extent, and loss of life and damage to property can be reduced if the investments in disaster preparedness and management are made prudently and sufficient preparatory work is done. Government of India has understood the importance of investment in disaster preparedness, especially for strengthening early warning system, and made appropriate investments for improving the same.

Our Hon'ble Prime Minister's address at the Asian Ministerial Conference on Disaster Risk Reduction, New Delhi, last year highlighted the need for developing a network of universities to work on disaster issues. Universities are the biggest platforms for generating sensitivity towards social responsibilities. All educational institutions especially the colleges and the universities in the Northeast India should work towards achieving a disaster-free development in this region.

I am happy to know that the State Disaster Management Authorities of Manipur, Nagaland, Meghalaya and Mizoram actively participated in the transdisciplinary Disaster Research Programme of Jawaharlal Nehru University (JNU), and the State Disaster Management Authority of West Bengal was cooperative to the ongoing research of JNU around the Kalimpong hills which hand hold the north-eastern geomorphology. I congratulate the Central University of Manipur for initiating this collaboration-based disaster research in the north-eastern states. It fills my heart with pride when I read the scholars from this region contributing such original research and grassroot voices for this volume of academic work.

I am confident that the meaningful research that has been presented in the book *Development and Disaster Management: A Study of the*

Northeastern States of India will help us in preventing damages and losses through disasters in the Northeast India.

New Delhi, India
June 2017

Kiren Rijiju
Minister of State for Home Affairs
Government of India

PREFACE

SPECIAL CENTRE FOR DISASTER RESEARCH SERIES

This volume is part of a series of books that the Special Centre for Disaster Research, JNU, is planning to publish under Springer Nature's Palgrave imprint. The series editor for these works is Prof. Amita Singh, Chairperson of the Centre.

The transdisciplinary centre has been working upon a social science approach to disaster research and has undertaken three concerns very genuinely—first, the lack of non-western, local and indigenous literature in India and in the Asia; second, the gulf existing due to the language of law which prevents appropriate implementation of DRR laws; and third, an inherently patriarchal and ill-coordinated structures of disaster governance.

The philosophical pretence in a culturing of human beings is much provoked and stimulated from the particularities of relationships in any society. Research which enables the discovery of an idea of responsibility towards nature, which naturalizes laws and streamlines the structures of governance in accordance with the geospatial warnings, has greater possibility to reconnect decision-makers to the ecological processes. Nature and natural processes are beyond human contrivance, but to prepare communities and decision-makers for an eventuality when a natural process turns into a disaster is embedded in a social science perspective of understanding this universe of relationships in a habitat. One such perspective of social sciences called the reconstructive postmodern approach highlights communication, language and discourse generation as an inherent part of any structure of governance. Is there any possibility that Sendai Framework and its focus on resilience building can ever be achieved with a 'hazard based' approach? These approaches of social sciences are fundamental to resilience building yet go completely missing in a hagemonic universe of science and technology. The undertaking of this book series is to create a temporal model of institutional framework and decision-making which could at best be a guideline to those who wish to study disasters beyond existing engineering approaches. The pedagogy of disaster studies would most appropriately focus upon the deontology or the moral rightness of doing one's duty towards the social system and communitarian traditions which we live in. This series attempts to assemble experience, spawn community initiatives and highlight the gaps in the implementation processes which throw communities into recurrent disasters. Some of the forthcoming volumes in the series starting with the present one are: 'Development and Disaster Management', and 'Rural Communities, Agriculture and Disasters'.

New Delhi, India	Amita Singh
New Delhi, India	Milap Punia
New Delhi, India	Nivedita P. Haran
Imphal, India	Thiyam Bharat Singh

ACKNOWLEDGEMENTS

The editors of this book acknowledge three invincible pillars which provided the direction, pertinacity and continuity required in every social science-based research. First pillar of course is Shri. Kiren Rijiju, the deeply perceptive Hon'ble Minister of State (MoS) at the Ministry of Home Affairs (MHA) who heads the disaster management in India. His commitment to disaster risk reduction is unmatched, and he and his office extended all support to carry forward this daunting task in the vulnerable zones of the north-eastern states over the Himalayan region.

The second illustrious support came from the former Minister in the Manipur State Cabinet Shri. Hemochandra Singh from the Singjamei constituency, who generously threw open his powerful ground support in coordinating local institutions and officials. Many intricacies of approaching the unapproachable terrain and the anthropological demands were discussed at the culturally reinvigorating Vaishnavite community dinner for the whole research team at his residence.

The third empowering initiative was taken by the Vice Chancellor of JNU, Prof. M. Jagadesh Kumar, who recognized the need for transdisciplinary research in social sciences and also about bridging its gap with natural sciences. One can recollect that providential moment of history when he tabled the proposal and cleared it through the noisy and menacing Academic Council of 2016.

A young scholar in disaster research Ms. Sukhreet Bajwa who is currently steering through the Gujarat Disaster Management Authority at Gandhi Nagar deserves attention and appreciation from all authors in this book. She found time even during the multifaceted government responsibilities to bring convergence on a common thematic platform through constant communication with demanding editors on one hand and passionate authors on the other. The book could not have inspired so many local authors without her commitment to the cause. She was also supported by others in the Disaster Research Programme's young team such as Priyaanka Jha, Homolata Borah, Binod Kumar and Langthianmung Vualzong to whom the editors gratefully express their appreciation. Lastly, the fly-wheeling at the administrative desks was addressed and resolved by none other than Jhuman Yadav, the sole office attendant at DRP, whose multitasking is much remembered and applauded.

Amita Singh
Milap Punia
Nivedita P. Haran
Thiyam Bharat Singh

CONTENTS

Part I   Disasters, Management and Development

India's Northeast: Disasters, Development
and Community Resilience   3
Amita Singh

Natural Disasters in the North-East: Crucial Role
of Academia in Saving Lives   25
P. P. Shrivastav

Regional Climate Changes Over Northeast India:
Present and Future   41
A. P. Dimri, D. Kumar and M. Srivastava

Manipur Flood 2015: Preparedness and Disaster
Risk Reduction Strategies   65
Bupinder Zutshi

Lessons from Manipur: Managing Earthquake Risk
in a Changing Built Environment   83
Kamal Kishore and Chandan Ghosh

Varying Disasters—Problems and Challenges
for Development in Manipur
Jason Shimray

95

Manipur's Tryst with Disaster Management Act, 2005
Nivedita P. Haran

113

Reviving 'Public Policy' and Triggering 'Good
Governance': A Step Towards Sustainability
Pankaj Choudhury

123

Role of Heavy Machinery in Disaster Management
Keshav Sud

135

Part II Vulnérabilité Studies

Co-Seismic Slip Observed from Continuous Measurements:
From the 2016 Manipur Earthquake (Mw = 6.7)
Arun Kumar and L. Sunil Singh

149

Post-Earthquake Geodisaster of 6.7 Mw Manipur
Earthquake 2016
P. S. Ningthoujam, L. K. Lolee, C. S. Dubey, Z. P. Gonmei,
L. Thoithoi and T. Oinam

161

Landslide Zonation in Manipur Using Remote
Sensing and GIS Technologies
R. K. Chingkhei

169

Building a Resilient Community Against Forest Fire
Disasters in the Northeast India
P. K. Joshi, Anusheema Chakraborty and Roopam Shukla

187

The Khuga Dam—A Case Study
Langthianmung Vualzong, Aashita Dawer and Nivedita P. Haran

201

Dams and Disasters in the Northeast India:
A Collateral 'Ethical' Relationship 215
Prasenjit Biswas

Achieving Last Mile Delivery: Overcoming
the Challenges in Manipur 229
Sukhreet Bajwa

Part III Accomplishing Community Resilience

A Study of Socio-economic Community Resilience
in Manipur 241
Thiyam Bharat Singh

'Living with Floods': An Analysis of Floods Adaptation
of Mising Community—A Case Study of Jiadhal River 259
Tapan Pegu

Role of CBOs in Resilience Building: Good Practices
and Challenges 281
Thongkholal Haokip

Floods, Ecology and Cultural Adaptation
in Lakhimpur District, Assam 301
Ngamjahao Kipgen and Dhiraj Pegu

Disaster and Resilience Building among Women in Manipur 319
Mondira Dutta

Community Resilience Building and the Role
of Paitei Tribe of Churachandpur in Manipur 331
M. Pauminsang Guite and Langthianmung Vualzong

Urban Risks to Hazards: A Study of the Imphal
Urban Area 349
Sylvia Yambem

xvi CONTENTS

Community Awareness on Landslide Disaster:
'Experience of Nagaland'   361
Supongsenla Jamir and N. U. Khan

Risk Perception and Disaster Preparedness: A Case Study
of Noney, Tamenglong District   373
Homolata Borah

Institutions of Faith in Resilience Building During
Disasters: Church and Communities in Lamka
Vis-à-Vis Churachandpur   383
Priyanka Jha and G. V. C. Naidu

Relating Resource Extraction Pattern with Disasters:
Political Ecology of Barbhag Block, Nalbari   393
Snehashish Mitra

Communities and Disasters in Nagaland: Landslides
and the Cost of Development   409
Imkongmeren

Glossary   419

Index   421

EDITORS AND CONTRIBUTORS

About the Editors

Amita Singh Professor at the Special Centre for Law and Governance and Founding Chairperson, Special Centre for Disaster Research, Jawaharlal Nehru University, India. She is also editor of *Disaster Law* (Routledge 2017).

She has been the longest serving Secretary General of NAPSIPAG (Network of Asia Pacific Schools and Institutes of Public Administration and Governance) initiated by ADB 2004 at INTAN Malaysia. She is Member Secretary of the Institutional Ethics Review Board and Member of the Indian Council of Social Sciences Research, Delhi. She has a wide research experience of evaluating best governance practices and working with the government (DARPG, India) and the Global Innovators Network, at the John F. Kennedy School of Government, Harvard University. She was awarded the Australia–India Council Fellowship (2006–2007) for academic research in nine Australian Universities and was again awarded the Australian Monash Fellowship in 2017. She has been closely associated with the International Women's Association at Hunter College SUNY USA in 1990 to prepare for the Beijing Declaration in 1995. Her other authored publications include *A Critical Impulse to e-Governance in the Asia Pacific* (Springer 2014). Part of her jurisprudential research on Western Ghats has been published as a special paper *Coastal Ballads and Conservation Ironic: Understanding*

xvii

Implementation Slippages of the CRZ Laws (EPW February 2016). She received the Bangladesh National Award of 'Nawab Bahadur Syed Nawab Ali Chowdhury National Award' 2014 for women's empowerment. Prof. Singh is an ardent activist of the 'Rights of Nonhuman Species'. Her work can be reached at http://www.jnu.ac.in/Faculty/asingh.

Milap Punia Professor, Centre for the Study of Regional Development, Director of Admissions, Jawaharlal Nehru University, India.

Milap Punia is Professor of Geography at the Centre for the Study of Regional Development, School of Social Sciences, Jawaharlal Nehru University, New Delhi, India. He is specialized in Remote Sensing and Geographic Information System (GIS). He worked as Scientist in Photogrammetry and Remote Sensing Division, Indian Institute of Remote Sensing-Indian Space Research Organization (IIRS-ISRO) and adjunct faculty to Centre for Space Science and Technology Education in Asia and the Pacific (Affiliated to the UN), Dehradun, India. He was awarded C. V. Raman fellowship to carry out post-doctoral research work on urban studies from School of Forestry and Environment Studies, Yale University, USA, in the year 2013–2014. His areas of research include land use change, climate change, spatial data analysis for urban planning and regional development; spatial aspects of governance, disaster management, sustainability studies and to bridge research gap between social sciences and geospatial discipline. He had completed many international research projects, namely Subaltern Urbanisation funded by ANR France, Risks & Responses to Urban Futures funded by ESPA, UK and Remapping of Global South funded by DAAD; and national projects, namely Snow Spectral Properties and Regional Climate funded by DST, GoI, Spatial Governance of Decentralised Local bodies funded by ICSSR and Geovisualisation of landscape and e-learning course funded by DST, GoI. He had published an edited book *Marginalization in Globalizing Delhi: Issues of Land, Livelihoods and Health* and has more than 50 research papers in the reputed national and international journals. He is currently holding administrative position of Director Admissions, Jawaharlal Nehru University, New Delhi.

Nivedita P. Haran Indian Administrative Service (Rtd.), Former Additional Chief Secretary, Government of Kerala who spearheaded the first Kerala State Disaster Management Authority.

Dr. Haran, a postgraduate in Philosophy from Jadavpur University, Kolkata, in 1976 in I Class and a Ph.D. in Sociology from IIT, Delhi, on the topic *Bureaucracy in India*, was selected to the Indian Administrative Service (IAS) in 1980. She has more than three decades of rich professional experience with the Indian Administrative Services in India and in the State of Kerala where she served in several senior positions of leadership and decision-making. She held crucially important positions as a District Planning Officer, as a Head of Revenue Administration, Land Administration, Land Records Management, Renewable Energy in Public Offices, coping Climate Change Strategies and post-Tsunami Rehabilitation project. She also held the position of Deputy Secretary in the Department of Administrative Reforms and Public Grievances, Ministry of Personnel, Government of India, New Delhi. As the Home Secretary, she brought some meaningful innovations such as the digitization of police records, simplification of procedures, bringing transparency and accountability through the use of new cost-effective technological innovations such as video conferencing and other ICT applications. She has also been the Director of The Centre for Innovations in Public Systems at Hyderabad. Her most passionate project with the NAPSIPAG (Network of Asia Pacific Schools and Institutes of Public Administration and Governance, JNU) was the creation of NYSAF (Network of Young Scholars and Administrators Forum) by bringing academic research closer to administrators and enabled them to work together for the country's development.

Thiyam Bharat Singh Reader, Centre for Study of Social Exclusion and Inclusive Policy (CSSEIP), Manipur University (MU), India.

Dr. Thiyam Bharat Singh is Reader in the Centre for Study of Social Exclusion and Inclusive Policy (CSSEIP), Manipur University (MU). He is currently Editor of Working Paper (ISSN No: 2393) of CSSEIP, MU. He is also editor of a book entitled *Social Exclusion, Marginalisation and Deprivation in Manipur*, published by the Akansha Publishing House, New Delhi. He has co-edited a book on *Social Exclusion and Inclusive Development in the North East*. Dr. Thiyam Bharat Singh was one of the Review Committee Members of Dictionary of Economics for Schools published by NCERT, New Delhi. He is currently undertaking a research project on 'Estimation of Poverty in Manipur', under the sponsorship of Planning Department, Government of Manipur. Apart from this, he has been able to complete two major research projects under the financial assistance of UGC and ICSSR, New Delhi. His articles have been published in Journals such as *Economic and Political Weekly* (EPW), *Indian Economic Journal* (IEJ), *MARGIN* (Sage publication) *NCAER* and *Journal of Exclusion Studies*. He has organized several seminars, workshops and conferences (both international and national) on various issues as a Convenor. Apart from that, he has also presented papers in various international and national seminars. His area of interest is development economics, trade, poverty, tourism, gender, socio-economic resilience and disaster reduction and social exclusion.

Contributors

Anusheema Chakraborty Department of Natural Resources, TERI University, New Delhi, India

Sukhreet Bajwa Gujarat Disaster Management Authority, Gandhinagar, India

Prasenjit Biswas North-Eastern Hill University, Shillong, India

Homolata Borah Centre for the Study of Regional Development, Jawaharlal Nehru University, New Delhi, Delhi, India

R. K. Chingkhei Department of Forestry and Environmental Science, Manipur University, Imphal, Manipur, India

Pankaj Choudhury Jawaharlal Nehru University, New Delhi, India

Aashita Dawer O.P. Jindal Global University, Sonipat, India

A. P. Dimri School of Environmental Sciences, Jawaharlal Nehru University, New Delhi, India

C. S. Dubey Department of Geology, University of Delhi, New Delhi, India

Mondira Dutta Jawaharlal Nehru University, New Delhi, India

Chandan Ghosh National Institute of Disaster Management, Ministry of Home Affairs, Government of India, New Delhi, India

Z. P. Gonmei Department of Geology, University of Delhi, New Delhi, India

M. Pauminsang Guite Jawaharlal Nehru University, New Delhi, India

Thongkholal Haokip Centre for the Study of Law and Governance, Jawaharlal Nehru University, New Delhi, India

Imkongmeren Centre for the Study of Law and Governance, Jawaharlal Nehru University, New Delhi, India

Supongsenla Jamir Jamia Millia Islamia, New Delhi, India

Priyanka Jha Jawaharlal Nehru University, New Delhi, India

P. K. Joshi School of Environmental Sciences, Jawaharlal Nehru University, New Delhi, India

N. U. Khan Jamia Millia Islamia, New Delhi, India

Ngamjahao Kipgen Department of Humanities and Social Sciences, Indian Institute of Technology Guwahati, Kamrup, Assam, India

Kamal Kishore National Disaster Management Authority, Government of India, New Delhi, India

Arun Kumar Department of Earth Sciences, Manipur University, Imphal, India

D. Kumar School of Environmental Sciences, Jawaharlal Nehru University, New Delhi, India

L. K. Lolee Department of Geology, University of Delhi, New Delhi, India

Snehashish Mitra School of Social Sciences, National Institute of Advanced Studies, Bengaluru, India

G. V. C. Naidu Jawaharlal Nehru University, New Delhi, India

P. S. Ningthoujam Department of Geology, University of Delhi, New Delhi, India

Nivedita P. Haran Disaster Research Programme (DRP), Jawaharlal Nehru University, New Delhi, Delhi, India

T. Oinam Department of Geology, University of Delhi, New Delhi, India

Dhiraj Pegu Department of Humanities and Social Sciences, Indian Institute of Technology Guwahati, Kamrup, Assam, India

Tapan Pegu Jiadhal College, Jiadhal, Assam, India

Jason Shimray Manipur State Disaster Management Authority, Imphal, India

P. P. Shrivastav Formerly, Member, North Eastern Council, Shillong, India

Roopam Shukla Department of Natural Resources, TERI University, New Delhi, India

Amita Singh Jawaharlal Nehru University, New Delhi, India

Thiyam Bharat Singh Centre for Study of Social Exclusion & Inclusive Policy (CSSEIP), Manipur University, Imphal, India

M. Srivastava Department of Environmental Sciences, Central University of Rajasthan, Rajasthan, India

Keshav Sud Amazon Robotics, Boston, MA, USA

L. Sunil Singh Department of Earth Sciences, Manipur University, Imphal, India

L. Thoithoi Department of Geology, University of Delhi, New Delhi, India

Langthianmung Vualzong Centre for the Study of Law & Governance, Jawaharlal Nehru University, New Delhi, India

Sylvia Yambem Department of Political Science, Manipur University, Imphal, India

Bupinder Zutshi Jawaharlal Nehru University, New Delhi, India

LIST OF FIGURES

Natural Disasters in the North-East: Crucial Role of Academia in Saving Lives

Fig. 1	The N-E Region: Major EQs and Tectonic Features	26
Fig. 2	Eurasian Plate	27
Fig. 3	The seismic vulnerability of NER	29

Regional Climate Changes Over Northeast India: Present and Future

Fig. 1	Topography (m) over the Northeast Indian region	43
Fig. 2	Observed near-surface air temperature climatology (°C) during 1970–2005 over Northeast Indian region from APHROTEMP for DJF season (**a**), MAM season (**b**), JJAS season (**c**) and ON season (**d**)	47
Fig. 3	Near-surface air temperature climatology (°C) during 1970–2005 over Northeast Indian region from REMO regional model driven by MPI-ESM-LR global model under CORDEX South Asia experiment for DJF season (**a**), MAM season (**b**), JJAS season (**c**) and ON season (**d**)	48
Fig. 4	Trends of observed near-surface air temperature (°C/yr) during 1970–2005 over Northeast Indian region from APHROTEMP for DJF season (**a**), MAM season (**b**), JJAS season (**c**) and ON season (**d**)	48
Fig. 5	Changes in near-surface air temperature (°C) calculated as near future (2020–2049) minus present (1970–2005) for RCP2.6 over Northeast Indian region from REMO	

xxiii

xxiv LIST OF FIGURES

	regional model forced with MPI-ESM-LR global model for DJF season (**a**), MAM season (**b**), JJAS season (**c**) and ON season (**d**)	49
Fig. 6	Same as Fig. 5 but for far future (2070–2099) minus present (1970–2005) under RCP2.6	49
Fig. 7	Changes in near-surface air temperature (°C) calculated as near future (2020–2049) minus present (1970–2005) for RCP4.5 over Northeast Indian region from REMO regional model forced with MPI-ESM-LR global model for DJF (December, January, February) season (**a**), MAM season (**b**), JJAS season (**c**) and ON season (**d**)	50
Fig. 8	Same as Fig. 7 but for far future (2070–2099) minus present (1970–2005) under RCP4.5	50
Fig. 9	Changes in near-surface air temperature (°C) calculated as near future (2020–2049) minus present (1970–2005) for RCP8.5 over Northeast Indian region from REMO regional model forced with MPI-ESM-LR global model for DJF (December, January, February) season (**a**), MAM season (**b**), JJAS season (**c**) and ON season (**d**)	51
Fig. 10	Same as Fig. 9 but for far future (2070–2099) minus present (1970–2005) under RCP8.5	51
Fig. 11	Trends of near-surface air temperature (°C/yr) over the period 1970–2099 under RCP2.6 over Northeast Indian region as simulated from REMO model under CORDEX South Asia experiment for DJF (**a**), MAM (**b**), JJAS (**c**) and ON seasons (**d**)	52
Fig. 12	Same as Fig. 11 but for RCP4.5	52
Fig. 13	Same as Fig. 11 but for RCP8.5	53
Fig. 14	Time series of near-surface air temperature (°C) averaged over the Northeast Indian region for the period 1970–2099 for DJF (December, January, February) season (**a**), MAM season (**b**), JJAS season (**c**) and ON (October, November) season (**d**) from REMO regional model forced with MPI-ESM-LR global model. The series in grey, green, dark blue and red corresponds to present, RCP2.6, RCP4.5 and RCP8.5 warming scenarios, respectively. The broken lines represent the baseline, as mean ±1 standard deviation of the historical climate	54
Fig. 15	Observed daily seasonal precipitation (mm/day) during 1970–2005 over Northeast Indian region from APHRODITE for DJF (December, January, February)	

	season (a), MAM season (b), JJAS season (c) and ON season (d)	55
Fig. 16	Daily seasonal precipitation (mm/day) during 1970–2005 over Northeast Indian region from REMO regional model driven from MPI-ESM-LR global model under CORDEX South Asia experiments for DJF (December, January, February) season (a), MAM season (b), JJAS season (c) and ON season (d)	55
Fig. 17	Observed trends of daily seasonal precipitation (mm/day/yr) during 1970–2005 over Northeast Indian region from APHRODITE for DJF (December, January, February) season (a), MAM season (b), JJAS season (c) and ON season (d)	56
Fig. 18	Changes in daily precipitation calculated as percentage change of near future (2020–2049) minus present (1970–2005) for RCP2.6 over Northeast Indian region from REMO regional model forced with MPI-ESM-LR global model for DJF (December, January, February) season (a), MAM season (b), JJAS season (c) and ON season (d)	56
Fig. 19	Same as Fig. 18 but for far future minus present under RCP2.6	57
Fig. 20	Changes in daily precipitation calculated as percentage change of near future (2020–2049) minus present (1970–2005) for RCP4.5 over Northeast Indian region from REMO regional model forced with MPI-ESM-LR global model for DJF (December, January, February) season (a), MAM season (b), JJAS season (c) and ON season (d)	57
Fig. 21	Same as Fig. 20 but for far future minus present under RCP4.5	57
Fig. 22	Changes in daily precipitation calculated as percentage change of near future (2020–2049) minus present (1970–2005) for RCP8.5 over Northeast Indian region from REMO regional model forced with MPI-ESM-LR global model for DJF (December, January, February) season (a), MAM season (b), JJAS season (c) and ON season (d)	58
Fig. 23	Same as Fig. 22 but for far future minus present under RCP8.5	58
Fig. 24	Trends of daily precipitation (mm/day/yr) for the period 1970–2099 under RCP2.6 over Northeast Indian region from REMO regional model forced with MPI-ESM-LR	

xxvi   LIST OF FIGURES

	global model for DJF (December, January, February) season (a), MAM season (b), JJAS season (c) and ON season (d)	59
Fig. 25	Same as Fig. 24 but for RCP4.5	59
Fig. 26	Same as Fig. 24 but for RCP8.5	59
Fig. 27	Time series of total seasonal precipitation (mm) averaged over the Northeast Indian region for the period 1970–2099 for DJF (December, January, February) season (a), MAM season (b), JJAS season (c) and ON season (d) from REMO regional model forced with MPI-ESM-LR global model. The series in grey, green, dark blue and red corresponds to present, RCP2.6, RCP4.5 and RCP8.5 warming scenarios, respectively, and the one in cyan represents the observation. The dotted black line shows the baseline, as mean ±1 standard deviation of the historical climate	60

Manipur Flood 2015: Preparedness and Disaster Risk Reduction Strategies

Fig. 1	Flood July–August 2015 in Manipur Valley	73
Fig. 2	Encroachment of river channel (Thoubal River)	74

Lessons from Manipur: Managing Earthquake Risk in a Changing Built Environment

Fig. 1	Did You Feel It (DYFI) response map of 4 January 2016 Manipur earthquake, India (*Source* USGS)	84
Fig. 2	Ima Market in Imphal in April 2016 and partial damage of the same due to the M6.7 earthquake of 4 January 2016 (*Source* Chandan Ghosh and www.nelive.in)	85
Fig. 3	Collapsed frame of a soft-storeyed RCC structure (women vendors' market) in Saikul	86
Fig. 4	A traditional house (wooden frame, with bamboo mat infill walls and CGI sheet roofing) with no noticeable damage in Noney, one of the nearest locality from earthquake epicentre	87
Fig. 5	Interim market space for women vendors built after the earthquake in Saikul	88
Fig. 6	Propping damaged RCC structures by newly constructed columns not connected to the beam above	89

Varying Disasters—Problems and Challenges for Development in Manipur

Fig. 1	Zone Mapping: Northeastern region most vulnerable in India (*Source* http://delhi.gov.in/DoIT/DOIT_DM/seismap.gif)	97

Fig. 2	Seismo-tectonic map of Manipur	98
Fig. 3	Photographs of the recent floods and landslides in Manipur during July–August 2015	101
Fig. 4	Photographs of the recent hailstorm/thunderstorm during the month of April 2016	105

Role of Heavy Machinery in Disaster Management

Fig. 1	Major roles of heavy machinery in disaster management	137
Fig. 2	Landslide in village of Malin on 30 July 2014 in Pune, Maharashtra	138
Fig. 3	Landslide in village Phamla, Tawang district of Arunachal Pradesh on 21/22 April 2016 (*Source* NDMA site)	140

Co-Seismic Slip Observed from Continuous Measurements: From the 2016 Manipur Earthquake (Mw = 6.7)

Fig. 1	General tectonics in the Indo-Burmese region and location of GPS permanent station	151
Fig. 2	Map showing plot of seismicity in Northeast India within a window of and 20–29°N latitude and 89–98°E longitude during the period 1964–2015	152
Fig. 3	GPS stations in Manipur	154
Fig. 4	Time series at IMPH in ITRF 2008. **a** North component (20.9 ± 0.5 mm/yr). **b** East component (29.7 ± 0.5 mm/yr). **c** Up component (0 ± 19 mm/yr)	155
Fig. 5	Time series at IMPH showing coseismic offset due to 2016 Manipur earthquake	156
Fig. 6	Time series at SOMP showing coseismic offset due to 2016 Manipur earthquake	157
Fig. 7	Preliminary results showing normal stress change along the fault plane	158

Landslide Zonation in Manipur Using Remote Sensing and GIS Technologies

Fig. 1	Map showing the study area. The red polygon indicates the 2 km buffer area along the national highway where study has been conducted	172
Fig. 2	Flow chart of the steps involved in the present study	174
Fig. 3	Landslide Hazard Zonation (LHZ) map of 83 H/9 along NH-37	178
Fig. 4	Landslide Hazard Management (LHM) map of 83 H/9 along NH-37	179
Fig. 5	Landslide Hazard Zonation (LHZ) map of 83-K/3 along NH-2	180

xxviii LIST OF FIGURES

Fig. 6	Landslide Hazard Management (LHM) map of 83-K/3 along NH-2	181
Fig. 7	Landslide Hazard Zonation (LHZ) map of 83-K/7 along NH-202	182
Fig. 8	Landslide Hazard Management (LHM) map of 83-K/7 along NH-202	183

Building a Resilient Community Against Forest Fire Disasters in the Northeast India

Fig. 1	Forest fire locations as per Fire Information for Resource Management System (FIRMS) over Northeast India (**a**) 01-07-2015 to 31-12-2015, (**b**) 01-01-2016 to 29-02-2016, (**c**) 01-03-2016 to 30-04-2016, (**d**) 01-05-2016 to 30-06-2016 (Adapted from Satendra and Kaushik, A. D. (2014): Forest Fire Disaster Management. National Institute of Disaster Management, Ministry of Home Affairs, New Delhi)	192

The Khuga Dam—A Case Study

Fig. 1	The Khuga/Mata Dam	203
Fig. 2	Cracks in the dam	209
Fig. 3	Map showing sites of irrigation projects (supra note 1)	213

Dams and Disasters in the Northeast India: A Collateral 'Ethical' Relationship

Fig. 1	Channel Migrating Zones (CMZ) of the river Barak	221

Achieving Last Mile Delivery: Overcoming the Challenges in Manipur

Fig. 1	Disaster management cycle and last mile delivery activities	232
Fig. 2	Challenges to last mile delivery in Manipur	234

'Living with Floods': An Analysis of Floods Adaptation of Mising Community—A Case Study of Jiadhal River

Fig. 1	Map of district Dhemaji scale: NTS (*Source* http://dhemaji.nic.in)	264
Fig. 2	Traditional Mising village set up in floodplains (*Source* Researcher's own findings [map not in scale])	266
Fig. 3	Crop cultivation (*Source* Researcher's own findings)	267
Fig. 4	Use of material in house making (*Source* Researcher's own findings)	270

LIST OF FIGURES xxix

Role of CBOs in Resilience Building: Good Practices and Challenges

Fig. 1	Hail damaged galvanised iron sheet roofings	287
Fig. 2	Distribution of galvanised iron sheet roofings	288
Fig. 3	Joumol village after the landslide (Photo: Michael Lunminthang)	289
Fig. 4	Collapsed women market in Saikul	289
Fig. 5	Pillars of the office of the Saikul Hill Town developing cracks after 4th January Manipur earthquake	290
Fig. 6	An earthquake affected house in Noney	290
Fig. 7	A new spring developed after the 4th January Earthquake at Noney	291
Fig. 8	Rescue efforts by a joint community based organisations at Joumol	291
Fig. 9	Relief donation centre and collection of relief materials	292
Fig. 10	Transportation and distribution of relief materials	292
Fig. 11	Distribution of relief materials under the banner of Natural Disaster Relief Committee	293

Floods, Ecology and Cultural Adaptation in Lakhimpur District, Assam

Fig. 1	Map of Lakhimpur district, site of the field study (*Source* http://lakhimpur.nic.in/)	305
Fig. 2	A Mising woman seen here is engaged in weaving cloths at Matmora	308
Fig. 3	Stilt houses (*chaang ghar*) in Laiphulia	309
Fig. 4	The ongoing construction of the new embankment in Matmora where geotube technology is being used for the first time in India	314

Disaster and Resilience Building among Women in Manipur

Fig. 1	Ima Market, Imphal, Manipur, 9 April 2016	325
Fig. 2	Ima Market, Imphal, Manipur, 9 April 2016	326

Community Resilience Building and the Role of Paitei Tribe of Churachandpur in Manipur

Fig. 1	Manipur administrative map 2011 (http://www.censusindia.gov.in/Data/Census_2011/Map/Manipur/00_Manipur.pdf)	333
Fig. 2	Location map of Lamka town in Churachandpur district, Manipur (merge of various maps from Census of India 2011)	334
Fig. 3	One of the many houses in Khominthang area of Lamka town damaged by hailstorm in May 2013	337

xxx   LIST OF FIGURES

Fig. 4   Rescue works at Joumol village in Chandel district
by Assam Rifles in the first week of August 2015
which was hit by a major landslide on 1 August 2015
(*Picture Credit* PRO, HQ IGAR(S), Imphal. Retrieved
from http://www.e-pao.net/epGallery.asp?id=5&src=
News_Related/Calamities_News_Gallery/Joumoul20150814)   338
Fig. 5   A market shed of New Lamka bazar was flooded in July 2015   339
Fig. 6   A footbridge damaged by Lanva River in July 2015   339
Fig. 7   YPA relief team for Joumol landslide flag-off by the Deputy
Commissioner of Churachandpur on 12 August 2015
(Retrieved from http://www.virthli.in/2015/08/
hmasawnna-thar-13-august-2015.html)   342
Fig. 8   Members of the Paitei community on their altruistic activities
at Khominthang area, Lamka that was affected by hailstorm
in May 2013   345

Urban Risks to Hazards: A Study of the Imphal Urban Area
Fig. 1   Imphal urban area (*Source* Directorate of Census
Operations, Manipur)   354

**Relating Resource Extraction Pattern with Disasters: Political
Ecology of Barbhag Block, Nalbari**
Fig. 1   Locating Barbhag Block, Nalbari district
(Top: *Source* http://www.mapsofindia.com/india/
where-is-nalbari.html. Accessed December 8, 2016)
(Bottom: *Source* http://www.districtnalbari.com/Index.aspx.
Accessed December 9, 2016)   394
Fig. 2   'Diversion Based Irrigation' canals constructed
by Gramya Vikash Mancha (GVM) with the labour
participation of the villagers in Nagrijuli Block, Baksa
district (*Source* Photograph taken by the author
during fieldwork)   404

LIST OF TABLES

India's Northeast: Disasters, Development and Community Resilience

Table 1   Comparative economic indicators (*Source* Economic Statistical Organisation Punjab Central Statistical Organisation, New Delhi STATE WISE DATA as on 29 February 2016, http://www.esopb.gov.in/static/PDF/GSDP/Statewise-Data/statewisedata.pdf. Report of Fifth Annual Employment-Unemployment Survey (2015–2016), Ministry of Labour & Employment, p. 120. Retrieved 24 November 2016 and The *Central Electricity Authority* 2016, *Executive Summary*. Available at http://www.cseindia.org/userfiles/factsheet-north-east-india.pdf. Retrieved November 2016)   5

Table 2   North-east states: land and population dividend (*Source* Census of India 2011, GoI, New Delhi)   6

Table 3   Comparative infrastructure in higher and professional education (*Source* Key results of AISHE, Department of Higher Education, Ministry of Higher Education, New Delhi 2013, and GPI data from report of McKinsey Global Institute, 'The power of parity, how advancing women's equality can add $12 trillion to global growth', September 2015)   8

Manipur Flood 2015: Preparedness and Disaster Risk Reduction Strategies

Table 1   Comparative status of north-east states and India (*Source* Census of India, general population tables)   67

xxxi

xxxii LIST OF TABLES

Table 2	Major floods in Barak River Basins 1990–2015 (*Source* Manipur State Disaster Management Plan, Vol I and reported by Earth Sciences Department, Manipur University, Canchipur, Imphal)	69
Table 3	Frequency of rainfall (*Source* Deka et al. 2013)	70
Table 4	Sample survey respondent characteristics (*Source* Field survey by research team at Thoubal and Chandel District of Manipur April–June 2016)	78

Lessons from Manipur: Managing Earthquake Risk in a Changing Built Environment

Table 1	Changes in the built environment (housing) of Manipur over 2001–2011 (*Data Source* Census of India 2011, and BMTPC, Vulnerability Atlas of India 2006)	90

Varying Disasters—Problems and Challenges for Development in Manipur

Table 1	Major earthquake recorded in Manipur since 1926	99
Table 2	Major floods recorded in Manipur since 1916	102

Co-Seismic Slip Observed from Continuous Measurements: From the 2016 Manipur Earthquake (Mw = 6.7)

Table 1	Average co-seismic displacement at permanent GPS stations using GPS data of December 2015 and January 2016	156

Landslide Zonation in Manipur Using Remote Sensing and GIS Technologies

Table 1	Parameters used in the preparation of LHZ and LHM maps	175
Table 2	Landslide management scheme in LIS	184

Building a Resilient Community Against Forest Fire Disasters in the Northeast India

Table 1	Forest cover in north-eastern states of India (*Source* India State of Forest Report 2015)	188
Table 2	Major forest management context in Northeast India (*Source* Barik et al. (2006). *Forest sector review of Northeast India*. Background Paper No. 12. Community Forestry International, Santa Barbara, CA, USA)	190
Table 3	Forest fire alters for year 2014 and 2015 and the fire season in north-eastern states of India (*Source* India State of Forest Report 2015)	193
Table 4	State-wise causes and mitigation option in case of forest fire (Adapted from Satendra and Kaushik, A. D. (2014). Forest Fire Disaster Management, National Institute	

LIST OF TABLES   xxxiii

| | of Disaster Management, Ministry of Home Affairs, New Delhi) | 195 |
| Table 5 | Framework for promoting community participation in fire management | 199 |

The Khuga Dam—A Case Study

| Table 1 | Pre- and post-occupation of people | 207 |
| Table 2 | Income after displacement (ibid.) | 207 |

A Study of Socio-economic Community Resilience in Manipur

Table 1	District-wise area, population, sex ratio and density of population (2011) (*Source* Economic Survey of Manipur 2013–2014)	243
Table 2	Share in the disaster relief fund for Manipur, 2010–2011 to 2014–2015 (*Source* Economic Survey of Manipur 2013–2014)	244
Table 3	District-wise people at risk during natural disaster in Manipur (*Source* Manipur State Disaster Management Plan Volume I & II (2013))	244
Table 4	Distribution of amenities of Imphal West district (*Source* Census of India (2011), Transport & District Elementary Education Report Card, 2013–2014)	246
Table 5	Distribution of amenities in inhabited villages of Senapati district (*Source* Census of India, 2001 & 2011[a])	248
Table 6	Selected socio-economic indicators for two districts of Manipur (*Source* Data compiled from various sources including Economic Survey Manipur, Planning Department, Government and Census of India (2011))	250

'Living with Floods': An Analysis of Floods Adaptation of Mising Community—A Case Study of Jiadhal River

| Table 1 | List of the fish availability (*Source* Researcher's own findings) | 272 |
| Table 2 | Time spent for fishing (daily consumption) (*Source* Researcher's own findings) | 274 |

Floods, Ecology and Cultural Adaptation in Lakhimpur District, Assam

Table 1	Damage caused by riverbank erosion due to flood (Government of Assam)[a]	302
Table 2	Educational profile of Matmora and Laiphulia	306
Table 3	Primary occupation of Matmora and Laiphulia villages	307

xxxiv LIST OF TABLES

Community Resilience Building and the Role of Paitei Tribe of Churachandpur in Manipur

Table 1   Relief materials collected by YPA Lamka Block
           for Joumol landslide and flash flood, Dated:
           8 August 2015 (*Source* Retrieved from YPA general
           headquarters' record file handwritten in Paitei)   343
Table 2   Details of contribution to Joumol landslide
           (*Source* Record of YPA general headquarters)   343

Urban Risks to Hazards: A Study of the Imphal Urban Area

Table 1   Imphal urban area (*Source* Directorate of Census
           Operations, Manipur)   351

Risk Perception and Disaster Preparedness: A Case Study of Noney, Tamenglong District

Table 1   To show the various episodes of earthquake in Manipur
           (*Source* Manipur, NIDM, National disaster risk reduction
           portal) http://asc-india.org/seismi/seis-manipur.htm   375
Table 2   The socio-economic profile of Noney village, Manipur
           (*Source* Computed from census, 2011)   379

PART I

Disasters, Management and Development

India's Northeast: Disasters, Development and Community Resilience

Amita Singh

THE DEVELOPMENTAL DEFICIT AND THE WEAKENING COMMUNITY RESILIENCE TO DISASTERS

Nature will remain pristinely transparent about its cyclic regularity and would continue to behave the way it should from time immemorial. If man could be God, nature would be his equipment to be used at will. But, this could only be true in sheer romanticism of Percy Bysshe Shelley's 'The desire of the moth for the star, of the night for the morrow', so the policy maker's world is an encounter with realities, an exploration of causes and a search for solutions. The luxury of poetry and an extended bereavement of destruction are counterproductive to the idea of governance.

Therefore, the coming of excess rainfall, floods, landslides, hurricanes, earthquakes or tsunamis is simply natural occurrences, but when these natural processes damage and destroy life and property, they

---

Professor of Law and Governance at JNU and PI of Disaster Research Programme (now a Special Centre for Disaster Research at JNU).

A. Singh (✉)
Jawaharlal Nehru University, New Delhi, India

© The Author(s) 2018
A. Singh et al. (eds.), *Development and Disaster Management*,
https://Doi.org/10.1007/978-981-10-8485-0_1

become 'disasters'. The task of the government is to prevent such hazardous natural processes from turning into disasters. This suggests that proper scientific studies should identify vulnerable zones, share research with policy makers and help in appropriate housing, roads, medical preparedness, availability of ambulances for timely evacuation, rescue and relief. The ability of local administration to help the community bounce back to normalcy at the earliest is successful disaster management. Since appropriate houses, roads, medical support, community training in relief and rescue, and administrative accountability cannot occur during disasters, it is an indispensable imperative of the pre-disaster preparedness stage. It is therefore necessary to explore if the North-Eastern Region is disaster prepared? A comparison with other states could bring the truth out more appropriately.

In a comparative framework, one can safely say that land is the only resource which the north-eastern communities have as industries, institutions and knowledge hubs are quite invisible or absent. Thus, the land area of the north-east is taken as a primary unit of study over which its population dividend or its human resource and its agriculture-based livelihood suggest the nature of developmental initiatives required in the region. How much of this land is forest and what percentage is being used for agriculture suggest the developmental challenges which this region encounters. A comparison is being made with the adjoining West Bengal and Bihar on one side and highly forested, quarried and dammed states like Chhattisgarh, Odisha, and Himachal Pradesh on the other hand. The data from Delhi are only brought into show the truth about an age-old 'Delhi Centric Development' in India. This paper does not take account of rapid population increase in some states due to cross-border infiltration, intrusion or migration as this would divert the focus from disaster management. Sikkim has also not been included in a comparative frame since the focus has been on vulnerable regions where floods, landslides and earthquakes converge most frequently.

Of the total north-eastern land area of seven states, 101,248 square miles which is almost 7% of India's land area (2011 Census) has only 3.7% of the country's population and the smallest road network in the country of 8480 kms only (see Table 1). It is hardly understandable that Delhi with a land area of mere 573 share miles. has a road network of 28,508 kms. This explains the lack of supplies of both information, infrastructure and governance support to

Table 1 Comparative economic indicators (*Source* Economic Statistical Organisation Punjab Central Statistical Organisation, New Delhi STATE WISE DATA as on 29 February 2016, http://www.esopb.gov.in/static/PDF/GSDP/Statewise-Data/statewisedata.pdf. Report of Fifth Annual Employment-Unemployment Survey (2015–2016), Ministry of Labour & Employment, p. 120. Retrieved 24 November 2016 and The *Central Electricity Authority* 2016, *Executive Summary*. Available at http://www.cseindia.org/userfiles/factsheet-north-east-india.pdf. Retrieved November 2016)

S. No.	Eco. indicators	Bihar	Odisha	WB	Delhi	HP	AP	Assam	Manipur	Meghalaya	Mizoram	Nagaland	Tripura
1	GDP lakh crores INR	6.32 LCr	4.12 LCr	10.94 LCr	6.22 LCr	1.24 LCr	19,492 Cr	2.58 LCr	14,324 Cr	27,305 Cr.	11,458 Cr.	17,727 Cr.	26,810 Cr.
2	GSDP per capita income INR/income rank in 33 of all states/UT	33,954 rank 33 (lowest)	71,184 rank 27	87,672 rank 21	275,174 rank 2	147,330 rank 10	110,217 rank 17	60,621 rank 30	58,442 rank 31	75,156 rank 26	97,687 rank 19	89,607 rank 20	77,358 rank 25
3	Unemploy. rank 33 of all states/UT	14	17	25	–	04	06	13	16	19	24	07	1 (worst in employment rate)
4	Road network in kms.	19,928	23,8034	92,023	28,508	–	1992	2836	959	810	927	494	400
5	Per capita electricity consumption in kilowatts per hour	117.48	837.55	506.13	1447.72	1144.94	718.57	240.28	352.86	690.20	469.38	268.49	296.05

Table 2 North-east states: land and population dividend (*Source* Census of India 2011, GoI, New Delhi)

S. No.	States	Land area sq. km	Forest area (%)	Pop. 2011 census	Pop. increase (%)	Sex ratio 2011 census
1	AP	83,743	80.59	1,383,727	26.03	938
2	Assam	78,438	35.28	31,205,576	17.07	958
3	Manipur	22,327	76.10	2,855,794	24.50	985
4	Meghalaya	22,429	77.08	2,966,889	27.95	989
5	Mizoram	21,081	90.38	1,097,206	23.48	976
6	Nagaland	16,579	78.68	1,978,502	0.58	931
7	Tripura	10,486	75.01	3,673,917	14.84	960

the larger north-east habitats which continue to drag much insulated community support to encounter disasters. This also explains why the two most participative and vibrant states of the north-east, i.e. Assam and Manipur which are one of the biggest human resource providers to the rest of India, are also the most backward states in the country with an NSDP capita INR ranking of 31 and 30, respectively, ironically just above the last rank 32 of Bihar. Tripura which is undergoing a population explosion has the worst and the highest unemployment rate.

The fragile land of the north-eastern states is mostly hills, forests, valley and plateaus (see Table 2). The population increase over this land is comparable to the rest of the mainland states. Yet, land is the most contestable issue over its alleged misuse by policy planners. There are more than eight dams already built over its many rivers. More than 84.34% of its population resides in villages practicing various forms of agricultural practices like Jhum (shifting cultivation) which covers an area of 386,300 ha/year, and more than five lakh families depend upon this cultivation (Patel 2013). Besides Jhum, the rural north-east communities also undertake wet rice and Aji systems in which they grow rice and millets with fish in deep waters. They have also been practicing an agri-horti-silvipastoral system unique for every community over the hills. The government tried to popularize the terrace land cultivation, but the support from the government in terms of input costs of soil testing, fertilizers and crop rotation did not follow as a result of which this terrace cultivation remains almost an undersupplied livelihood arrangement of the small and marginal farmers who constitute more than 78.92% of agricultural communities. With all these shortcomings of an apathetic governance and a complete dependence on agriculture, the north-east is able to produce only 1.5% of the country's food grains which is not even sufficient enough to feed its own self. Their major livelihood comes from agriculture, and almost the whole of its farming is organic yet no government has ever tried to seek certification for these farmers to give them access to international markets. Women are the most vibrant and deserving community leaders in the rural countryside of the north-east. The sustenance of Manipur on Ema Markets (or Mothers' Market) which is one of the most vibrant women led market places, an enduring location for local farm and kitchen produce besides a hang out location for all motivated women to showcase their creative art products, designs and home based inventions. This exponential resilience demonstrated in the Manipur's local culture is subsequently reflected in their relatively better sex ratio, a higher number of women teachers, medical practititoners as well as agricultural workers. Ironically, most agricultural practices which they master are getting distanced from them due to lack of micro-credit and other support systems. The national and state level IITs and agricultural training institutes have not been able to connect with north-eastern communities and empower their skills and mutually understand and learn from their indigenous knowledge (Table 3).

The comparable mainland states in the neighbourhood of north-east have a much lower enrolment of women in higher education than the states of north-east. The GER in higher education is startling in some of these states especially in Manipur, Mizoram and Nagaland. Similarly, the GPI in all north-east states barring Nagaland and Tripura is higher than most mainstream comparable states. The GPI for the SC/ST is much higher in at least four states of the north-east, i.e. Assam, Manipur, Meghalaya and Mizoram, where it exceeds Bihar, Odisha and even MP. Most north-eastern states have a better sex ratio and the Gender Parity Index than the country's average of 940 and GPI score of 48, respectively. While higher GPI is considered a population dividend,[1] yet these

---

[1] GPI is a socio-economic index which measures relative access to education of males and females. It has been established in research that improving gender parity can reap rich rewards of economic growth and a latest McKinsey report establishes that 'it can boost GDP by USD 7 trillion in 2025 … or 1.4% per year of incremental GDP growth for India'.

8   A. SINGH

Table 3 Comparative infrastructure in higher and professional education (*Source* Key results of AISHE, Department of Higher Education, Ministry of Higher Education, New Delhi 2013, and GPI data from report of McKinsey Global Institute, 'The power of parity, how advancing women's equality can add \$12 trillion to global growth', September 2015)

S. No.	States	Colleges/lakh pop.	GER in higher ed. M/F/total	GPI in higher ed. all cat./SC/ST	Total universities central/state
1	Bihar	649/6	14.7/11.21/13.1	0.76/0.57/0.76	01/14
2	Odisha	1089/23	18.4/14.3/16.3	0.78/0.79/0.79	01/12
3	West Bengal	899/8	14.7/10.7/12.8	0.73/0.71/0.64	01/19
4	Delhi	184/9	35.7/33.6/34.8	0.94/0.79/–	04/05
5	HP	296/38	25.7/24.2/25	0.94/0.90/0.94	01/04
6	Arunachal Pradesh	26/16	36.9/24.9/30.9	0.67/–/0.63	01/nil
7	Assam	485/13	14.5/14.2/14.4	0.98/0.96/0.90	02/03
8	Manipur	79/26	32.3/34.4/33.4	1.07/0.94/0.83	02/nil
9	Meghalaya	61/17	14.3/18.3/16.4	1.28/0.49/1.56	01/nil
10	Mizoram	29/22	21.6/19.6/20.6	0.91/1.31/0.89	01/nil
11	Nagaland	57/22	22/13.7/17.9	0.62/–/0.62	01/nil
12	Tripura	39/9	14.2/9.1/11.6	0.64/0.63/0.53	01/nil

states have not been able to reap this benefit due to lack of higher educational, specialist and vocational training institutions. This region lacks in institutions of higher education. The Department of Higher Education (2013) data suggest that this region lacks school of engineering, architecture, palliative care, physiotherapy, nursing, hospitality and hotel management institutes, medical education, veterinary sciences, music, dance, journalism and media studies, fine arts and culture. It even lacks a sports university despite the fact that the north-east schools have some of the best soccer players, marathon athletes and other sports activity excellence records. The question is about what prevents the government from building community resilience against disasters? The region is loosing its land[2] to floods, landslides and preventable gradual erosion, and the community is loosing trust in governance making development more problematic every day.

---

[2] Majuli island has shrunk from 1250 sq. kms 10 years ago to one-third of its original size. This process has been happening from the Kalimpong (West Bengal) and Darjeeling side of the north-eastern hills to the Manipur and beyond. 'The slow death of India's Majuli Island' by Das (2013).

MANAGING FRAGILE HIMALAYAN ECOLOGY

The north-east states of the country have enormous diversity and a highly inconsistent and fragile terrain spread over the most precarious 'Himalayan' mountain belt. This multi-coloured, multi-cuisine and culturally sensitive communities camouflage this highly erratic, harsh and callous nature of loose Himalayan rocks which have a ruthless history of cold-blooded destruction of habitats, its people and animals. 'Himalaya' in Sanskrit suggests 'an abode of snow' which brings an unmatched water and food security to people over these hills as well as those in the downhill regions. Himalayas have the third largest deposit of ice and snow in the world after the Antarctic and the Arctic. In the midst of more than 15,000 glaciers covering the Himalayas, the strategic Siachen Glacier is one of the largest besides the other strategically located Baltoro, Biafo, Nubra, and Hispur. So much is our dependence upon Himalayas that just one Siachen Glacier which is situated over the Karakoram range extends to more than 48 miles or 72 kms and has been the largest single source of freshwater on the Indian subcontinent. It also feeds the Pakistan side of the Punjab plains through the Nubra River. Siachen is not just ecologically important as a food security to communities in the north-east but is a major safety gadget to India's strategic security too. The impact of climate change, neglect of development and misuse may also bring other threats as the Saltoro ridge which runs from the north-west to the south-east and bypasses the Gilgit and Baltistan in the west is a watchman against access from Skardu in the Gilgit–Baltistan area to the Karakoram pass. Any presence of Pakistan in Karakoram makes Ladakh in India more vulnerable since China continues to remain in the occupation of Aksai Chin since 1962. Therefore, Siachen Glacier which is also called the third pole of the world becomes the highest battleground of the world, where Indian Army has to be posted at an altitude of 21,000 ft. sea level. Under these vulnerable situations, it would be imprudent to have unhappy and resentful people in the north-east.

With a fragile placement of tectonic plates within the Himalayan mountain range, they remain incessantly in motion and collision with a relatively higher occurrence of tremors, earthquakes and landslides. This fragility over the hills and concomitant glacier dynamics affects water circulation channels, agri-systems, food security, ecosystem conservation, roads building efforts which in turn severely disrupt livelihood opportunities. Ironically, while this region requires the most committed, coordinated and mission-driven mitigation and risk reduction initiatives, the state institutional functioning has failed to even make a modest effort towards achieving it. The laggard approach on disaster mitigation in this region is further exposed when Murthy et al. highlight the advancements which have taken place in the Himalayan disaster mitigation strategies such as 'the emergence and adoption of near-real time systems, Unmanned Aerial Vehicles (UAV), board-scale citizen science (crowd-sourcing), mobile services and mapping, and cloud computing (which) have paved the way towards developing automated environmental monitoring systems, enhanced scientific understanding of geophysical and biophysical processes, coupled management of socio-ecological systems and community based adaptation models tailored to mountain specific environment'.

One of the major problems for disaster management in the states of the north-east is to adopt a disaster tested developmental strategies. The hill communities of the north-east are highly dependent upon natural resources, and this provides them the food and a secure habitat which no government has a substitute for, but development has not only been inconsistent but also in denial of natural resource-based sustainable mountain development strategies. This has completely prevented and discouraged communities from participating in any government effort which affects their areas. In the absence of confidence-building measures, the most understandable reaction from them to any developmental initiative is to oppose government which would steal their resources away. The government should develop a transparent data analytics on the collection, management and corroboration of data on cultural communities, indigenous and customary practices, their basic resources including their consumption patterns, infrastructural capacities and local community-based regulations to the use of land and their natural common areas. Much of this has to seep into state mitigation planning to disaster management. Modern advances in earth sciences have made much of this possible and expanding education, and use of English as a language of education has simplified the availability of locally placed technical support. The educational institutions and the universities in the north-east should now be catering to this demand of reviving and rejuvenating community institutions and local people on consultative developmental planning as a strategy to disaster risk reduction.

Need for Transdisciplinarity in Planning and Enforcing Disaster Management

The disaster research centre at JNU has adopted a transdisciplinary approach in managing disasters. Earth scientists have been coming up with important scientific findings suitable in preparedness policies for disaster management, but these researches have mostly been within academic group or individually project driven to raise the impact factor of their publications in international journals. Most of these scientists have remained distanced to the DMA 2005 framework, and the local institutional framework which makes their work less understood to a policy implementer. Therefore, administrators are left with no option but to focus on post-disaster rescue and rehabilitation programmes. The last decade of DMA's 2005 implementation graph indicates the following concerns:

1. Weak performance of State Disaster Management Authorities and completely naïve district disaster management authorities
2. Greater visibility of NDRF and the Army
3. Limited dissemination (early warning capacity) of meteorological, geological and remote sensing authorities of government
4. Continued government focus on post-disaster relief, rescue and rehabilitation
5. Absence of community resilience building through mitigation and preparedness programmes.

Transdisciplinarity unlocks scientific data and makes it a public property which could be tested, accepted and put on trial as well. It brings forth a participatory approach to disaster risk reduction which draws its rationale from the data on carrying capacity but its legitimacy rests on the constitutional rights to life, property and safety which are derivatives of intra- and inter-generational equity and justice. The poor is a bigger victim to disasters, not due to location but due to fragile houses, unprotected and unauthorized dwellings and locations and due to disinterested state in providing timely warnings to the poor and making arrangements for them as much as they do for the rich. Data would demarcate boundaries of settlements and prohibited activities such as mining, damming and constructions in vulnerable areas. Mining and damming could be structured more appropriately rather than throwing the developmental policies open to information deficient governments and people. ICIMOD brings together the concerns of co-working to suggest a 'mainstreaming earth observation and geospatial technologies to its programs and enhancing inter and transdisciplinary approaches to solve mountain development problems' (Murthy et al. 2014). This approach has also been applied in other regional and international disaster mitigation organizations. The Terrestrial Observation Panel for Climate (TOPC) conducts a balanced and integrated system of in situ and satellite observations of the terrestrial systems. It is sponsored by the Global Climate Observing System (GCOS) and the World Climate Research Programme (WCRP). There are other specialized global agencies working for data collection and corroboration under United Nations such as the Global Climate Observing System (GCOS), Global Ocean Observing System (GOOS), Global Terrestrial Observing System (GTOS) and the United Nation's Framework Convention on Climate Change (UNFCCC) Subsidiary Body for Scientific and Technological Advice (SBSTA). The Brussels-based Centre for Research on the Epidemiology of Disasters (CRED) launched Emergency Events Database (EM-DAT) to keep record of essential core data of over 22,000 mass disasters from 1900 to the present day. NatCat and Sigma are highly sophisticated databases managed by the two largest insurance companies Munich Re and Swiss Re, respectively.

COMMUNITY RESILIENCE BUILDING AND GOVERNMENT'S PROTECTIVE COVER TO HOUSEHOLDS

India's north-east is a collage of many hill and tribal communities. No task of disaster management is completed without substantial knowledge of these communities which are highly vulnerable to disasters. The ability of these communities to withstand disasters is highly linked to the government's timely enforcement of mitigation plans in the region. The state and district disaster management authorities responsible for mitigation plans are important for the design and dimensions of developmental planning yet they remain distanced, unvoiced and disjuncted from these communities. An intimidating presence of state institutions bulldozing their will over local communities as if they were all errant and non-cooperative has been proved wrong by the team of researchers coordinated by the disaster research centre at JNU. The steel frame of laws which control land, community initiatives and the ability of local institutions to share responsibilities has been overlooked. Disaster management in the region demands scientific findings to strengthen accountability, lack of preparedness as suggested by the DMA 2005 and an appropriate intervention strategy to save lives and property.

Most underdeveloped regions do not have insured households, and one disaster can destroy their total life even if they are able to save themselves from being killed. However, information on some really meaningful innovations by insurance companies functioning like the micro-credit rural companies is scarce to the NDMA and the SDMAs. One such company, DesInventar, managed by a coalition of non-governmental actors, covers 16 countries in Latin America, the Caribbean including the USA, Brazil, Colombia, South Africa, and India. It tends to record higher number of people affected than other databases. These databases are important foundations to any disaster management policy and are vital for identifying trends in the impacts of disaster and tracking relationships between development and disaster risk.

Community resilience in the north-east is based on community participation in governance and an insurance cover to each household against disasters. However, implementation of resilience building has the danger of getting lost in political rhetoric and parochial community leadership. The way out is to bring north-east region immediately within a data management company to base policy rationale in scientific data, with information on different thematic areas such as land cover mapping, landscape planning, biodiversity conservation, agriculture monitoring including drought forecasting, food security, cropping systems, forest cover mapping and monitoring, rangeland mapping, disaster allocation and snow & glacier studies. This information should more appropriately form the basis for developmental planning rather than the demands of the market and politics of the state. Such a development would make communities and systems more resilient to disasters in a sustainable manner which even the best and most efficient government may not be able to achieve.

THEME OF THIS BOOK

The concern of this book unlike any other conventional disaster research publication is to link disasters to development and to highlight the fact that simply by taking serious action towards disaster mitigation, the government can allow the region to bloom in prosperity. This is brought out under three major sections in which the book is divided. Some of the main authors are local people who feel and understand their homeland as no specialist can. The knowledge providers on disaster zoning through GPS mapping are also from the region. The perceptions and governance challenges are an outcome of an extended brainstorming debates and discussions among different tribes, hill vs. valley people, administrators–academia–politicians' mixed groups. The effort is to make the book as much realistic as possible to that the State Disaster Management Authorities of the north-eastern states could consult its readings as they venture out in those regions with their policy plans. The three sections of the book are:

1. Governance institutions and community support programmes
2. Vulnerability of communities living in disaster zones
3. Community resilience and ability to withstand disasters

The second Chapter, 'Natural Disasters in the North-East: Crucial Role of Academia in Saving Lives' by *P. P. Shrivastav* IAS Rtd., a former member of the North-Eastern Hill Council, sets the platform where others join in the later chapters. He identified this region as one of the six most earthquake-prone areas of the world, the other five being Mexico, Taiwan, California, Japan and Turkey (Honolulu Workshop: May 1978). The reason for such high seismic vulnerability lies in the turbulent geological past and the continuing tectonic plate movements in the Himalayan belt. The paper attempts to look into the geophysical history of this broad region to identify the main factors that have led to such high incidence of natural disasters in the NER and suggest effective remedial measures. The third Chapter, 'Regional Climate Changes Over Northeast India: Present and Future' by *A. P. Dimri, D. Kumar* and *M. Srivastava* from the School of Environmental Sciences, Jawaharlal Nehru University, suggests that in the light of many diverse cultural and ethnic groups, it is noteworthy that the ecosystem services provided by the forest and other components in this region should not be overlooked. This mountain ecosystem is quite sensitive to the changes in the climatic pattern due to its geo-ecological fragile nature, its location in the Himalayan landscape, transboundary hydrological regime and its inherent socio-economic instabilities. This paper helps to design and implement appropriate adaptation and mitigation measures. The fourth Chapter, 'Manipur Flood 2015: Preparedness and Disaster Risk Reduction Strategies' by from the Centre for the Study of Regional Development at the Jawaharlal Nehru University, investigates the most devastating 2015 floods ever experienced in the areas of Manipur State during the last 200 years. It affected the entire Thoubal district and parts of Chandel district. The chapter examines the level of prevention and preparedness of State Government and local communities, recovery and response mechanisms of State Government, civil society and local communities to propose developing a disaster risk reduction measures which are based on community's indigenous experiences. The Chapter five, 'Lessons from Manipur: Managing Earthquake Risk in a Changing Built Environment' by *Kamal Kishore* Member, National Disaster Management Authority (NDMA) and *Chandan Ghosh* from National Institute of Disaster Management (NIDM), made an in-depth analysis of the Manipur earthquake of 4 January 2016, to remind about the high level of earthquake risk in Manipur as well as other north-eastern states of India. The epicentre of the earthquake being Tamenglong district of the state which was also one of the most backward had never discussed the high damage was related to the designs of structures. In Manipur, a gradual and inevitable shift from traditional forms of construction to RCC buildings affected the seismic safety in the state. Chapter six on the 'Varying Disasters—Problems and Challenges for Development in Manipur' by *Jason Shimray*, Secretary and Head of the Manipur State Disaster Management Authority, powerfully adds the intricate cultural and ethnic mosaic facing challenges of sociocultural identity resulting in ethnic conflict coupled with factors such as geographical location, difficult terrain and poor road connectivity, poor infrastructure, problem of law and order due to insurgency, etc. These have been roadblocks to overcoming challenges for ensuring a responsive administration and governance. The argument of the paper is about a major insight into disaster management in the valley which cannot be addressed without incorporating the entwined cultural perceptions of various communities inhabiting the region. The Chapter seven, 'Manipur's Tryst with Disaster Management Act, 2005' by *Nivedita Haran*, a former Addl. Chief Secretary of Kerala who writes with her experience of having set up the first State Disaster Management Authority in Kerala after the DMA 2005, has some concrete solutions at palms length. The paper suggests that since the DM Act has completed its 10th anniversary, an assessment of its impact in the field and at the grassroot level has become necessary. The paper brings out the different ways in which benefits of the DM Act and rules can reach the people, how the funds can be utilized more effectively, ensuring better transparency and accountability and some novel methods of making the citizens disaster resilient. Manipur can show the way to the other north-eastern states, and a very welcoming suggestion comes in the end that the newly established special centre for disaster research at JNU which is aptly placed in the midst of all vibrant social science and natural sciences research departments should coordinate the task to fill this gap.

Chapter eight on 'Reviving 'Public Policy' and Triggering 'Good Governance': A Step Towards Sustainability' by *Pankaj Choudhury* from the Centre for the Study of Law and Governance at Jawaharlal Nehru University, is clearly addressing the techno-centricity and commodity-fetishism as the leading cause of fragility and vulnerability over the hills. This chapter seeks to address the significance of transparency and good governance in relation to the state disaster reduction programmes in India and its relationship to the rising number of hydroelectric projects in the State of Arunachal Pradesh (AP). The Chapter nine on 'Role of Heavy Machinery in Disaster Management' by *Keshav Sud,* from the Robotics at Amazon, Boston, USA, is about the prospective but urgent need for the disaster management agencies to adapt to machines and bring their operations close to communities and young people in the hills to save and protect their people and property in time. The paper analytically suggests that the disasters in India and much of the South Asian subcontinent have been the work of Samaritans and martyrs brought in from the Army or more recently from an army like battalion such as the National Disaster Relief Force (NDRF) or Central Industrial Security Force (CISF). Huge amount of funds are pumped in post-disaster relief and recovery despite the fact that much of this would have been saved through right planning and training available to people in the vulnerable zones.

The Chapter ten of the book on the 'Co-seismic Slip Observed from Continuous GPS Measurements: For the 2016 Manipur Earthquake (Mw = 6.7)' by *Arun Kumar* and *L. Sunil Singh* from the Department of Earth sciences, Manipur University Imphal, suggests that the use of GPS to study crustal movement and explore geodynamic phenomenon in the world and its application in geodesy, geophysics, geology explores an understanding of global tectonic plates, as the cause of an earthquake. The paper attempts to simplify dissemination of this knowledge to policy makers on the north-east areas. The Chapter eleven on the 'Post-Earthquake Geodisaster of 6.7 Mw Manipur earthquake 2016' is a research finding made by *Ningthoujam, Lolee, Dubey, Gonmei, Thoithoi* and *Oinam T* from the Department of Geology, University of Delhi. This study focuses on the analysis of post-earthquake geological and geomorphic changes in parts of Imphal and Tamenglong districts of Manipur to understand the causes of the earthquake, nature of the movement of crust and any possible future disaster. This earthquake with its epicentre located in a remote village near Noney in western Manipur throws some light on the role of institutions such as the Geological Society of India (GSI) which categorized Manipur as Zone V in their Seismic Zoning Map of India (Bureau of Indian Standard). Chapter twelve, 'Landslide Zonation in Manipur Using Remote Sensing and GIS Technologies', by *R. K. Chingkhei* from the Department of Forestry and Environmental Science, suggests action around the landslide vulnerable zones along the national highways of Manipur that are severely affected. The Landslide Hazards Zonation (LHZ) map and a landslide hazard management (LHM) map using remote sensing data and Geographical Information System (GIS) techniques with data-sets consisting of fourteen parameters have been generated in GIS platform from the topographic maps, satellite imageries and field visits. The resulting landslide hazard zonation map delineates the area into different zones of six classes of landslide hazard zones, i.e. severe, very high, high, moderate, low and very low. The result of LHZ map is also validated using the field data. Chapter thirteen, 'Building a Resilient Community Against Forest Fire Disasters in the Northeast India' by *P. K. Joshi, Anusheema Chakraborty* and *Roopam Shukla* from the School of Environmental Sciences, Jawaharlal Nehru University and the Department of Natural Resources, TERI University, is the only chapter that addresses forest fires. Fire events are major cause of change in forest structure and function, challenging the supply of ecosystem goods and services. Among the different floristic regions, the northeast India suffers the maximum due to spread of forest fires from age-old practice of shifting cultivation, Jhum fields. Therefore, the authors advocate the use of satellite remote sensing inputs and GIS that can assist in the identification of fire-prone areas. With this background, the focus of this article is on the broader aspects of forest vulnerability to fire, current methods used in mitigating its impacts and the need for more participative forest fire management plans in north-east India.

In Chapter fourteen, 'The Khuga Dam—A Case Study' by *Langthianmung Vualzong* and *Aashita Dawer* from the Centre for the Study of Law and Governance, Jawaharlal Nehru University, and

Nivedita Haran, former Addl. Chief Secretary of Kerala, highlighted that since water has the highest potential for danger that can cause a disaster the nature of technical interventions most suitable in managing water, the optimal size of a dam, the best ways to retain the organic character of a river, and similar should be made more open and visible in disaster management policies. Intervention of state in managing the water resources is thus considered to be of prime importance. Ideally, dams are expected to bring about development in the surrounding areas, thereby increasing the overall economic benefit to the nation. Dams that are engineered structures need the right location, optimal design and correct execution. The story of the Khuga Dam in Manipur reveals many inputs to the design of law and governance in Manipur and surrounding states.

Chapter fifteen, 'Dams and Disasters in the Northeast India: A Collateral 'Ethical' Relationship' by *Prasenjit Biswas* from the North-Eastern Hill University (NEHU) questions the sustainability of Central Government plans of constructing a mega dam with huge private investment over every river that flows into the region. Particularly, the proposed Tipaimukh project is proposed on the confluence of the Barak and Tuivai Rivers in Manipur's south-western area, and there will be an earthen/rock-filled 178-metre high dam. In case of any breach, it has the possibility of not only inundating certain parts of Manipur's districts like Senapati, Tamenglong, Churachandpur and Jiribam, but also the whole of the Barak Valley of southern Assam especially Silchar town. This paper suggests that such unwarranted consequences of dam-led development in a geologically and ecologically fragile region need to be taken into consideration in the aftermath of what happened in Uttarakhand. Chapter sixteenth on 'Achieving Last Mile Delivery: Overcoming the Challenges in Manipur' by *Sukhreet Bajwa,* one of the key researchers from the erstwhile transdisciplinary Disaster Research Programme at JNU and currently at the Gujarat Disaster Management Authority, Gandhinagar, suggests an imperative need for preventing the exclusion of large number of communities from basic services during preparedness discourses. The click of a button approach to identify the unserved is what matters in a governance policy which would rightfully focus on preparedness. This paper intends to explore the challenges arising to achieve the last mile delivery in Manipur.

The seventeen Chapter of the book on 'A Study of Socio-economic Community Resilience in Manipur' by *Thiyam Bharat Singh* at the Central University of Manipur. The paper brings out a report by the

United Nations ESCAP to highlight that half of the world's disasters in 2014 occurred in Asia and the Pacific, affecting 80 million people and incurring nearly $60 billion in economic losses. The average number of natural disasters in the Middle East and North Africa (MENA) almost tripled since the 1980s, resulting in average annual losses of over $1 billion. The objective of this present paper is to examine resilience of local community (socio-economic resilience) during disaster by selecting two districts of Manipur. The two districts are Imphal West (valley) and Senapati (hill district). Chapter eighteen on the "Living with Floods': An Analysis of Floods Adaptation of Mising Community—A Case Study of Jiadhal River' by *Tapan Pegu*, from the Jiadhal College at Jiadhal, Assam, brings out the plight of the Mising people who are known as 'river loving' people. After many years of settlement in the river bank of Brahmaputra and its tributaries, the community has webbed deep relations with river ecosystem. The study reveals that in adaptation with floods, the community have developed adjustments mechanism to cope with flood. The house architecture, crops, land use, use of natural resource, etc., are all based on traditional ecological knowledge which support in mitigating the adverse effect of natural hazard like flood. Anthropogenic factors depleting the natural resource base have further reduced the resilience capacity of the community. The flood hazard of Jiadhal River has now become the cause for great sorrow for the Mising community. Chapter nineteen on the 'Role of CBOs in Resilience Building: Good Practices and Challenges' by *Thongkholal Haokip* from the Centre for the Study of Law and Governance, Jawaharlal Nehru University, examines the role of community-based organizations in resilience building within the north-eastern States of India. There is a plethora of indigenous practices and value systems in the largely egalitarian societies of the north-east that help them in facing disasters without much assistance from the state. The value systems as well as a practice among the Kukis known as tomngaina and khankho will be taken as a reference point to show how CBOs are obliged under these value systems to assist anyone within the community in the face of any kinds of disasters. The study assesses the historical past as well as from 2015 Manipur landslide, the 2016 Manipur Earthquake and hailstorms, particularly to prove or disprove the role played by community's cultural value systems and practices in building resilience. The next twentieth Chapter, 'Floods, Ecology and Cultural Adaptation in Lakhimpur District, Assam' by *Ngamjahao Kipgen* and *Dhiraj Pegu* from the

Department of Humanities and Social Sciences, Indian Institute of Technology, Guwahati, Kamrup, Assam, identifies and analyses the determinants of the problems and captures the vulnerability of the locals, adaption mechanism and sources of livelihood of the Mising community in Lakhimpur district, Assam. Through this study, it was found that most of the people affected by the flood have been resettled and relocated over six to seven times from their original location. This empirical study was conducted in Matmora and Laiphulia villages during November 2014 to April 2015. The following Chapter twenty one, 'Role of Women in Disaster Risk Reduction—Perspectives from Manipur, India' by *Mondira Dutta* from the Centre for Inner Asian Studies, School of International Studies, Jawaharlal Nehru University, is indicative of the plight of girls and women. Women and girls have been found to be at a much greater risk in terms of reproductive health problems, sexual abuse, forced marriages and other forms of gender-based violence including death. The region of Northeast India and the neighbouring parts of Myanmar, China, Bhutan and Bangladesh encompass one of the most active seismic regions of the world. The paper makes visible many of the earthquakes in this region being among the largest in the world impacting large areas nearby tend to be most severe is perpetrating violence against women. Chapter twenty two on the 'Community Resilience Building and the Role of Paitei Tribe of Churchanpur in Manipur' by *M. Pauminsang Guite and Langthianmung Vualzong* studies how a tribal community namely Paitei tribe in Manipur engaged to tackle natural disasters in their own approach. The Paiteis living in Manipur have found solutions to the recurrent flood and hailstorm that caused damages to lives and property. This paper also highlights the disorganized and ill-approach of the government in reduction of disaster risk and the much needed for social intervention on natural disasters. The local community with its strong bonding on ethnic line is prominent even in disaster relief management. Chapter twenty three on the 'Urban Risks to Hazards: A Study of The Imphal Urban Area' by *Sylvia Yambem* from the Central University of Manipur in Imphal identifies the linkage of bad planning and aggravated vulnerability of Manipur people. She highlights the inability of state agencies to institutionalize an effective structure of disaster risk reduction in the state even though disasters are most frequent and recurrent in the region. The effect of disasters such as floods, earthquakes and landslides in the urban areas not only present a deeper social, economic and environmental challenge but also exposes a lack of inclusive development. This chapter finds it imperative to integrate disaster risk reduction strategies in urban planning procedures such as land use policies, building codes, standards, rehabilitation and reconstruction practices, etc., which have also been emphasized in the Hyogo Framework for Action 2005–2015. Chapter twenty four, 'Community Awareness on Landslide Disaster: 'Experience of Nagaland'' by *Supongsenla Jamir* and *N. U. Khan* from the Nagaland State Disaster Management Authority has some important policy solutions. The Nagaland State is prone to all kind of natural disasters like earthquake, flash floods, landslide and forest fire, etc. Every year the state faces huge losses of property due to natural disaster. The geographical location places a major role on the occurrence of natural disaster. Both districts, Kohima and Mokokchung, have been experiencing a steady growth towards urbanization and attracting more population from rural area and from outside state. This paper asserts that landslide awareness consideration is very important in assessing the quality of life since the environment impact is directly on the well-being and life of the residents. Chapter twenty five, 'Risk Perception and Disaster Preparedness: A Case Study of Noney, Tamenglong District' by *Homolata Borah* from the Kamala Nehru College and a researcher at the erstwhile Disaster Research Programme at Jawaharlal Nehru University, relates the pillars of Sendai Framework of Disaster Risk Reduction (SFDRR 2015–2030) in the context of disaster mitigation work in the new district Noney of Manipur. It aimed to reduce mortality and protecting livelihood by understanding risks, making investment in disaster risk reduction, building better governance and to 'build back better'. This chapter attempts to observe risk perception before and after the occurrence of the earthquake in relation to the preparedness strategies as laid down and practiced by the government and to find associations if preparedness and mitigation strategies are available in a high risk perceived situation. The next Chapter twenty six, 'Institutions of Faith in Resilience Building During Disasters: Church and Communities in Lamka Vis-à-Vis Churachandpur' by *Priyanka Jha* and *G. V. C. Naidu* from the Centre for Political Studies at Jawaharlal Nehru University makes visible the phenomenal role of institutions of faith in the midst of destruction and extreme melancholy. Church in Churachandpur is an example of sustaining faith in times of crisis and destruction. Its role has only expanded as a security cover to local communities because the district continues to be the country's most impoverished districts which is surviving since 2007 on the Backward Regions Grants Fund. Lamka vis-à-vis Churachandpur are ancient historic areas which have got one name of Churachandpur around the Second World War. It is alleged by some local people that merging Lamka into Churachandpur was part of a strategy of the Meiti tribe to control the smaller tribes such as Kuki, Paitei and Zo. So with the minimum support and help coming in from the institutions of the state, the communities over a period of time have started drawing help from community-based organizations such as the church in coping with disasters. This paper asserts an invocation and recognition of such institutions of faith in community resilience building. The last Chapter twenty seven, 'Relating Resource Extraction Pattern with Disasters: Political Ecology of Barbhag Block, Nalbari' by *Snehashish Mitra*' from the Calcutta Research Group highlights abundant natural resources and their management as a key to disaster mitigation. The resources have provided sustenance to both indigenous and migrant communities, while serving as objects of extraction for British and Indian states. Tea, timber and crude oil were the main resources which went into colonial accumulation, while post-independence focus has gradually moved to additional resources like water (for hydroelectricity), uranium, coal, rubber plantation and jatropha plantation. North-east India is a high seismic zone area, and therein it is essential to factor in the nature of resource exploitation that holds a direct impact on the cause and probability of disasters affecting especially the marginalized and vulnerable communities of the region.

CONCLUDING REMARKS

This chapter has raised issues which are fundamental in considering disaster mitigation and risk reduction in the north-east of India which inhabits a very large population of hill tribes over a high vulnerability zone. The chapter highlights three main requirements of building community resilience, i.e. availability of complete community resource data, participatory and collaborative transdisciplinary governance structures and government's protective cover such as insurance and basic services to all communities. The chapter also gives a brief survey of the twenty-six chapters in the book which are classified under three sections, governance institutions and community support programmes, vulnerability of communities living in disaster zones and Community resilience and ability to withstand disasters.

REFERENCES

Census (2011). Office of the Registrar General and Census Commissioner, India. Ministry of home Affairs, Government of India. Available at http://censusindia.gov.in

Das, Bijoyeta. (2013). 'The Slow Death of India's Majuli Island' by Retrieved from http://www.aljazeera.com/indepth/features/2013/08/20138894413586685.html.

Disaster Management Act. (2005). Available at website National Disaster Management Authority, Government of India.

Economic Statistical Organisation Punjab Central Statistical Organisation, New Delhi STATE WISE DATA as on 29 February 2016. http://www.esopb.gov.in/static/PDF/GSDP/Statewise-Data/statewisedata.pdf.

Key Results of AISHE, Department of Higher Education, Ministry of Higher Education. (2013). New Delhi.

Murthy, M. S. R., Bajracharya, Birendra, Pradhan, Sudip, Shrestha, Basanta, Bajracharya, Rajan, Shakya, Kiran, Wesselman, Sebastian, Ali, Mostafa, Bajracharya, Sameer, and Pradhan, Suyesh. (2014). 'Adoption of Geospatial Systems Towards Evolving Sustainable Himalayan Mountain Development', *The International Archives of the Photogrammetry, Remote Sensing and Spatial Information Sciences*, Volume XL-8, ISPRS Technical Commission VIII Symposium, 9–12 December 2014, Hyderabad, India (pp. 1319–1324).

Patel, Amrit. (2013). 'Harnessing Agricultural Potential in North Eastern Region of India', 25 March. Available at http://indiamicrofinance.com/agricultural-in-north-east-india.html.

Report of McKinsey Global Institute, 'The Power of Parity, How Advancing Women's Equality Can Add $12 Trillion to Global Growth', September 2015.

Report of Fifth Annual Employment-Unemployment Survey (2015–2016), Ministry of Labour & Employment, p. 120. Retrieved 24 November 2016.

The Central Electricity Authority. (2016). Executive Summary. Available at http://www.cseindia.org/userfiles/factsheet-north-east-india.pdf. Retrieved November 2016.

Natural Disasters in the North-East: Crucial Role of Academia in Saving Lives

P. P. Shrivastav

Northeast India has been identified as *one of the six most earthquake-prone areas of the world*, the other five being Mexico, Taiwan, California, Japan and Turkey (*Honolulu Workshop: May 1978*). In fact, seven states in our North-Eastern Region (NER) namely, Arunachal Pradesh, Assam, Manipur, Meghalaya, Mizoram, Nagaland and Tripura, fall in the maximum seismic vulnerability Zone V: only Sikkim (included in NER in the year 2005) falls in Zone-IV.

As many as 17 large earthquakes (Eqs) of magnitude 7 and above on the Richter scale had occurred in this general area over the 100 years starting from the Cachar EQ of 1869 (magnitude 7.8). These include the Great Shillong EQ of 1897 (M > 8) followed 53 years later by another Great EQ of similar intensity in 1950 just across our borders (near Rima in Tibet), not to speak of several hundred small and micro-earthquakes spread all over. Sixty-six years have gone by since then. Apprehensions of another Great EQ hang heavy in the air (Fig. 1).

P. P. Shrivastav (✉)
Formerly, Member, North Eastern Council, Shillong, India

© The Author(s) 2018
A. Singh et al. (eds.), *Development and Disaster Management*,
https://doi.org/10.1007/978-981-10-8485-0_2

THE N-E REGION: MAJOR EQs AND TECTONIC FEATURES

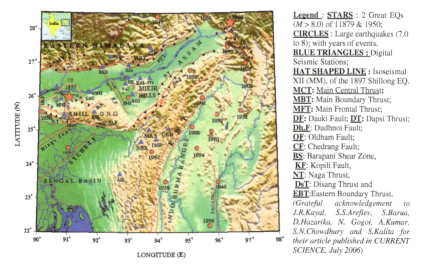

Fig. 1 The N-E Region: Major EQs and Tectonic Features

WHAT MAKES NER SO VULNERABLE TO EARTHQUAKES (EQS)

The reason for such high seismic vulnerability lies in the turbulent geological past and the continuing tectonic plate movements in the Himalayan belt. A brief overview of the geophysical history of this broad region will help identify the main factors that have led to such high incidence of natural disasters in the NER. That will also give us the needed insight that would help in devising effective remedial measures—both preventive and curative—for saving lives and damage to property.

Geological Past of India's NER

Strange as it may appear, practically the entire NER was a seabed around 65 million years back and remained so for about 25–30 million yrs. Then, the gradual drift of the Indian Plate, and its collision against and below the Eurasian Plate in the north and the Burmese Plate in the east (as shown below) resulted in upward thrusts giving birth to the Himalayan range in the North and the Axial Ridge along the Indo-Myanmar border in the East (Fig. 2).

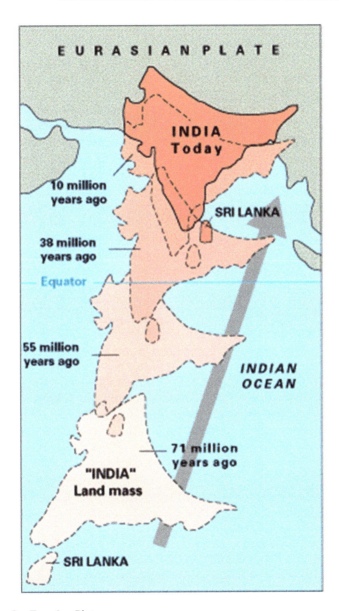

Fig. 2 Eurasian Plate

As the Himalayan range continued to rise, lots of sediments started flowing downhill south-westwards and getting deposited in the lower reaches raising the ground level and resulting in delta formation. The sea started receding southwards and the rivers, flowing down with increasing velocity, started lengthening their courses in N-S and later in E-W directions. This deltaic environment persisted for around 10 million years resulting in heavy sedimental deposits in what is now known as the Upper Brahmaputra Valley. Around 15 million years back, some seismic activity caused re-incursion of sea for a couple of million years but with further rise of the Himalayan range and emergence of Shillong Plateau and the Mikir Plateau towards the south, the sea receded fully into the Bay of Bengal and the fertile Assam Valley was formed. In course of time, the plant, animal and mineral material that had got buried under the heavy sedimental deposits are transformed into the rich coal, oil and mineral deposits of the present times. The rivers earlier draining directly into the sea waters became tributaries of the Siang which emerged as the main channel, now called Brahmaputra, which took its shape according to the contours of the valley.

Another consequence of the rise of the mighty Himalayan range has been the heavy precipitation that occurs when the moisture-laden monsoon winds from the Bay of Bengal and beyond strike against the Patkoi hills and the Himalayan system. Combination of all these geophysical and geoclimatic upheavals over millions of years has shaped the area that is now our fascinating NER and its neighbourhood, with waves of high hills (<50 to >5000 m above MSL) spread over 70% of the 2.62 lakh sq. km of its geographical area (that is slightly below 8% of the country's area) rising from the evergreen plains of the mighty Brahmaputra and reaching out to the crimson sky of the rising sun. Rich in natural resources, fertile valley lands, luxuriant forests and substantial mineral wealth, the NE Region has been the meeting point of great civilizations, inhabited by 4.56 crores of people (3.76% of the country's population as in 2011) with unmatched cultural vivacity.

The continuing north-eastward thrust of the Indian Plate below and against the two neighbouring plates has made the Himalayan belt and NER so highly prone to natural disasters and will continue to do for millennia to come. Fortunately for us, the deposit of soft belt of sediment along the Indo-Myanmar border on the Nagaland–Manipur side functions somewhat as shock absorber that lessens the damage potential of the intense seismic activity and numerous earthquakes that keep on occurring on the Myanmar side. While precise estimation of the rate of

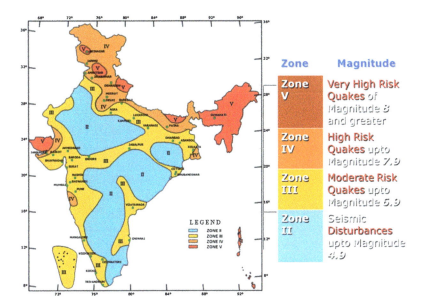

Fig. 3 The seismic vulnerability of NER

this movement is not very clear, differing as it is from source to source, Wikipedia describes it in the following terms:

> ...The Indian plate is still moving at 67 mm per year,... and over the next 10 million years it will travel about 1,500 km into Asia. About 20 mm per year of the India-Asia convergence is absorbed by thrusting along the Himalaya southern front. This leads to the Himalayas rising by about 5 mm per year, making them geologically active....

The seismic vulnerability of NER as a consequence of this background may be seen in the map shown in Fig. 3.

Natural Disasters in NER

The main natural disasters that NER has been facing include *high frequency of earthquake, annual occurrence of floods, cyclonic disturbances* and *cloudbursts* causing *flash floods, riverine erosions* and very large number of *landslides* including those triggered by earthquakes, floods, cloudbursts, etc. Let us now identify the specific factors that result in loss of life and destruction of material assets and try to find or devise

30 P. P. SHRIVASTAV

effective preventive and curative remedies for each of them, including credible Early Warning Systems (EWS) for such acts of nature prevention of which is beyond the frontiers of our present-day knowledge and capability.

EARTHQUAKES

The last decade (2000–2009) witnessed 290 EQs and the resulting landslides, tsunamis and floods claimed *453,553 lives* and caused *damage worth US $183,425 million*, far more than any other type of natural disaster (World Disasters Report 2010). Interestingly, earthquake by itself hardly kills directly. Death, injury and damage result mainly from collapse of buildings/structures by tremors, by induced landslides and flooding arising from bursts of dams or blockage of waterways. These killer-factors can be prevented and neutralized through foresight and advance action and casualties minimized. This process was initiated with the enactment of the DM Act, setting up DMAs, implementation of its provisions and training of the people. Now it is the turn of the people to come forward willingly to cooperate, get trained and prepared to save life and property.

The Great Shillong Earthquake of 1897, one of the world's largest earthquakes, devastated the thriving Shillong city, razing to the ground all 'modern' (pucca) structures. But the traditional light timber/bamboo structures could withstand the shaking and remained intact (Ref: Oldham: 'Great Earthquake of 1897'), a tribute to the traditional wisdom. The next great earthquake (M > 8) occurred in 1950, just across our border with Tibet in the north-east. It caused landslides which temporarily blocked the big Lohit River flowing down the hills of Arunachal Pradesh into Assam. Pressure of the fast accumulating waters finally broke through, and the resultant deluge swept away the thriving market town of Sadiya. Dibrugarh city further downstream was also seriously damaged by the maddening force of the rushing waters.

The basic lesson learnt is that earthquakes hardly kill people directly; casualties result mainly from collapse of man-made buildings and structures. Havoc is thus the heaviest in newer settlements where traditional wisdom has been ignored and overcrowding at a fast pace has resulted in jungles of non-engineered constructions, obstruction of natural drainage, in absence and/or violation of town planning and building norms. Such unplanned, hasty and largely illegal developments suffer very high damage and casualties in disasters.

EARLY WARNING OF EARTHQUAKES (EWS-EQ)

In the Haicheng EQ (M-7.3) in China, the authorities, after a few initial cautionary warnings, noticed widespread unusual animal behaviour and finally on the morning of the 4 February 1975 ordered the local people (about a million) to move away fast. The EQ struck in the evening and casualties were limited to 2041 dead and >27,000 injured. Strangely however, it has not yet led to development of any reliable short-term EWS-EQ of practical utility. EQ-warnings based on differential velocity of P and S waves gives too short a notice (of seconds generally) to be of practical value in Indian conditions where distances are vast and fast access to the common man is lacking. It needs hours for officials to disseminate warning and for the people to move to save themselves.

Fortunately, an encouraging development of late is that Dr Arun Bapat, our distinguished Pune-based Indian Seismologist and a keen researcher, has achieved a remarkable breakthrough as a result of serious studies and critical analysis of the pattern of occurrence of certain phenomena that precede big EQs in a systematic manner. These include hydrological, thermal, seismo-electro-magnetic and also animal and even human precursors, as was seen in a number of prominent recent major EQs including those at **Latur M-6.2** (1993), **Bhuj M-7.9** (2001), tsunami affecting **Andaman M-9.3** (2004) and **Kashmir M-7.9** (2005) in India. The factual data stands authenticated by the routinely maintained official records. 'Articles by Dr Arun Bapat' clicked on Google Search will give access to this fascinating novel concept for a new EWS-EQ with tremendous potential for saving lives.

The CSIR-NEIST (N-E Institute of Science & Technology) has collected a lot of data on EQs with special focus on those that occur in NER under imaginative guidance of Dr Saurabh Baruah. It is now ready for critical analysis to identify precursors that may be useful in forecasting major EQs.

Both these issues have been brought to the notice of National Disaster Management Authority (NDMA) and let us hope something positive emerges from there soon.

LANDSLIDES (LS)

Landslides, triggered by earthquakes, slope failures, heavy precipitation or other causes, are in fact, real big killers of men and animals and destroyers of property. According to NDMA, more than 15% of India's total land territory, in all 22 states and 2 union territories, are affected by LSs of varying types, magnitudes and frequency. A study (2009) by Dr Surya Parkash (Asst Professor in NIDM and acknowledged Geohazards Expert) revealed a total of 10,930 LS sites in the country, bulk of them being in the Western and Eastern sectors (NER) of the Himalayan belt.

Treatment of LS so far has left much to be desired. Effective prevention and mitigation call for a well-coordinated multidisciplinary and multi-agency approach. Inter-agency coordination not being a strong point in the colonial system that we have inherited, treatment of landslide menace has been highly fragmented and hence grossly ineffective, leading to heavy casualties which could have been prevented. This state of affairs led us to organize a series of three multi-stakeholder sessions of *Indian Landslide Congress* (the first at Lucknow in 2010; the second at Guwahati on 15–16 September 2011 and the third at Port Blair in 2013). The initiative for these came from Dr R. K. Avasthy, a retired but highly committed and motivated GSI officer who had done commendable work in NER. Dr Manisha Ramesh HOD Amrita University, Kollam (Kerala) was invited to the Guwahati Conference to give a presentation on their EWS for Landslides (EWS-LS) that had actually saved lives in a risky LS site in the Western Ghats (as learnt in our earlier meeting with Prof P. Venkat Rangan, Vice Chancellor and Dr Manisha Ramesh in the office of Prof Vinod K. Menon, the then Member of NDMA). The outcome was the installation of an EWS-LS at the highly vulnerable Chandmari locality of Gangtok (Sikkim) which was made operational in 2015.

The latest on this subject is the recent mega-initiative of the internationally acclaimed expert on natural disasters, Dr R. K. Bhandari. He organized (under the aegis of *Indian National Academy of Engineering*) a series of interactive meetings involving top-level experts from all disciplines and agencies concerned with LS which culminated in two-day long Roundtable Meetings on *Engineering Interventions in Landslide Risk Reduction* at CSIR-CRRI, New Delhi on the 11 May and 4 November 2015. There was unanimity in the assertion of the top national-level participants that:

NATURAL DISASTERS IN THE NORTH-EAST: CRUCIAL ROLE ... 33

'....unlike earthquakes and tsunamis, *landslides are predictable, preventable and controllable*'; and that 'there is a need to strengthen, amplify and accelerate the pace of scientific studies on occurrence, distribution, processes, frequency and impact of landslides and link these studies organically and intimately with landslide management in the multi-hazard context'. The 18 unanimous recommendations, duly signed by the participants, 'are all aimed at

(a) establishing fully operational, well-coordinated, result-oriented and accountable institutional mechanisms;
(b) inculcating scientific temper in landslide investigation and management; and
(c) underscoring the importance of multi-disciplinary team-building, education and training to be able to measure up to the growing challenges posed by the relentless pressure of population, urbanization, unregulated constructions and extreme weather events'.

These have been forwarded to NDMA and Government of India for consideration, adoption and speedy implementation by the dealing ministries, institutions and agencies with the hope and expectation that policy makers will take a fresh, holistic look at the landslide risk reduction strategies, mainstream landslide risk reduction with the development planning, and proactively convert every disaster into a possible opportunity to test policies, learn lessons and effectively use those lessons to build a strong culture of safety against landslides.

Let us hope that MHA, NDMA, SDMAs, DDMAs and all concerned departments/agencies/institutions will give due regard to the unanimous recommendations of top national-level scientists of the country and that landslides are given the holistic multidisciplinary attention and we look forward to sharp decline in the casualties and damage caused by Landslides.

Floods

Cherrapunji (Sohra) in Meghalaya holds two Guinness World Records, one for the maximum rainfall in one year (of $26{,}471$ mm $= 1042.2$ inches between August 1860 and July 1861), and the other for the maximum rainfall in a single month (of 9300 mm $= 370$ inches in July 1861). Now this distinction has shifted to the nearby Mawsynram, where average rainfall is 11,873 mm (467.4 inches). No wonder, the entire region gets heavy monsoon downpour mostly between May and June to September, and that flood is one natural disaster that recurs with clockwork regularity in NER. The larger river valleys of Brahmaputra and Barak in Assam and their tributaries take a heavy toll of life, crops, private property and public assets, besides causing large-scale displacement of people, year after year. What is worse is the accumulation of heavy silt load that these rivers carry down from the young Himalayan hills to the plains below. Mounds get formed in the river beds, and these cause the rivers to change course leading to breaches in embankments, causing flooding, prolonged inundations and to serious riverine erosions in the thickly populated plains. Raising the heights of embankments and protective structures helps only for a short while. Real solution would perhaps lie in effective desilting by holding the waters en route in depressions in the hills/foothills/plains at appropriate locations, where damage to ecology and environment would be the least. Strings of such multi-purpose dams could substantially reduce the silt load in the plains. Besides, these water bodies could be used for fishery, irrigation, generation of hydropower, etc. The idea of inter-linking of the country's river systems that would enable utilization of surplus fresh potable water in one system to the others facing deficit, instead of letting the surplus to continue flooding and merging into the sea unutilized. Arunachal Pradesh has on hold many such dams that could contribute hydropower (in the prevailing power-deficit scenario) and to reduce floods to an extent, but doubts are being expressed that these might inundate the plains of Assam. This calls for high-level technical review and mass awareness of unbiased facts to allay local apprehensions being created and/or exploited by politicians to mobilize a following for themselves.

FLOOD FORECASTING AND ADVANCE WARNING

FLEWS (Flood Early Warning System) was conceived and initiated by us in May 09 on a pilot basis in Lakhimpur District and its neighbourhood, the worst flood-affected area of Assam, by synergising the expertise and experience of IMD, CWC, ISRO and Assam SDMA, with NESAC (a joint venture of ISRO and NEC) playing the coordinating and operational role. Motivation for free sharing of relevant information, data, expertise, knowledge, skill and unreserved cooperation and coordination as members of one team working together for a laudable cause to save life could be achieved by having meetings with the heads of these agencies and key persons at personal level. Dr Radhakrishnan, the then Chairman of ISRO, readily made available to NESAC on real-time basis the relevant data collected by our satellites circling our skies. He also sanctioned a dedicated multidisciplinary team of scientists to FLEWS. That enabled credible warning messages with maps of precise villages/areas likely to be inundated, to be given 12 to 24 hours in advance of the actual inundation. That gives adequate time to the local officials and the potential victims to save themselves, their cattle and belongings and move to safer sites. FLEWS proved very successful and was extended, at Assam SDMA's request, to give coverage to the entire Brahmaputra and Barak Valleys.

Combined with other measures, the results have been very encouraging, with far less casualties despite higher rainfall, as would be seen from the following chart taken from a report of the Government of Assam.

Year	Rainfall from March to September (mm)	Population affected (lakh)	Human lives lost	Crop area affected (lakh hectares)
2007	1999.1	108.67	134	6.74
2010	2067.4	22.41	12	1.16

FLEWS was discussed in two meetings of experts convened by NDMA, and the view that emerged was that it would be useful to emulate this initiative and set up appropriate early-warning mechanism in other flood-affected areas of the country.

CYCLONIC DISTURBANCES

Mizoram, Tripura and Western Meghalaya areas are prone to cyclonic disturbances. Cyclone warnings are, therefore, being sent as add-on to FLEWS warning messages.

RIVERINE EROSION

Riverine erosion caused by the fast-flowing rivers is the most serious problem in the valley areas of Assam. According to the state authorities, the loss of land on this account has been of the order of 8000 ha/yr. The most serious casualty has been *Majuli*—the world's largest riverine island, as shown in the following report of CISR-NEIST (NE Institute of Science and Technology, Jorhat).

36 P. P. SHRIVASTAV

Year	Land area (sq. km)	Remarks (sq. km)	Average annual rate of erosion (sq. km)
1920	735.01	Reduced by 509.99[a]	–
1975	613.63	Reduced by 121.39	2.21
1998	577.65	Reduced by 35.96	1.56

[a]As per historical records, originally the area of Majuli Island was 1245 sq. km

Efforts are being made by the State Government to protect the banks and preserve this heritage site. Let us hope that the efforts bear fruit before it is too late.

LEGAL AND ADMINISTRATIVE MACHINERY FOR DISASTER MANAGEMENT

Enactment of the Disaster Management Act 2005 and setting up of the NDMA chaired by PM, State DMAs under CMs and District DMAs under DMs, have brought about a paradigm shift in Disaster Management (DM) strategy. The colonial-time emphasis on post-disaster Relief & Rehabilitation has now been replaced by *Pre-Disaster Risk Reduction and Mitigation strategy*. Each Ministry/Department of the Central Government is now mandated by the DM Act to *prepare disaster management plans* (sec 37), and the State Governments have to integrate into ... development plans and projects, the measures for prevention of disaster and mitigation; and to allocate funds for the purpose. The ground realities and the ways these are being implemented in practice vary from place to place and are well known to the local people. Without spending more time and space on these, let us move on to the three crucial issues, namely (i) what needs to be done; (ii) how best to do that, and (iii) who are the best equipped to do it.

Action Points of the Disaster Prevention and Mitigation Strategy:

(i) What needs to be done; and
(ii) How best to do that

What needs to be done is known to the experts and the SDMA/ DDMAs and to some extent to the better-informed sections of society: others will also not find it difficult to understand if someone takes interest in motivating them and making them aware of the natural disasters to which their area is prone and how best to protect and save themselves. How the different natural disasters arise and the way they endanger life and property has been presented earlier in a manner that the way to prevention and mitigation becomes self-evident. For example, traditional wisdom of using light resilient locally available building material (like timber or bamboo) is well known to all. Those who want 'pucca' masonry/stone/RCC structures will get easily convinced if told by someone whom they trust that EQ-vulnerability of the area demands that they go in for EQ-resilient structures only. They will then not mind the extra cost involved for the safety of their family. Likewise, advice to keep emergency kits ready and handy at home all the time will be thankfully accepted, but gratitude will follow if they are advised about the basics of EQ-safety in households (like removing heavy boxes stored on top of cupboards near the bed; the duck-hold-crouch drill if caught indoors when EQ-strikes, etc.).

Likewise, everyone will appreciate what Dr R. K. Bhandari, our top-level scientist and humane specialist in disaster management seeks to convey on landslides, when he writes in his eloquent style in one of his published articles:

> Today, we have the knowledge, tools and experience we need to predict and avert most, if not all, landslides. By tapping the phenomenal power of geotechnology, instrumentation, remote sensing, integrated GPS and information communication systems, we can monitor unstable areas in real time even during unfavourable weather conditions. It is time therefore to launch selected mission-mode projects to initially cover timely prediction of (a) possible reactivation of major old, dormant and seasonal landslides, (b) landslides and floods due to bursting of glacial lakes, (c) flash floods due to bursting of landslide dams, (d) first-time landslides in urban and strategically important areas falling in the zone of exceptional landslide hazard and (e) rockfalls.

The case of flood disaster is no different in principle. Here, it has to be recognized that FLEWS though very successful is only the first step of giving warning in time for saving life, and not the total strategy for mitigation of this Disaster. The intensity of floods will continue to increase unless an effective solution is found to the problem of continued heavy silting of river beds, as mentioned earlier. This task has to be taken on by the governments. The task for the community is to build pressure on the governments to commence action, commission experts to make comprehensive plans at the national, state and river basin levels and execute them in time-bound manner. A national plan for linking of the river systems was envisaged earlier. That could conserve clean potable water and divert surplus from one system to make up the deficit in others. Of course, it has various environmental and other implications, which have to be gone into in-depth. Informed public pressure has to be built up to take this concept to its logical conclusion. Meanwhile desilting has to continue, and accumulation of fresh silt in river beds has to be effectively prevented otherwise the problem will continue to multiply.

What has to be done in preventing loss of life and property from natural disasters and how best it can be done is thus now well understood: the crucial issue that still remains is who is best equipped to take on the task. Let us consider this issue now.

Action Points of the Disaster Prevention and Mitigation Strategy:

(iii) Who are the best equipped to do it.

Life is invariably the one thing that is valued the most by all and the next in priority are one's material possessions. Both are threatened by natural disasters, and hence, the responsibility for saving their own life and property lies primarily with the people themselves. Unfortunately, the common man, though classed as '*Master*' in our democratic set-up, remains ignorant of modern DM techniques, as also reluctant to open up before the officials ('*Servant*' of the people) who have the knowledge, skill, resources, authority and responsibility of DM. This considerable psychological distance between the two, inherited from the long colonial past, will take its own time to disappear. It is, therefore, imperative to locate an intermediary close to both 'Master' and 'Servant' (Community and the government) that can seamlessly bridge the gap between the two.

The academic Community would appear to be the ideal and perhaps the only choice available. It is ideally placed to provide seamless organic linkage with the government on the one hand and on the other with the Community, rural and urban, and with all the different sections of society across the country. All institutions, universities, colleges, schools are centres of learning with ever-advancing frontiers of knowledge and skills relevant to DM. The faculty, scholars and teachers command respect and confidence of all and can easily liaise symbiotically with DMAs at all levels. Students n2u (nursery to university) coming from all sections are easily accepted by the entire community, being their own. With their youthful energy and enthusiasm, they would love to actively disseminate DM-awareness/knowledge/skills among the people, who will love to listen to their own children, relish being trained by them and enjoy the mock-drills with them. A person may not be willing to spend extra money to build EQ-resilient house, but if his grandson and his friends convince him that their life would otherwise be in danger, he will have no heart to be miserly. Can we imagine a more effective trainer than our own near-and-dear ones, whom we love and treasure more than our own lives?

Against this backdrop, the Academic Community may like to take on this role and:

(i) Volunteer to disseminate DM-awareness and mobilize/prepare the community to face the challenge of natural disasters with fortitude through knowledge, skill and regular practice;

(ii) Liaise symbiotically with different organs of the State/District DMA-machinery, especially NDRF to gain all general and technical know-how needed for action on the point (i) above;

(iii) Liaise likewise with local self-government institutions like Gram Sabhas/Panchayats in villages and Municipalities, Resident Welfare Associations, Traders Associations, Youth and Women's Clubs, etc., to motivate and involve them in DM-awareness, capacity building and prepare them to face disasters successfully;

(iv) Draw up area-wise plans to cover all educational institutions right down to primary schools for spreading DM-awareness and training programmes and making these activities as fascinating and playful as possible for the different age groups and motivate them to adopt DM as a benign voluntary community service to their people;

(v) Make the fullest use of EDUSAT for wide dissemination of DM-related programmes in NER. Studio facilities are available at NESAC (North-Eastern Space Applications Centre) at Umiam (Barapani) near Shillong, for recording imaginative television programmes on DM for NER;

(vi) Inspire the University faculty to take the initiative to organize motivation-cum-DM-awareness/training for the teachers in colleges and schools in their respective areas, and function as voluntary resource persons for their students and make this movement self-propelling;

(vii) Motivate and encourage day-scholars to organize voluntary groups in their residential localities who may be enabled to prepare local safety plans at the household, neighbourhood, locality levels and integrate them with village/town DM-Plans in coordination with the relevant DDMAs. Hostellers may be encouraged to take up similar activities in their hometowns/villages during vacations in tune with the local DM authorities;

(viii) Inspire the Lady faculty members and lady students to organize motivation/awareness/training programmes among women organizations/club/groups in their respective areas giving special attention to safety in households and the neighbourhood. Ladies are likely to take keen interest in DM since safety and welfare of their family members is always their prime concern;

(ix) Set up Apex-level Committees at University level to (a) monitor and guide the DM-related activities of the academic Community in their areas at state/district/institution Levels as may be appropriate. (b) Remain in close touch with the relevant S/D DMAs and inter alia press for their preparation/periodic updating of area DM plans, preparation of microzonation maps for vulnerable area/localities and (c) integrate DM in their social/cultural programmes; and

(x) Create strong public opinion against non-compliance of law/regulations. Town planning/zoning/building regulations that may lead to enhanced vulnerability and higher casualties.

With the degree of dedication and efficiency witnessed at the Imphal Conference (9–11 April 2016), one looks forward to the day when every person living in Zones 5–4 in the north-east will be aware of his duties and social obligation to follow all laws, rules and regulations to minimize risk from disasters and will be aware of what to do and how best to save himself, his family and neighbours in times of crises. In this, the example of the Japanese people is worthy of emulation.

Regional Climate Changes Over Northeast India: Present and Future

A. P. Dimri, D. Kumar and M. Srivastava

INTRODUCTION

The easternmost part of the India is considered as Northeast (NE) India. It consists of the 'seven sister states' (Arunachal Pradesh, Assam, Meghalaya, Mizoram, Manipur, Nagaland and Tripura) along with Sikkim under the Indian province. The region shares its international boundary with China (in north), Myanmar (east), Bangladesh (south-west) and Bhutan (north-west). The region can be categorized into different physiographic regions across the states such as the Eastern Himalaya, north-eastern hills and the plain areas including Brahmaputra and Barak Valley plains. The climate of the region is characterized by subtropical type, which is influenced by the different relief features as well as the monsoonal currents during south-west as well as the north-east monsoon seasons.

A. P. Dimri (✉) · D. Kumar
School of Environmental Sciences, Jawaharlal Nehru
University, New Delhi, India

M. Srivastava
Department of Environmental Sciences, Central University
of Rajasthan, Rajasthan, India

© The Author(s) 2018   41
A. Singh et al. (eds.), *Development and Disaster Management*,
https://doi.org/10.1007/978-981-10-8485-0_3

The climate of the region has been characterized as subtropical humid climate as per the Koppen's classification of climate (Oliver and Wilson 1987). Due to its distinct geographical setting, the climate is influenced by the mighty Himalayas in the north, plateau of the Meghalaya in the south and hilly regions of the Manipur, Mizoram and Nagaland in the east. The climate is also influenced due to the vicinity of the Bay of Bengal. Overall, the distinct features in the topography as well as unique geographical setting result into a complex interaction of the land, sea and atmosphere, which results into peculiar climatic traits over the area. These special traits are also due to the presence of special features such as orography, the alternating pressure cells of NE India and that of the Bay of Bengal (BoB), the maritime tropical air masses from the BoB followed by South Indian Ocean, the wandering periodic western disturbances and the localized mountain and valley winds (Barthakur 2004). One of the wettest places over the earth lies in this region; with these particular traits of climate and topography, a distribution of wide variety of flora and fauna is found over this region. Many of the endemic species of plants, animals and birds are found in the North-Eastern Region leading to a very distinct floral and faunal biodiversity. A number of forest reserves and national parks are situated over the NE India region, which consists of most of the types of forest cover found in India ranging from tropical moist deciduous forest to tropical semi- as well as evergreen forest along with alpine forest in the higher elevation regions. The economy of the region is mostly agrarian while being non-settled and localized, and preferably Jhum cultivation (with slash and burn approach) is practised. The region is richest in terms of the freshwater resource, with a number of rivers, lakes, streams and other natural reservoirs of freshwater. The inaccessible terrain and some other local logistic problems have made it difficult for the industrialization to take place.

The states situated in the higher elevations (e.g. Arunachal Pradesh and Sikkim) represent the alpine type of climate with typically mild summers and cold and snowy winter seasons (Dash et al. 2007). Figure 1 shows the distribution of the topography over the study region. Most of the area across the region is characterized by moderate elevation ranging up to 1000–1500 or below, except some of the higher reaches of Himalayan region in the provincial area of Arunachal Pradesh and Sikkim.

The NE Indian region is not only important from the point of view of its location and importance in the international strategy, but also abode

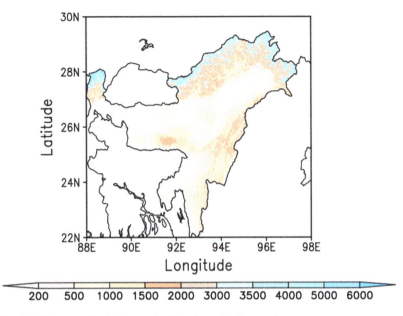

Fig. 1 Topography (m) over the Northeast Indian region

of diverse cultural and ethnic groups. It is noteworthy that the ecosystem services provided by the forest and other components in this region are also of prime importance. This mountain ecosystem is quite sensitive to the changes in the climatic pattern due to its geo-ecological fragile nature, its location in the Himalayan landscape, transboundary hydrological regime and its inherent socio-economic instabilities; thus, it becomes imperative to investigate and understand the behaviour of the climatic regime in the past and also to produce accurate climate projections for the future carbon-intensive scenarios which are coherent to the development pace of the current human civilization. This will not only help in appraising the past changes over the study area, but also help design and implement appropriate adaptation and mitigation measures.

It is now well established that the warming of the earth system as a response to the gradual rise in the concentration of the greenhouse gases (GHG) emission through anthropogenic activities is unequivocal (IPCC 2007). This warming of the earth system is leading to the generation of some of the abrupt weather and climate systems across the globe including erratic patterns of rain, temperature events and increase in the frequency of the extreme events. Following the increase in the global mean surface temperatures, it was suggested that the increased moisture content in response to global warming (Trenberth et al. 2007) might lead to increasing global mean precipitation. However, no clear or significant trend has been observed in the all India annual rainfall or monsoonal rainfall pattern during last 100 years, but significant long-term rainfall changes have been reported at different spatio-temporal scales (Dash et al. 2007; Kumar et al. 2010; Guhathakurta et al. 2011). At the same time, a study carried out by Dash et al. (2007) suggests that for the period 1871–2002, a decreasing trend in monsoonal rainfall and increasing trend in the pre-monsoon and post-monsoon seasons have occurred. In the same study, a decreasing trend over Northeast India, east Madhya Pradesh and adjoining areas along with the parts of Gujarat and Kerala has been reported. Further, in a study based on the period 1871–2010, Kulkarni (2012) observed a falling trend of rainfall during the last three decades of the twentieth century. Besides this, while talking particularly of the Northeast India and west coast region of India, Goswami et al. (2006) have reported that Northeast India is a region of the high mean and high variability of the rainfall events attributed to the strong influence of the local orography on the rainfall over both the regions. The region has experienced several climatic and hydrological extremes in recent times which include the changes in the thermal environment, series of extreme precipitation events and spells of severe droughts. A study based on station data over Agartala suggests that there has been decrease in the total pan evapotranspiration and increase in the relative humidity (Jhajharia et al. 2007) with simultaneous rising trend of mean air temperature at the rate of 0.32 °C per decade. Same study also claims that for the period (1969–2000), a declining trend of the annual evaporation has also been observed. Moreover, same authors have concluded a declining trend of the total annual rainfall over the period (1953–2000) with the magnitude of the order of 2.4 mm/year. Another study by Jhajharia et al. (2012) over some stations in Assam reports a significant decreasing trend of precipitation during the June and July months. Another interesting study carried out by Rupa Kumar et al. (1992) suggests that the north-east peninsula, Northeast India and north-west peninsular region experienced a declining trend (-6 to -8% of the normal per 100 years) for the monsoonal precipitation. A significant decreasing trend of precipitation with a rate of 3.2 mm/day was observed over the central Northeast India during the period 1889–2008 (Subash et al. 2011). Studies have revealed that, in general, the frequency of the more intense rainfall over many parts of Asia has increased with a simultaneous decrease in the total number of rainy days and total annual precipitation across the region (Dash et al. 2007; Kumar and Jain 2011).

For the period 1901–1982, an increasing trend in the mean annual temperature over many parts of India including the North-Eastern Region was observed by Hingane et al. (1985). Again, a change in the diurnal temperature range of the maximum and minimum air temperature over several places in India was reported by Rupa Kumar et al. (1994), which was attributed to the increasing trends of the maximum temperature with no changes in the trend of minimum temperature pattern. In addition to these, some of the modelling efforts have also been carried out in order to quantify the trends of temperature and precipitation over the Northeast Indian region. Dash et al. (2012) used a regional climate model (RegCM3) to show that the model was able to capture the trends of annual mean temperature over the Northeast Indian region correctly. While simulating the precipitation over the region, the regional model showed an inherent wet bias, which led to the overestimation of the precipitation over the study region. In the same study, the authors also projected a consistent rise in the annual mean temperature, and precipitation with more frequent occurence of warm events as compared to the cold events. However, a number of studies have been carried out in order to assess the present climate changes over the Northeast India but very few efforts have been undertaken in order to incorporate the information available from the climate models. Again, in order to plan for the appropriate adaptation and mitigation measures under the changing climate scenario, generating accurate projection of the climate system is necessary following the different available climate scenarios.

To overcome this gap, in the current study we have tried to investigate and summarize the trends of the climate in the present as well as the future projected climate. The regional climate models have been found very instrumental in providing the regional-scale information pertaining to climate change due to their high resolution and representation of the local scale features such as orography and clouds in simulating the climate over the limited area.

MATERIALS AND METHODS

The regional climate model data from the simulation over South Asia Domain at the spatial resolution of 50 km have been obtained for the regional model (REMO) (Jacob 2001) driven by MPI-ESM-LR global model under the COordinated Regional Downscaling Experiment (CORDEX) framework. The CORDEX programme is supported by World Climate Research Programme (WCRP) and aims at exploring and generating the regional climate information worldwide through collaborative efforts. The meteorological variables viz. 2 m air temperature (near-surface air temperature) and precipitation have been analysed for the present historical (1970–2005) climate. Moreover, the changes in the temperature and precipitation patterns over the study area have been calculated for three different emission scenarios called as representative concentration pathways (RCP) (Moss et al. 2010), namely RCP2.6, RCP4.5 and RCP8.5 as prescribed under the IPCC AR5. These RCPs are characterized by the changes in the radiative forcing of the atmosphere to the pre-industrial values by the amount (W/m^2) prefixed to each RCP. In order to calculate the changes in the future, the data of temperature and precipitation have been used for each RCP for the period of 2006–2100. In order to quantify the changes in the immediate future as well as the late future, the period 2006–2100 has been time sliced into two different periods such as near future (2020–2049) and far future (2070–2099). The mean climatology of temperature and precipitation and changes in the temperature climatology under the different carbon-intensive scenarios for different seasons, namely DJF (December-January-February), MAM (March-April-May), JJAS (June-July-August-September), ON (October-November) has been calculated in order to assess the changes in the climate regime under different seasons for each RCP. The percentage change in the precipitation in the near future as well as for far future has been calculated for each season mentioned above. In addition to this, long-term linear trend of temperature as well as the precipitation has been calculated in order to investigate the rate of change. In order to compare the spatial patterns of model-simulated output, observations at a horizontal resolution of 50 km have been used as Asian Precipitation–Highly-Resolved temperature (APHROTEMP) data set (Yasutomi et al. 2011) and Asian Precipitation–Highly-Resolved Observational Data Integration Towards Evaluation of Water Resources (APHRODITE) (Yatagai et al. 2012) for temperature and precipitation, respectively. The results have been discussed in the following sections.

Results and Discussion

Near-Surface Air Temperature

The climatology of the near-surface air temperature has been computed for each season (DJF, MAM, JJAS and ON) from the observed as well as the model data sets for the period 1970–2005. Figure 2 shows the climatology of the observed near-surface air temperature (2 m air temperature) for different seasons from APHROTEMP data set. The magnitude of temperature climatology has been found to be higher in the JJAS month as compared to other months. The contribution to the higher magnitude of JJAS temperature might be due to the warmer climatology in the month of June or the late onset of the monsoon over the region.

Figure 3 shows the climatology of the near-surface air temperature over the study region from the REMO model experiment. The model is able to capture the spatial patterns of the temperature climatology. However, cold bias in the temperature could be noticed in all the seasons, as the magnitude of the temperature is comparatively less in the model-simulated outputs.

The trends of the observed near-surface air temperature for the period 1970–2005 are shown in Fig. 4 for different seasons. An increasing trend in the temperature has occurred during the present climate period over major parts of the study region in all seasons except the MAM.

However, an interestingly falling trend of temperature has been noticed over the some of the upper reaches of the Arunachal Pradesh in all the seasons, which might be due to the localized effect of the higher

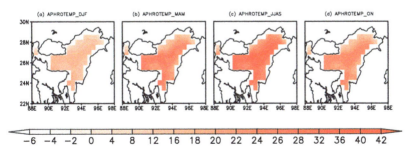

Fig. 2 Observed near-surface air temperature climatology (°C) during 1970–2005 over Northeast Indian region from APHROTEMP for DJF season (**a**), MAM season (**b**), JJAS season (**c**) and ON season (**d**)

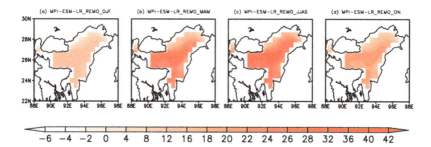

Fig. 3 Near-surface air temperature climatology (°C) during 1970–2005 over Northeast Indian region from REMO regional model driven by MPI-ESM-LR global model under CORDEX South Asia experiment for DJF season (**a**), MAM season (**b**), JJAS season (**c**) and ON season (**d**)

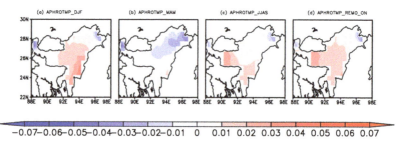

Fig. 4 Trends of observed near-surface air temperature (°C/yr) during 1970–2005 over Northeast Indian region from APHROTEMP for DJF season (**a**), MAM season (**b**), JJAS season (**c**) and ON season (**d**)

elevation orography over the region. The magnitude of warming trend in many parts is as high as 0.03 °C/yr. The decreasing trend in the JJAS season might be attributed to the role of atmospheric moisture during the monsoonal season, which needs to be investigated carefully. In the model outputs (figure not presented), almost all the seasons show rising trend of the temperature of the region, which is consistent with the trends of the observations.

Changes in the temperature climatology for the near future and far future under RCP2.6 are shown in Figs. 5 and 6, respectively. It is projected that in the near future the winter season will experience more warming over the study region as compared to other seasons. The upper

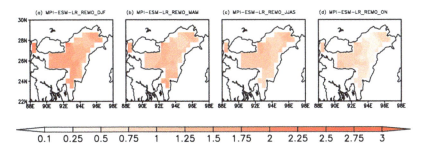

Fig. 5 Changes in near-surface air temperature (°C) calculated as near future (2020–2049) minus present (1970–2005) for RCP2.6 over Northeast Indian region from REMO regional model forced with MPI-ESM-LR global model for DJF season (**a**), MAM season (**b**), JJAS season (**c**) and ON season (**d**)

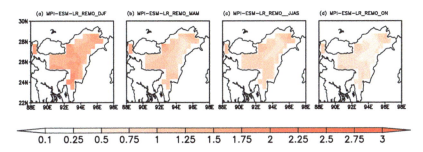

Fig. 6 Same as Fig. 5 but for far future (2070–2099) minus present (1970–2005) under RCP2.6

reaches of Sikkim will experience elevated temperature with the magnitude as high as 3 °C to the average temperature.

The pattern of temperature changes in far future (2070–2099) is similar to the near future changes; however, in the MAM season, more pronounced changes are evident in near future. And the magnitude of changes is relatively smaller for the far future, which is due to the fact that the climate system stabilizes itself as a response to the increased GHG concentration in the latter half of the twenty-first century.

Under RCP4.5, a warmer winter season (DJF) in the near future climate projection is shown in Fig. 7, with the magnitude of temperature changes of the order of 2–2.5 °C on an average.

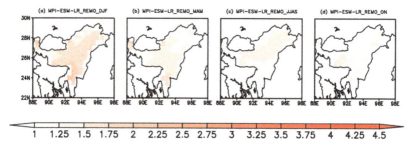

Fig. 7 Changes in near-surface air temperature (°C) calculated as near future (2020–2049) minus present (1970–2005) for RCP4.5 over Northeast Indian region from REMO regional model forced with MPI-ESM-LR global model for DJF (December, January, February) season (**a**), MAM season (**b**), JJAS season (**c**) and ON season (**d**)

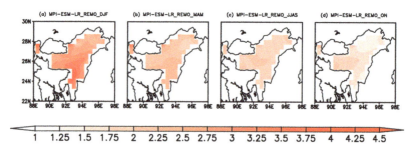

Fig. 8 Same as Fig. 7 but for far future (2070–2099) minus present (1970–2005) under RCP4.5

Figure 8 shows the changes for the far future under RCP8.5, with a considerable change of magnitude by 4–4.5 °C; a warmer winter along with warmer pre-monsoon (MAM) temperature could occur over the study region.

Under RCP8.5, which is the strongest warming scenario with change in radiative forcing over the globe by 8.5 W/m^2, the warming of the North-Eastern Region is highly pronounced in the winter season in the upper as well as lower reaches of the area. The far future shows changes in the temperature climatology, with a magnitude as high as 6–7 °C (Fig. 10).

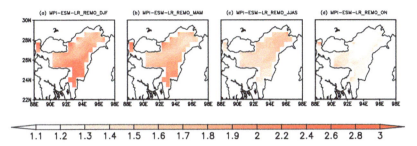

Fig. 9 Changes in near-surface air temperature (°C) calculated as near future (2020–2049) minus present (1970–2005) for RCP8.5 over Northeast Indian region from REMO regional model forced with MPI-ESM-LR global model for DJF (December, January, February) season (**a**), MAM season (**b**), JJAS season (**c**) and ON season (**d**)

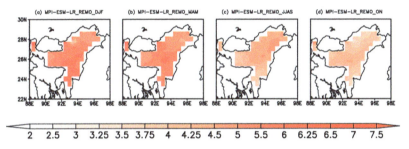

Fig. 10 Same as Fig. 9 but for far future (2070–2099) minus present (1970–2005) under RCP8.5

In order to assess the changes in the temperature over the region, the linear trend is also calculated. The rate of change of temperature (°C/yr) in terms of the linear long-term trend for the period 1970–2099 has been presented in the following figures.

The trends of near-surface air temperature over 1970–2099 for RCP2.6 are shown in Fig. 11. All the regions show a warming trend of the temperature, with more pronounced warming trend in the winter season. The rate of increase in temperature is as high as 0.03 °C/yr at places.

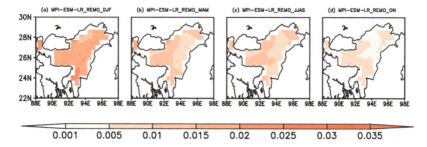

Fig. 11 Trends of near-surface air temperature (°C/yr) over the period 1970–2099 under RCP2.6 over Northeast Indian region as simulated from REMO model under CORDEX South Asia experiment for DJF (**a**), MAM (**b**), JJAS (**c**) and ON seasons (**d**)

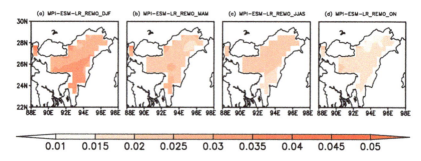

Fig. 12 Same as Fig. 11 but for RCP4.5

The trends for RCP4.5 are shown in Fig. 12; a consistent increase in temperature with magnitude as high as 0.04 °C/yr has been projected in the model-simulated climate. Again, in the strongest warming scenario (i.e. RCP8.5), the warming gets more pronounced in both winter and spring (MAM) seasons (Fig. 13).

The area-averaged time series of the near-surface air temperature is shown in Fig. 14, for the period 1970–2099 for different seasons. The area-averaged temperature profile shows that although the model is able to capture the spatial patterns of the temperature patterns, but a higher variability in year-to-year temperatures has been shown. Moreover, the model shows a similar behaviour in projecting the climate for RCP2.6 and RCP4.5, especially for the 2 m air temperature. During the JJAS

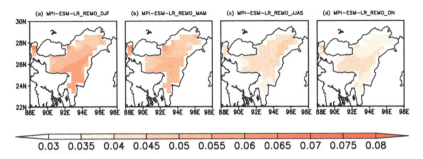

Fig. 13 Same as Fig. 11 but for RCP8.5

season, less year-to-year variability can be observed in the projected climate, which may be attributed to the role of moisture in redistributing the heat over the region.

Precipitation

The climate of the north-east region is influenced by the monsoonal precipitation. Also, the pre-monsoonal showers and convective precipitation are also dominant over many parts across the region. The observed daily seasonal precipitation climatology (mm/day) is shown in Fig. 15. Daily precipitation during the JJAS season has been found to be of the order of 12–15 mm/day on an average.

In the lower parts of Meghalaya, the higher intensity of precipitation is observed during pre-monsoonal and monsoon season, which corroborates the wet characteristic of precipitation over Meghalaya.

The model-simulated climatology for the present climate is shown in Fig. 16. The spatial pattern of precipitation is well captured by the model; however, the wet climatology over the Meghalaya could not be captured in the model simulations (Fig. 16).

The observed trends of precipitation for the period 1970–2005 suggest a decrease in the monsoonal precipitation over most of the regions except the Meghalaya and southern part of Assam and Mizoram. Again, the rainfall over Sikkim peaks has been found to be increasing along with increasing trend of the pre-monsoonal precipitation (MAM season). Also, a mixed trend for the JJAS season can be seen overall across the study region (Fig. 17).

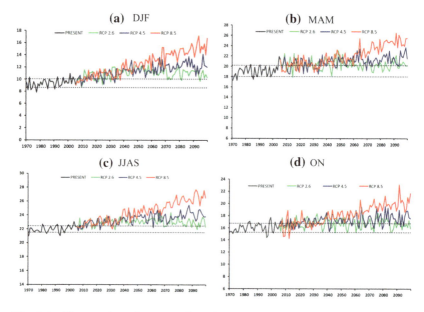

Fig. 14 Time series of near-surface air temperature (°C) averaged over the Northeast Indian region for the period 1970–2099 for DJF (December, January, February) season (**a**), MAM season (**b**), JJAS season (**c**) and ON (October, November) season (**d**) from REMO regional model forced with MPI-ESM-LR global model. The series in grey, green, dark blue and red corresponds to present, RCP2.6, RCP4.5 and RCP8.5 warming scenarios, respectively. The broken lines represent the baseline, as mean ± 1 standard deviation of the historical climate

The overall changes in the precipitation climatology in the future climate have been calculated. Figures 18 and 19 show the percentage changes in the daily seasonal precipitation climatology under the future projected climate in RCP2.6 scenarios for near future and far future, respectively. It is clearly noticed that, during the monsoon season, most parts of the study area shall receive lesser amount of daily precipitation, with a reduction as greater as 5–10%. However, in the southern part of the Assam, 5–10% increase in the daily precipitation can be noticed for the near future climate. Such changes are projected for the least carbon-intensive scenario, which suggest a slight change in the radiative forcing

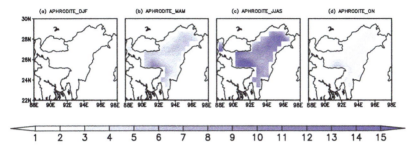

Fig. 15 Observed daily seasonal precipitation (mm/day) during 1970–2005 over Northeast Indian region from APHRODITE for DJF (December, January, February) season (**a**), MAM season (**b**), JJAS season (**c**) and ON season (**d**)

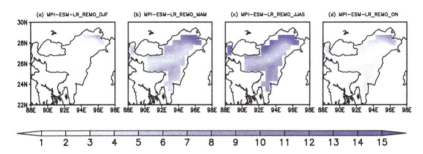

Fig. 16 Daily seasonal precipitation (mm/day) during 1970–2005 over Northeast Indian region from REMO regional model driven from MPI-ESM-LR global model under CORDEX South Asia experiments for DJF (December, January, February) season (**a**), MAM season (**b**), JJAS season (**c**) and ON season (**d**)

(2.6 W/m^2) by the end of 21st century. Again in the near future itself, a substantial increase in the post-monsoonal precipitation and a consistent decrease in the wintertime precipitation have been projected by the model. On the other hand, for far future positive change has been projected for all the seasons except for DJF and JJAS seasons.

The magnitude of changes is negative for the near future under RCP4.5 in monsoonal season (Fig. 20), but an increase in the post-monsoonal precipitation has been projected similar to the RCP2.6 (Fig. 21).

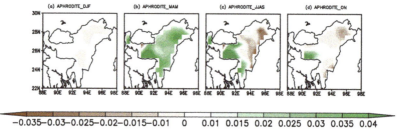

Fig. 17 Observed trends of daily seasonal precipitation (mm/day/yr) during 1970–2005 over Northeast Indian region from APHRODITE for DJF (December, January, February) season (**a**), MAM season (**b**), JJAS season (**c**) and ON season (**d**)

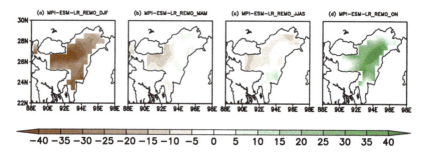

Fig. 18 Changes in daily precipitation calculated as percentage change of near future (2020–2049) minus present (1970–2005) for RCP2.6 over Northeast Indian region from REMO regional model forced with MPI-ESM-LR global model for DJF (December, January, February) season (**a**), MAM season (**b**), JJAS season (**c**) and ON season (**d**)

A stronger deficit in total daily precipitation for the winter time has been projected under both RCP4.5 and RCP8.5 (Figs. 22 and 23). JJAS season is not affected with a decrease in greater magnitude.

The changes become even stronger in the latter half of the twenty-first century under RCP8.5.

The linear trends of the precipitation for the period of 1971–2099 under RCP2.6 show that the daily precipitation may experience a considerable declining trend during winter and monsoon season, while for the other two seasons it projects an increasing trend across the region except few areas.

REGIONAL CLIMATE CHANGES ... 57

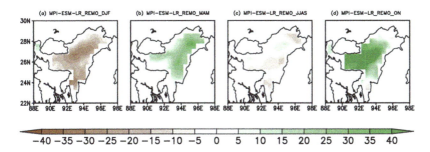

Fig. 19 Same as Fig. 18 but for far future minus present under RCP2.6

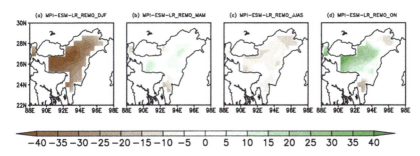

Fig. 20 Changes in daily precipitation calculated as percentage change of near future (2020–2049) minus present (1970–2005) for RCP4.5 over Northeast Indian region from REMO regional model forced with MPI-ESM-LR global model for DJF (December, January, February) season (**a**), MAM season (**b**), JJAS season (**c**) and ON season (**d**)

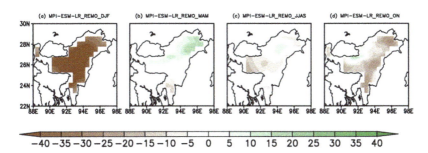

Fig. 21 Same as Fig. 20 but for far future minus present under RCP4.5

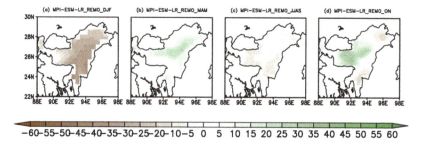

Fig. 22 Changes in daily precipitation calculated as percentage change of near future (2020–2049) minus present (1970–2005) for RCP8.5 over Northeast Indian region from REMO regional model forced with MPI-ESM-LR global model for DJF (December, January, February) season (**a**), MAM season (**b**), JJAS season (**c**) and ON season (**d**)

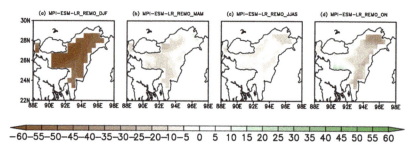

Fig. 23 Same as Fig. 22 but for far future minus present under RCP8.5

Conversely, RCP4.5 scenario projects an overall falling trend in the daily precipitation across the study area in all the seasons and RCP8.5 projects a general mixed decreasing trend (refer to Figs. 24, 25 and 26). The area-averaged time series of the total seasonal precipitation for each season is shown in Fig. 27. A large inter-annual variability in the total seasonal precipitation is observed for all seasons except monsoonal season (JJAS). Moreover, a consistent wet bias has been simulated in the model simulations.

REGIONAL CLIMATE CHANGES ... 59

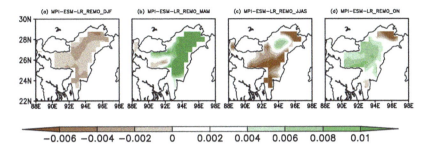

Fig. 24 Trends of daily precipitation (mm/day/yr) for the period 1970–2099 under RCP2.6 over Northeast Indian region from REMO regional model forced with MPI-ESM-LR global model for DJF (December, January, February) season (**a**), MAM season (**b**), JJAS season (**c**) and ON season (**d**)

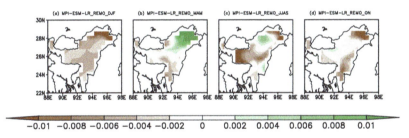

Fig. 25 Same as Fig. 24 but for RCP4.5

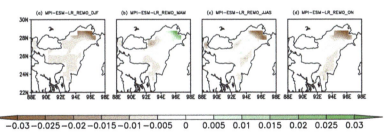

Fig. 26 Same as Fig. 24 but for RCP8.5

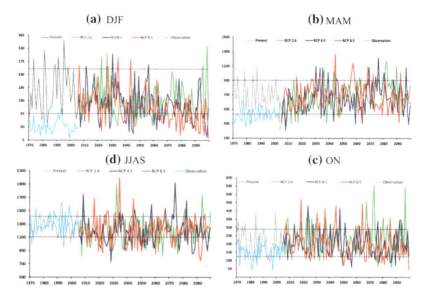

Fig. 27 Time series of total seasonal precipitation (mm) averaged over the Northeast Indian region for the period 1970–2099 for DJF (December, January, February) season (**a**), MAM season (**b**), JJAS season (**c**) and ON season (**d**) from REMO regional model forced with MPI-ESM-LR global model. The series in grey, green, dark blue and red corresponds to present, RCP2.6, RCP4.5 and RCP8.5 warming scenarios, respectively, and the one in cyan represents the observation. The dotted black line shows the baseline, as mean ± 1 standard deviation of the historical climate

Summary and Conclusions

The Northeast Indian region is characterized by its distinct geographical setting giving rise to sublime landscape beauty and also proving shelter to a wide variety of flora and fauna. The complex topographical settings accompanied by the extreme variations in the elevation and aspect give rise to many peculiar climatic regimes from east to west across the region which behave differently. There has been a lot of discussion regarding the changes in the climate over the region; in this respect, an effort has been made to summarize the recent changes in the precipitation and temperature regime across the area followed by the changes in the projected future climate using regional climate models.

The climatology for the precipitation and temperature suggests that the model is able to capture the spatial patterns similar to the observation. In the present climate (1970–2005), a warming trend for all seasons except the MAM (pre-monsoon) season has been observed. In the future warming scenario of RCP2.6, more pronounced changes in the near-surface air temperature would be experienced, especially during the winter season in the immediate future. The warming of DJF (winter) season has also been noticed from RCP4.5 and RCP8.5 simulations. The warming of the winter season is also confirmed in the long-term (1970–2099) future trends. This may lead to a warmer winter, which may have severe implications for the regional-scale climate and associated agricultural practices over the region. In this respect, it would be interesting to assess, whether the warming trend of the DJF season influences the diurnal temperature range over the region.

A general increase in the pre-monsoonal precipitation (MAM) has been observed for the period 1970–2005, which is again followed by the decrease in monsoonal precipitation. This may be attributed to the changes in the pre-monsoonal convective activity, driven by the evaporation and evapotranspiration changes. The contribution of the changes in evaporation and evapotranspiration to the pre-monsoonal precipitation needs to be confirmed. For the future projected climate, it is expected that the daily seasonal precipitation would increase in all the seasons except the winter. For RCP4.5 and RCP8.5, the precipitation is expected to decrease except for the pre-monsoon season. This suggests that for the future warmer climate, a temporal shift in the monsoonal precipitation may occur with lesser rain in monsoon followed by higher precipitation in pre-monsoon as well as post-monsoon seasons. This behaviour needs further investigation, using theoretical mechanism to explain such changes. The RCP8.5, which is the strongest warming scenario, shows positive changes in pre-monsoon as well as post-monsoon seasons in the near future, but it projects a drier precipitation regime in the latter half of the twenty-first century. This fact has been confirmed by the increasing trends of the precipitation in the pre-monsoon as well as the post-monsoon seasons correspondingly.

These changes in precipitation and temperature provide an essence of the changing climate over the region which may have greater influence on the population residing in such difficult circumstances. The changing climatic pattern may affect the agriculture, water resources, flora and fauna of the region. Also, these changes may introduce extreme weather

62   A. P. DIMRI ET AL.

events in future, which need to be assessed in further studies. However, large uncertainty may be associated with the single regional model projections, and for more accurate projection of the future climate, an ensemble projection may be used but single regional model experiment gives a baseline information on the changing climate across the region. Further devoted studies are required to confirm the signatures of the climate change over the north-east region; this will not only help develop the understanding of the possible drivers of the climatic changes in the region but will also help plan and execute appropriate mitigation and adaptation measures.

References

Barthakur, M. (2004). Weather and Climate. In V. P. Singh, et al. (Eds.), *The Brahmaputra Basin Water Resources* (p. 22). Dordrecht: Kluwer Academic Publishers.

Dash, S. K., Jenamani, R. K., Kalsi, S. R., & Panda, S. K. (2007). Some Evidence of Climate Change in Twentieth-Century India. *Climatic Change, 85*(3–4), 299–321.

Dash, S. K., Sharma, N., Pattnayak, K. C., Gao, X. J., & Shi, Y. (2012). Temperature and Precipitation Changes in the North-East India and Their Future Projections. *Global and Planetary Change, 98,* 31–44.

Goswami, B. N., Venugopal, V., Sengupta, D., Madhusoodanan, M. S., & Xavier, P. K. (2006). Increasing Trend of Extreme Rain Events Over India in a Warming Environment. *Science, 314*(5804), 1442–1445.

Guhathakurta, P., Sreejith, O. P., & Menon, P. A. (2011). Impact of Climate Change on Extreme Rainfall Events and Flood Risk in India. *Journal of Earth System Science, 120*(3), 359–373.

Hingane, L. S., Rupa Kumar, K., & Ramana Murty, B. V. (1985). Long-term Trends of Surface Air Temperature in India. *Journal of Climatology, 5*(5), 521–528.

IPCC. (2007). *Climate Change 2007—The Physical Science Basis Contribution of Working Group I to the Fourth Assessment Report of the IPCC* (ISBN 978 0521 88009-I hardback; 0521 70596-7 paperback).

Jacob, D. (2001). A Note to the Simulation of the Annual and Inter-Annual Variability of the Water Budget Over the Baltic Sea Drainage Basin. *Meteorology and Atmospheric Physics, 77,* 61–73.

Jhajharia, D., Roy, S., & Ete, G. (2007). Climate and Its Variation: A Case Study of Agartala. *Journal of Soil and Water Conservation, 6*(1), 29–37.

Jhajharia, D., Yadav, B. K., Maske, S., Chattopadhyay, S., & Kar, A. K. (2012). Identification of Trends in Rainfall, Rainy Days and 24 h Maximum Rainfall

Over Subtropical Assam in Northeast India. *Comptes Rendus Geoscience, 344*(1), 1–13.

Kulkarni, A. (2012) Weakening of Indian Summer Monsoon Rainfall in Warming Environment. *Theoretical and Applied Climatology, 109*, 447–459.

Kumar, V., & Jain, S. K. (2010). Trends in Seasonal and Annual Rainfall and Rainy Days in Kashmir Valley in the Last Century. *Quaternary International, 212*(1), 64–69.

Kumar, V., & Jain, S. K. (2011). Trends in Rainfall Amount and Number of Rainy Days in River Basins of India (1951–2004). *Hydrology Research, 42*(4), 290–306.

Kumar, K. R., Kumar, K. K., & Pant, G. B. (1994). Diurnal Asymmetry of Surface Temperature Trends Over India. *Geophysical Research Letters, 21*(8), 677–680.

Kumar, K. R., Pant, G. B., Parthasarathy, B., & Sontakke, N. A. (1992). Spatial and Subseasonal Patterns of the Long-term Trends of Indian Summer Monsoon Rainfall. *International Journal of Climatology, 12*(3), 257–268.

Moss, R. H., Edmonds, J. A., Hibbard, K. A., Manning, M. R., Rose, S. K., Van Vuuren, D. P., & Meehl, G. A. (2010). The Next Generation of Scenarios for Climate Change Research and Assessment. *Nature, 463*(7282), 747–756.

Oliver, J. E., & Wilson, L. (1987). Climate Classification. In J. E. Oliver & R. W. Fairbridge (Eds.), *The Encyclopedia of Climatology* (pp. 221–236). New York: Van Nostrand Reinhold Company.

Subash, N., Sikka, A. K., & Mohan, H. R. (2011). An Investigation into Observational Characteristics of Rainfall and Temperature in Central Northeast India—A Historical Perspective 1889–2008. *Theoretical and Applied Climatology, 103*(3–4), 305–319.

Trenberth K. E., Jones, P. D., Ambenje, P., et al. (2007) Observations: Surface and Atmospheric Climate Change. In S. Solomon, D. Qin, & M. Manning (Eds.), *Climate Change 2007: The Physical Science Basis. Contribution of Working Group I to the Fourth Assessment Report of the Intergovernmental Panel on Climate Change.* Cambridge: Cambridge University Press.

Yasutomi, N., Hamada, A., & Yatagai, A. (2011). Development of a Long-term Daily Gridded Temperature Dataset and Its Application to Rain/Snow Discrimination of Daily Precipitation. *Global Environmental Research, 15*(2), 165–172.

Yatagai, A., Kamiguchi, K., Arakawa, O., Hamada, A., Yasutomi, N., & Kitoh, A. (2012). APHRODITE: Constructing a Long-term Daily Gridded Precipitation Dataset for Asia Based on a Dense Network of Rain Gauges. *Bulletin of the American Meteorological Society, 93*(9), 1401–1415.

Manipur Flood 2015: Preparedness and Disaster Risk Reduction Strategies

Bupinder Zutshi

INTRODUCTION

Floods of July–August 2015 are said to be worst flood in the plain areas of Manipur State during the last 200 years. The flood of 2015 affected entire Thoubal district and parts of Chandel district. According to the government reports, nearly 600 square kilometres of area with over 500,000 populations was affected in the Thoubal and Chandel districts. Majority of the population in the flood-affected area are farmers, whose paddy crop cultivation was destroyed and it also affected crop production for the next year due to acute shortage of seeds. Fisheries farms got submerged and people lost livelihood opportunities. Animal life also suffered due to inundation of inhabited plain areas for several days. The market and shops were closed, and many were submerged for at least one week. Since it was the season for agricultural plantation, the farmers also suffered from food security for the entire season. The present study examines causes and consequences that lead to the worst floods of July–August 2015 in the plains of Manipur State. An in-depth analysis of role of government policies and programmes and human activities and interventions

---

B. Zutshi (✉)
Jawaharlal Nehru University, New Delhi, India

© The Author(s) 2018
A. Singh et al. (eds.), *Development and Disaster Management*,
https://doi.org/10.1007/978-981-10-8485-0_4

65

has been studied to identify the causes and consequences of the disaster, based on field study experiences. The paper also examines the level of prevention and preparedness of State Government and local communities to face the disasters. It also examines the recovery and response mechanisms of State Government, civil society and local communities to face the challenges of disaster of such magnitude. The study also proposes recommendations for developing disaster risk reduction measures and community resilient approaches, based on community experiences.

OBJECTIVES AND METHODOLOGY

The present study examines causes and consequences that lead to the worst floods of July–August 2015 in the plains of Manipur State. The study evaluates the status of disaster preparedness, response and rehabilitation undertaken by State Disaster Management Authority of Manipur. The study also examines the community response and disaster risk reduction measures adopted by the state for future disaster eventuality.

The present study is based on observations and information collected during the field visits by the research team and information gathered during the seminar deliberations at Manipur University from 9 to 11 April 2016. The study has been conducted under the auspices of Disaster Research Programme (DRP)—a collaborative research programme of Jawaharlal Nehru University, New Delhi, and National Institute of Disaster Management (NIDM, GOI). DRP in collaboration with Manipur University and other academia from several research institutes, mass media, civil society organizations, research students and bureaucrats conducted a seminar on Disaster Management and Community Resilience in Northeast India from 9 to 11 April 2016. A field visit was also conducted on 10 April 2016 in District Thoubal and District Chandel. Several sites including Vanzing Tentha village in Thoubal District and Rungchang Chakpikarong village in Chandel district of Manipur were visited. These two villages were flooded during July–August 2015.

FLOOD HISTORY OF NORTHEAST INDIA REGION

India's North-Eastern Region consists of eight states, namely Arunachal Pradesh, Assam, Manipur, Meghalaya, Mizoram, Nagaland, Sikkim and Tripura. With a total population of 45.53 million (2011 census) and

Table 1 Comparative status of north-east states and India (*Source* Census of India, general population tables)

States	Population (as per census 2011)	Area (sq. km)	Percentage all India		Person (per sq. km)
			Population	Area	
Arunachal Pradesh	1,382,611	83,743	0.11	2.54	17
Assam	31,169,272	78,438	2.57	2.38	397
Manipur	2,721,756	22,327	0.22	0.67	122
Meghalaya	2,964,007	22,429	0.24	0.68	132
Mizoram	1,091,014	21,081	0.09	0.64	52
Nagaland	1,980,602	16,579	0.16	0.50	119
Sikkim	607,688	7096	0.05	0.21	86
Tripura	3,617,032	10,486	0.29	0.31	345
Total NE	45,533,982	262,179	3.07	7.97	174
All India	1,210,000,000	3,287,263			374

an area of 262,179 square kilometres, the North-Eastern Region is relatively sparsely populated compared with much of the rest of India. However, population density varies widely among the north-eastern states. Assam and Tripura are the most densely populated followed by Meghalaya and Manipur, while Arunachal Pradesh is the least densely populated (17 persons per square kilometre) (see Table 1).

DISASTERS AND VULNERABILITY IN NORTH-EAST STATES

Owing to its specific geoclimatic, geological and physical features, North-Eastern Division (comprising of 8 states) is vulnerable to all major natural hazards (drought, flood, cyclone, earthquake, landslides, fires, etc.). The north-east states being mountainous topography is also vulnerable to the occurrence of other technological/human-caused hazards such as transportation accidents, human-induced landslides and terror attacks. Human intervention in terms of development programmes without taking into account carrying capacity of the area, use of appropriate technologies in sync with the natural ecosystem and local requirements in this fragile ecosystem has made the region vulnerable to disasters.

Brahmaputra and Barak River Basin falling in the North-East Division of India are among the most flood-prone areas in the world.

The Brahmaputra River flows through Assam from east to west over a length of approximately 650 kilometres. Its main branch originates in the Tibetan plateau, flowing from west to east as the Tsangpo River, and then turns south through the eastern Himalaya as the Dihang River to enter Assam, where it is joined by other branches to form the Brahmaputra. The Barak River rises in the Indian State of Nagaland at an elevation of approximately 2300 meters and passes through the Manipur Hills of Manipur State over a river length of nearly 400 kilometres. It then flows generally westward from Lakhimpur through the Cachar Plains region of Assam over a river length of approximately 130 kilometres to enter Bangladesh near Bhanga. Flood in Assam, Manipur and Tripura (three states of North-East Division) are characterized by their extremely large magnitude, high frequency and extensive devastation. This natural hazard repeats itself every year and not only costs lives but also affects the economy of the state, which is largely agricultural. Flood is not a new phenomenon in these three states (refer Table 2). Frequency of occurrence of floods in Brahmaputra and Barak River Basins during last 100 years clearly shows that these two river basins have recorded excess water discharge and floods (refer Table 3).

FLOOD VULNERABILITY IN MANIPUR

About two-third population of Manipur State (1.79 million out of 2.72 million populations, as per Census 2011) is concentrated in the Manipur Valley surrounded by mountain hills (Manipur Valley constitutes only about 8.2% of the state area). Rivers from these mountain hills flow into valley and very often lead to flash floods and landslides during rainy seasons in the mountain areas and flood the plain areas. All the major river systems of Barak River in the state are vulnerable to flash floods/flooding during rainy season, as captured in the Vulnerability Atlas of Manipur State. Flood in Manipur Valley is primarily due to heavy rainfall in the upper catchment areas. Intensity of rainfalls is higher in the hilly region than in the plain region. Thus Manipur Valley has large upper catchment area where rainfall is normally high. These good amounts of rainfall feed many streams and rivers, which finally drain, through Manipur Valley. In the hilly region, very steep slopes occupy the major portion. The denuded land due to deforestation in the hilly areas enhances more erosion and run-off conditions. There are many vulnerable points along the riverbanks of the major rivers of Manipur Valley, which get breached

Table 2 Major floods in Barak River Basins 1990–2015 (*Source* Manipur State Disaster Management Plan, Vol I and reported by Earth Sciences Department, Manipur University, Canchipur, Imphal)

Date and year	Remarks
First week of August 2015	The state has witnessed moderate-to-heavy rainfall in the first week of August 2015, with several rivers flowing above the danger mark. The flood washed away bridges and national highways besides rendering thousands of people homeless in Manipur. Low-lying areas. The flood also inundated the capital city of Imphal and its outskirts
August 2002	Severe flood occurred in Manipur Valley in August 2002. Breach of embankment took place at 59 places. Due to incessant rain in the catchments, all the rivers flowing in and around Imphal, Thoubal and Bishnupur districts were rising from 11 August 2002. On 13 August 2002, the water levels in all major rivers/streams in Manipur Valley were rising alarmingly crossing the R.F.L on the same day. The flood mainly occurred in the south-eastern parts of Manipur Valley. About 10,000 houses and 20,000 hectares of paddy fields were affected
June–July 2001	Flood of low magnitude occurred in some parts of Manipur Valley. On 7 June, breach of embankment of Nambol River took place at Nambol, Kongkham: inundating Kongkham, Sabal Leikai, Maibam and Naorem
September 2000	Flood occurred in Manipur Valley in September 2000. Breaches of river embankment take place at 30 different places. Not less than 2400 houses and 7800 hectares of paddy field were affected. Breaches of river embankment take place at 11 places of Thoubal River, 6 places of Wangjing River, 2 places of Arong River, 2 places of Sekmai River and 3 places of Manipur River. The flood was of moderate magnitude
September 1999	There was incessant rainfall from 24 August to 3 September 1999. The flood mainly affected the southern parts of the valley. Not less than 7300 houses and 15,300 hectares of paddy fields were affected
July, August 1998	Flood occurred in the valley in July 1998 affecting some areas of Iroisemba. In August, breach of river embankment took place at one place of Wangjing River, as a result inundating the areas of Lamding Nashikhong, Lamding Laishram Leikai and some adjoining areas

(continued)

70   B. ZUTSHI

Table 2   (continued)

Date and year	Remarks
September 1997	Flood occurred in Manipur Valley in September 1997. All the rivers flowing through Manipur Valley were rising rapidly from 25 September 1997. Breaches of embankments took place at four different places of Nambul River, two places of Wangjing River, one place of Merakhong River, two places of Imphal River, two places of Thongjaorok River, one place of Khujairok River and one place of Khabi River. Due to the flood, damage caused to houses rose up to 4965 numbers. The flood was of high magnitude
October 1992	Due to the incessant rainfall in the upper catchment area of the major rivers of Manipur Valley, water level of all the rivers rose rapidly. The daily precipitation in the form of rainfall on 14, 15 and 16 October 1992 was very high and heavy discharge occurred in the rivers and caused breached, overtopping and piping at some of the places

Table 3   Frequency of rainfall (*Source* Deka et al. 2013)

Year	Brahmaputra River Basin		Barak River Basin	
	Excess	Deficient	Excess	Deficient
1901–1910	1	0	2	0
1911–1920	1	1	1	1
1921–1930	2	1	2	0
1931–1940	3	2	1	0
1941–1950	1	0	3	1
1951–1960	1	0	0	2
1961–1970	1	3	0	4
1971–1980	1	2	2	1
1981–1990	4	0	3	1
1991–2000	0	4	1	0
2001–2010	1	4	0	5

every year during rainy season. In these areas, erosion, sliding and slumping of the banks are common. All the major river systems in the state are vulnerable to flooding, as captured in the Vulnerability Atlas.[1] According to report of the Manipur State Disaster Management Plan (2014) and

[1] Ibid.

Earth Sciences Department, Manipur University, it is found that the high stream velocity of the Thoubal river causes breaching of river banks at Okram, Sabaltongba, Khekman, Ningombam, Leishangthem, Phoudel and Haokha. The confluence of the Imphal River with the Iril River at Lilong makes it voluminous and rapid, causing the breach of embankment in Chajing, Haoreibi, Samurou and Lilong. Khuga River flows through strong to steep slope. Hence, river velocity is moderate to high, creating flood problem in Kumbi and Ithai Village. The Chakpi River meets the Manipur River at a reverse direction causing flood in surrounding areas of Sugnu extending up to Wangoo, during rainy season. Most of the embankments are poorly maintained. Many vulnerable points in Imphal river, Thoubal River and Iril River were identified for prone to flooding due to river channel encroachment. The rapid increase in the valley's built-up areas is also an important factor for the recent increase in flash floods in urban areas. In addition to meteorological factors like the intensity and amount of rainfall and other forms of precipitation, during specific period, flood vulnerability is affected by the structural and non-structural parameters like the following.

Structural and non-structural measures include:

- Structural design and maintenance of embankments/banks, flood walls, flood levees,
- Channel improvement desilting/dredging of rivers, dams, reservoirs and other water storages,
- Drainage improvement, diversion of flood water, catchment area treatment/afforestation, anti-erosion works,
- Alignment, location, design and provision of waterway, i.e. vents, culverts, bridges and causeways in national highways, state highways, district and other roads, and
- Railways embankments and inspection, rehabilitation and maintenance, and
- River channels should be free from encroachments.

River flooding is a regular hazard faced by the state, but it has usually manifested into disaster during last few years, due to drainage failures in urban areas, increased run-off loads in hard surfaces, due to large-scale deforestation causing increasing erosion capacity of flowing water, illegal encroachment of river channels and engineering defective construction of dams and bridges, creating barriers in the free flow of excessive water during heavy rainfall in catchment areas. Other identified causes of floods in Manipur Valley includes poor urban drainage, deforestation, breaching of river banks, improper damming and their poor maintenance, sediment pollution, siltation, swallowing of river beds and lakes, changing of land use pattern, vanishing of traditional recharging structures and water bodies. Earlier various lakes of Manipur, mostly located in the southern part of the valley, served as effective reservoirs of excess run-off. Most of the lakes are severely degraded in quality to the extent of complete disappearance resulting in severe curtailment of their water-holding capacities.

MANIPUR STATE FLOOD DISASTER 2015

Manipur witnessed worst-ever floods in July–August 2015, as large area in Imphal, Thoubal and Chandel districts were inundated for nearly 10–15 days due to heavy rainfall from 29 July to 2 August 2015. According to GIS map, a large area of Imphal, Thoubal and Chandel districts was inundated for near 15–20 days. Surprisingly, neither the national media nor state media reported the worst flood that affected entire Thoubal district and parts of Chandel district according to affected people. This is said to be the worst flood in last 200 years. The Asian Highway No. 1, connecting Imphal and Moreh, Myanmar, was affected and remained cut-off for nearly 10–15 days. The old bridge at Pallel was damaged, and the lone and newly constructed Pallel Bridge, yet to open, had been affected. Chakpi River in southern Chandel district washed away the lone Chakpikarong Bridge. The Chakpi River flooded the entire Serou region in southern part of Thoubal district. The worse affected areas in Thoubal district were Kakching subdivisional areas, Wabagai-Hiyangalam and Sugnu and Serou area in Thoubal district. According to the government reports, nearly 600 square kilometres of area with over 500,000 populations was affected in the Thoubal and Chandel districts.

Majority of the population in the flood-affected area are farmers, whose paddy crop cultivation was destroyed and it also affected crop production for the next year, due to acute shortage of seeds. Fisheries farms got submerged and people lost livelihood, as animals, poultry and piggery in domestic homes are the sole source of livelihood for most

Fig. 1 Flood July–August 2015 in Manipur Valley

of the villagers, which were also destroyed. The market and shops were closed and many were submerged at least for one week. Since it was the season for agricultural plantation, the farmers had suffered food security for the entire season. The inundated area of Manipur in 2015 floods is also known as the rice bowl of Manipur that produces large quantity of agricultural products, not only for the region but also for the entire state. Hence, floods in 2015 created food scarcity condition for the state.

Major Causes of Flood Disaster

Failure to adhere to the basic tenants of both structural and non-structural measures mentioned above attributed to the floods of July–August 2015 in Manipur State (Fig. 1). The observations from the field visits and discussions with flood victims brought out the following major reason for flood devastation of July–August 2015, which could have been avoided, if appropriate measures and steps were taken by government, community and State Disaster Management Authorities.

Inadequate Carrying Capacity

All the major rivers of Manipur State after the confluence of the rivers from hilly areas meeting in the plain areas have inadequate carrying capacity in the plain areas. The capacity of the river channels in the plain areas has been reduced further from its original capacity owing to illegal encroachment of land for commercial agricultural activities (banana

Fig. 2 Encroachment of river channel (Thoubal River)

cultivation—see Fig. 2 which was taken during the field visit), construction of illegal buildings on the edge of river channels and continuous siltation of these rivers, raising the water levels of rivers. After the discussion with community and government staff, it was observed that de-siltation of river channels and maintenance of embankments have not been undertaken for a long time in Manipur. Instead community members have been allowed to encroach the river channels for commercial agricultural activities. These agricultural activities have created barriers in the free flow of water, during heavy rainfall season. All the three major rivers in Manipur proved insufficient to accommodate the enormous discharge of floodwater in July–August 2015. The floodwaters of such enormous volume led to the inundation of plain areas of Thoubal and Chandel districts in Manipur, due to breaches in the embankments as well as damming conditions created by land encroachments and defective engineering structures.

Defective Structural Design

Due to continuous rainfall for four days in the higher catchment area of surrounding mountain hills of Chakpikarong village, huge trees and wooden logs were brought down by the river Chapki from higher reaches. These trees, eroded soil and wooden logs brought down by the river Chapki were blocked at the old bridge of Chapki river due to defective structural design of the bridge (like huge and wide pillars closely spaced). This blockade created conducive damming condition. According to the flood victims, the new bridge was constructed 5 years earlier at the same point but due to State Governments lethargy, the old bridge was not dismantled, in spite of several repeated requests from the community. Owing to the poor maintenance of embankments, flood walls and flood levees of old bridge, the Chapki River was blocked at the old bridge for several days.

Unabated Urban Expansion

Imphal city has been expanding fast, converting many wetlands and water bodies into dwelling units. In this process of expansion, even the flood spill channels have been encroached upon and residential colonies have come up closer to water bodies. During the floods of July–August 2015, these areas were the worst affected. Floodwaters followed the natural path and inundated whatever came in their way.

Excessive and Unabated Deforestation

During last three decades, Manipur has been denuded from forest cover by human intervention on a large scale. Forest rights given to the tribals have often been misused for commercial felling of trees. The excessive deforestation in the hills and its tributary river basins has increased erosive capacity of rivers and increased siltation in river channels and water bodies, thereby increasing water levels in these river channels and water bodies. Unfortunately, no efforts were made towards channel improvement, desilting/dredging of rivers and other water bodies like wetlands and lakes.

Disaster Management and Risk Reduction Policy

Sendai Framework of Action has rightly endorsed disaster risk reduction management (DRM) through community resilience measures as a key to reduce the disaster-related affects. DRM is therefore the key to minimize the human fatalities, loss of building and other construction and infrastructure structures, agricultural crops, loss of animal and livestock. Mainstreaming disaster risk reduction management within the policies and programmes of different sectors ensures that the effects of disasters are minimized. At the same time, it enables governments to ensure that these policies and programmes do not put people at risk. According to the ESCAP-UNISDR (2009), DRM is

> "the systematic process of using administrative directives, organizations and operational skills and capacities to implement strategies, policies and improved coping capacities in order to lessen the adverse impacts of hazards and the possibility of disaster"

According to IPCC (2012), it is found that 'People are already experiencing the impacts of climate change through slow onset changes, for example sea level rise and greater variability in the seasonality of rainfall, and through extreme weather events, particularly extremes of heat, rainfall and coastal storm surges'. At the same time, economic damage from climate-related extreme events and disasters has increased dramatically in the last 50 years, with developing country economies being particularly badly hit. There has been significant rise in economic losses from disasters for the Asia and Pacific region.

While it is clear that climate and disaster-related shocks and stresses undermine economic growth and development, there are many actions that governments and other agencies can take to reduce the risks to lives, livelihoods and economies. Poor people suffer the most from disasters, as they lack the capacity and resources to effectively cope with disasters. Risk management in policies and programmes to reduce disaster risk reduction is vital for helping to ensure that the most vulnerable people can access the benefits of development (Mitchell and Tanner 2006). Some people see mainstreaming as a way to realize certain human rights, including the right to safety (Kent 2001). The impetus for mainstreaming risk in development can also be linked to a government's fiduciary responsibility. According to a study by Jackson, D. (2011), it has been pointed out that mainstreaming risk is a government's 'duty to their citizens to maximise the utility of the public resources disposable to them, similar to a private company's fiduciary duty to maximise value to shareholders'. Floods being the most common natural disaster, people have, out of experience, devised many ways of coping with them. On account of frequent occurrence of floods since times immemorial, people have learnt to live with them. They have generally set up settlements away from frequently flooded areas, which have been used for other activities such as agriculture and grazing of cattle.

FLOOD DISASTER MANAGEMENT IN MANIPUR

Manipur State has long history of flood management strategies through local traditions like widening of river channels and clearing river channels from wastes, but such measures have been stopped by the community to increase incomes through illegal encroachments of river channels. Although Manipur State had a historical record of flood management, yet surprisingly the state has remained a blind spot, in both the Central Water Commission and India Meteorological Department's monitoring and flood forecasting establishments. The Central Water Commission (CWC), India's premier water resources body that is responsible for flood forecasts and related advisories to states, had no forecast for any place in Manipur during July–August 2015 floods. This was despite the fact that the India Meteorological Department website clearly indicated heavy rainfall in the state (Thakkar Himanshu 2013). Unfortunately, the subject of flood control, unlike irrigation, does not figure as such in any of the three legislative lists included in the Constitution of India. The primary responsibility for flood control thus lies with the states. The flood disaster management policy of Manipur is completely dependent on Central Government. Manipur State Disaster Management Authority (SDMA) is still in its infancy stage unfortunately, the provisions of Disaster Management Act 2005 and Disaster Management Policy 2009 have not made progress in the state. The preparedness, response and recovery and rehabilitation to July–August 2015 flood disaster indicated that no efforts have been made to institutionalize Disaster Risk Reduction into governance, as envisaged in Disaster Management Act, 2005, and National Disaster Management Policy, 2009. Disaster Risk Reduction (DRR) has not been mainstreamed into development planning to build capacities and promote effective institutional mechanism and promote community-based DRR. Even the research has not been promoted for an effective, well-coordinated and timely preparedness

Table 4 Sample survey respondent characteristics (*Source* Field survey by research team at Thoubal and Chandel District of Manipur April–June 2016)

Areas/localities	Sex		Age group			Occupation		
	Male	Female	<20	20–40	40+	Service	Trade	Agriculture
Thoubal District	10	5	1	8	6	3	2	10
Chandel District	10	5	1	7	7	2	5	8
All areas	20	10	2	15	13	5	7	18

and responsive system. Table 4 depicts the characteristics of respondents selected for the survey. The survey was conducted among 20 male and 10 female respondents. These respondents represented diverse age and occupation groups.

KEY FINDINGS

The study found that Thoubal River has been encroached by villagers for banana cultivation which has hindered the free flow of discharge of rainwater. Discharge from dams located in the Manipur Valley due to structural defects also released water in the rivers in plain areas. Village Chakpikarong located at the foothill of the surrounding mountain ranges had never witnessed flooding in the past memory of the people. In fact, the village is located at higher elevation and is not prone to flooding. The village was flooded on 29 July 2015 midnight. The water entered the village from western side and on 30 July havoc was in store for the village. It rained continuously for 4 days. Debris brought down by the river from higher reaches (tree trunks, stones, boulders, mud and branches of tree) was blocked at Chapki old bridge creating damming condition. These man-made barriers in the river channel hindered the free flow of excessive water supply in the river because of continuous rainfall during July–August 2015 flooding. Unfortunately, neither government nor community has learnt any lessons from the past floods. Field observations and information collected in both villages of Chakpikarong, District Chandel and village Vanzing Tanthi of Thoubal district are tabulated, and major key findings are as follows:

- There was no early warning by government, regarding the floods. For one week, the inundated areas remained cut-off. There was no disaster preparedness for the floods of July–August 2015 in Thoubal and Chandel districts. The government was totally unprepared to face such disaster.
- Major cause of sudden flooding disaster of large areas in the plains of Manipur State was due to man-made actions like deforestation, encroachment of land for cultivation and other buildings in the existing river channels and defective construction practices of dams and bridges without any maintenance services. These human activities created artificial barriers in the way of free flow of river water.
- Government response to rescue people affected by floods was slow and weak. Communities, local civil societies and local clubs were more responsive to help people immediately after the flood disaster. People with support from local communities shifted to more secure areas, and some food was made available to them. Immediate relief in terms of makeshift shelter, food, drinking water and clothing was slow from government authorities. In the 'relief camps', people had nothing to sleep on except for a piece of cloth and men and women, boys and girls all cramped together.
- The roads damaged during 2015 floods were yet to be repaired completely. In fact government's pathetic actions towards stopping land encroachment in the river channels were a measure causes for sudden inundation of areas in both districts. In spite of community appeals to demolish the old bridge on river Chapki, no action was taken by government until our field visit.
- Response from government was weak, even after one month of the floods, inundating large habitable settlements. Community lost complete agricultural season without any agricultural activity, the main source of their livelihood for the whole agricultural season.
- The floods affected the subsistence and basic livelihood opportunities of affected people as growing agricultural season was disturbed due to water logging and sediments in the fields even during next agricultural season.
- Government support in terms of MGNERGA, food security scheme and IYA has been very slow due to weak response of conducting survey for the identification of victim's property, animal and agricultural losses. Even after nearly one year, no major rehabilitation package has been given to affected people.

80   B. ZUTSHI

- The government has not even identified any emergency operation sites for safety in case of emergency situation in the current year or in future years.
- However, people felt that they were better prepared to face any eventuality as there is a strong sense of solidarity among the community with support from civil society organizations.

CONCLUSION

In the light of the above, it may be concluded that strengthening the flood infrastructure in the Barak Basin is needed to cope up with the probability of next extreme flooding event of the magnitude observed in 2015. Dredging of the existing river channels, flood channels, wetlands and strengthening of breached and weak embankments are important. The management of the water bodies/lakes and wetlands in the Barak River Basin should be brought under one regulatory authority for their integrated management. The government, with the help of academia/research institutes, must consider undertaking a scoping study to assess the probability of flooding in immediate future, based on the understanding to be developed from the interactions of groundwater, surface water and the landslide-prone areas in the Barak River Basin. SDMA and district authorities must urgently operationalize, the Flood Early Warning System (FEWS) for Barak River Basin. NGOs and CBOs and local communities must be encouraged to launch awareness generation campaigns on disaster management and risk reduction measures converging both local and traditional knowledge with modern scientific knowledge. SDMA should formulate literature of do's and don'ts for building codes and disaster-related rescue in local/vernacular languages with help from experts and educate public in basic response measures.

REFERENCES

Annual Report. (2015). Prepared by Earth Sciences Department, Manipur University, Canchipur, Imphal.

Deka, R. L., Mahanta, C., Pathak, H., Nath, K. K., & Das, S. (2013, October). Trends and Fluctuations of Rainfall Regime in the Brahmaputra and Barak Basins of Assam, India. *Theoretical and Applied Climatology, 114*(1), 61–71.

Government of India. (2011). *Census of India 2011*. India: General Population Tables, The Office of the Registrar General & Census Commissioner.

Government of Manipur. (2014). *Manipur State Disaster Management Plan* (Vol. I). Manipur: Relief and Disaster Management.

IPCC. (2012). Summary for Policymakers. Managing the Risks of Extreme Events and Disasters to Advance Climate Change Adaptation. In C. B. Field, V. Barros, T. F. Stocker, D. Qin, D. J. Dokken, K. L. Ebi, M. D. Mastrandrea, K. J. Mach, G.-K. Plattner, S. K. Allen, M. Tignor, & P. M. Midgley (Eds.), *Special Report of Working Groups I and II of the Intergovernmental Panel on Climate Change*. Cambridge, UK, and New York: Cambridge University Press. www.ipccwg2.gov/SREX/images/uploads/SREX-SPMbrochure_FINAL.pdf.

Jackson, D. (2011). *Effective Financial Mechanisms at the National and Local Level for Disaster Risk Reduction*. Geneva: United Nations Office for Disaster Risk Reduction. www.unisdr.org/files/18197_202jackson.financialmechanismstosup.pdf, https://sandrp.wordpress.com/2013/06/25/central-water-commissions-flood-forecasting-pathetic-performance-in-uttarkhand-disaster/.

Kent, G. (2001). The Human Right to Disaster Mitigation and Relief. *Environmental Hazards, 3*(3), 137–138.

Mitchell, T., & Tanner, T. (2006). *Overcoming the Barriers: Mainstreaming Climate Change Adaptation in Developing Countries*. London: Tearfund.

Thakkar, H. (2013). *Central Water Commission's Flood Forecasting-Pathetic Performance in Uttarakhand Disaster*. Posted on June 25, 2013 by South Asia Network on Dams, Rivers and People (SANDRP).

Lessons from Manipur: Managing Earthquake Risk in a Changing Built Environment

Kamal Kishore and Chandan Ghosh

EARTHQUAKE OF 4 JANUARY 2016 IN MANIPUR

At least eight people were killed and more than 80 injured when a moderate M6.7 earthquake struck parts of Manipur, in the early hours of 4 January 2016. The earthquake caused widespread damage to buildings and in some locations split open roads across the highly seismic region. According to the India Meteorological Department, the epicentre of the earthquake was in Tamenglong district of the state. The focal depth of the earthquake was estimated to be 55 km by USGS. Four districts, Imphal (East), Imphal (West), Tamenglong and Senapati, reported the maximum property damage. Cracks also appeared in several buildings in Ukhrul where a church building collapsed. Outside of Manipur, the

---

K. Kishore (✉)
National Disaster Management Authority, Government of India, NDMA Bhawan, Safdarjung Enclave, New Delhi, India

C. Ghosh
National Institute of Disaster Management, Ministry of Home Affairs, Government of India, New Delhi, India

© The Author(s) 2018
A. Singh et al. (eds.), *Development and Disaster Management*,
https://doi.org/10.1007/978-981-10-8485-0_5

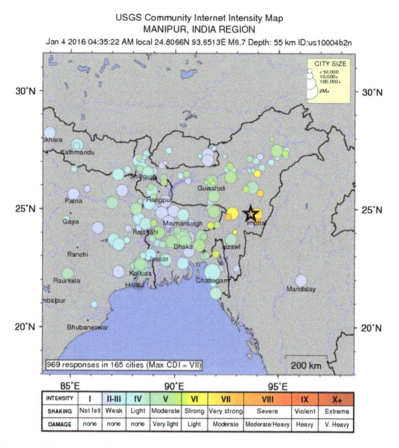

Fig. 1 Did You Feel It (DYFI) response map of 4 January 2016 Manipur earthquake, India (*Source* USGS)

quake was also felt in parts of Assam, Arunachal Pradesh, Meghalaya, Nagaland, Mizoram, Tripura, West Bengal as well as parts of Bangladesh. Minor damages were reported from areas as far as Guwahati and Silchar in Assam. Immediately after the quake, the Northeast Frontier Railway suspended train movement for two hours as a precautionary measure after minor cracks were reported along the railway line. Figure 1 shows the felt area map produced by USGS based on 969 response survey from 165 cities.

Fig. 2 Ima Market in Imphal in April 2016 and partial damage of the same due to the M6.7 earthquake of 4 January 2016 (*Source* Chandan Ghosh and www.nelive.in)

DAMAGE TO BUILDINGS AND INFRASTRUCTURE IN THE MANIPUR EARTHQUAKE

Most of the damage was observed in Reinforced Cement Concrete buildings built over the last two to five years. In Imphal, Women Vendors' market buildings (Fig. 2), particularly Laxmi Market and New Market buildings, suffered high level of damage. Neither of these buildings suffer catastrophic collapse. Given that a large number of families depend on the income from these markets,[1] it was a clear priority of the government to ensure seismic safety and restore functionality of these buildings. However, given the visible damage to structural elements

---

[1] Together Laxmi Market and New Market provide space for more than 2000 women vendors.

Fig. 3 Collapsed frame of a soft-storeyed RCC structure (women vendors' market) in Saikul

(columns and beams), the safety of these buildings for occupancy after the earthquake was highly suspect. A technical team[2] fielded by NDMA on 9–10 January recommended a 'detailed seismic safety assessment' of Laxmi Market and New Market buildings to help determine whether these buildings should be retrofitted or reconstructed. The team recommended a detailed assessment based on following information on construction of these buildings: design-basis report; soil investigation; structural design calculations; structural drawings (good for construction); construction-stage inputs: material tests, quality control and quality assurance steps; photographic evidences during construction; as-built drawings; modifications made in structure since construction; and details of subcontracting. Given that these buildings were built in recent years, it can be expected that this information is systematically stored and can be retrieved easily when required.

In Saikul, a women's market building (three-storeyed) had completely collapsed (Fig. 3) owing to soft-storey failure. The building had been

---

[2]The team comprised Prof. C. V. R. Murty (Director, IIT Jodhpur), Dr. Vineet K. Gahalaut (Director, National Center for Seismology, New Delhi), Dr. Ajay P. Chourasia (Principal Scientist, CSIR-CBRI, Roorkee) and a Senior Engineer of NBCC Limited, New Delhi.

Fig. 4 A traditional house (wooden frame, with bamboo mat infill walls and CGI sheet roofing) with no noticeable damage in Noney, one of the nearest locality from earthquake epicentre

built less than two years ago. In addition to the soft storey, a quick survey of the remaining frame on the site revealed several problems in the design of the structure and detailing of the reinforcement. For example, weak-column–strong-beam design; inadequate overlap in reinforcement bars at the column–beam junction; inappropriate stirrup details; and inappropriate spacing of reinforcement bars in the columns. With some basic technical inputs, all these designs and detailing errors could have been corrected at the time of construction. This clearly points to the need for training on a large scale for masons involved in RCC construction in rural areas.

The traditional buildings suffered little to no damage (Fig. 4).

Recovery Planning

Public Buildings in Rural Areas In case of public buildings in rural areas, the State Government was swift in making interim arrangements (Fig. 5). For example, in Saikul the government rebuilt a temporary marketplace for women. The quality of space of this interim marketplace is

Fig. 5 Interim market space for women vendors built after the earthquake in Saikul

good. However, the women in Saikul felt that the location of the interim market was not very good and has affected their earnings. The new market was occupied in mid-March. In the intervening two months (since early January), the women running the market had no source of income. They had to borrow from local moneylenders and subsequently, a large part of their earnings went towards servicing the loan.

Support to Private Housing/Community Buildings A vast majority of traditional houses sustained minimal or no damage. Most of the damage occurred to relatively recent RCC construction. The government announced a relief package for households whose houses were destroyed completely or damaged partially. However, in the absence of any technical input, people were repairing or retrofitting houses using inappropriate techniques thus rebuilding risk. Following is an example of a single-storeyed RCC structure in Saikul, where the columns have sustained serious damage. The structure has been propped up by constructing RCC columns right next to the original columns. There is hardly any connection between column and old beam. In a future earthquake, the ability of this structure to withstand any lateral force will be minimal as shown in Fig. 6.

Fig. 6 Propping damaged RCC structures by newly constructed columns not connected to the beam above

BUILDING EARTHQUAKE SAFETY IN MANIPUR

Changes in the Built Environment of Manipur

An analysis of changes in the built environment—focused only on housing—of Manipur is presented in Table 1. The Census of India provides data on roof and wall types of the state's housing stock. Although it does not provide figures for different roof–wall combinations, drawing on the experience of the 4 January 2016 earthquake and based on our preliminary understanding of housing typology in Manipur, we can deduce which roof type and which wall type are of greatest concern for earthquake safety. In general, houses with poorly executed heavy flat roofs and (unreinforced) brick and masonry walls are of greatest concern.

Over the period, 2001–2011, the number of houses with heavy roofs made of concrete, stone, slate or burnt brick grew by 82%. This is the most rapidly growing category of houses by roof type. When not built to standard, these houses are most vulnerable to severe damage or collapse

Table 1 Changes in the built environment (housing) of Manipur over 2001–2011 (*Data Source* Census of India 2011, and BMTPC, Vulnerability Atlas of India 2006)

	Roof type *Number and percentage of total (in parentheses)*			
Year	Grass/Thatch/Bamboo/Wood/ Plastic/Polythene/G.I.Metal/ Asbestos sheets	Handmade tiles Machine-made tiles	Concrete/Stone/Slate Burnt Brick/	Total Number of households
2001	467,775 (*95%*)	4059 (*0.9%*)	20,537 (*4.2%*)	492,371
2011	512,289 (*92%*)	5045 (*1%*)	37,379 (*7%*)	554,713
Growth 2001–2011 (%)	10	24	82	13

	Wall type *Number and percentage of total (in parentheses)*					
Year	Mud/Unburnt brick/ Stone not packed with mortar	Burnt brick/ Stone packed with mortar	Concrete	Wood	Grass/Thatch/Bamboo/ Plastic/Polythene/G.I./ Metal/Asbestos sheets/ other	Total Number of households
2001	212,826 (*43%*)	42,557 (*9%*)	1932 (*0%*)	46,000 (*9%*)	187,294 (*38%*)	492,371
2011	291,619 (*53%*)	64,961 (*12%*)	5659 (*1%*)	69,224 (*12%*)	123,250 (*22%*)	554,713
Growth over 2001–2011 (%)	37	53	193	50	−34	13%

causing death or serious injury to occupants. However, in 2011, these houses constituted only 7% of the overall stock. In 2001, the same proportion was even lower at 4.2%. It means that although there is a rapid shift to these types of roofs, this transition is at an early stage. This represents an opportunity. If capacities are built gradually but consistently to impart the right set of skills to design and execute housing construction involving heavy flat roofs, over a relatively short period it can be ensured that the transition does not lead to higher levels of risk. Capacity development of construction workers, contractors and engineers will need to be complemented by appropriate institutional arrangements to ensure compliance with seismic safety codes.

The trend observed in housing categories by wall type suggests even more rapid transition. Over the period 2001–2011, there is an increase of 53% in (unreinforced) masonry constructions using burnt brick or stone. Houses with mud, unburnt brick and stone (without mortar) walls have also seen a 37% increase over the same period. As of 2011, together these categories account for two-thirds of the total housing stock in the state. Both these categories of construction when not executed properly are highly susceptible to earthquake damage. The proportion of some of the traditional forms of wall construction that are quite earthquake resilient is declining. Enhancing capacities for earthquake-resistant masonry construction need to be a priority.

CHALLENGES AND OPPORTUNITIES FOR EARTHQUAKE RISK REDUCTION IN MANIPUR

Building on the lessons learned from the January earthquake, how do we work towards building a resilient future for Manipur? It would be prudent to take a multihazard approach. Manipur is a disaster-prone state. The entire state is in Seismic Zone V. Landslides are a common occurrence affecting not just towns and villages but also road transport including even the national highways. Flood affects all the low-lying areas of the state.

Within this context, three broad strategies are proposed for Manipur:

One, manage the transition of built environment to build resilience: Although the State of Manipur is highly earthquake-prone, according to the Vulnerability Atlas of India (2001), except in two districts—Imphal West and Imphal East—less than 4% of the houses are built using heavy-weight roofs.

The traditional forms of construction, particularly in the rural areas, are quite earthquake resilient. However, this is changing. There is a gradual and inevitable shift from traditional forms of construction to RCC buildings. In future, greater proportion of buildings in both rural and urban areas will be brick masonry and concrete roof. In most of the state, we are at an early stage of this transition. This transition needs to be supported by systematic transfer of both design and execution skills for RCC construction on the one hand and strengthening of institutional capacities for enforcement of building bye-laws on the other hand. The former includes training and capacity development of engineers, masons and other construction workers. The latter addresses compliance to building codes, earthquake safety guidelines, etc. It is important that all mechanisms are put in place for managing this transition—ranging from an appropriate techno-legal regime to training of people involved in construction industry so that they learn correct methods of building with new materials. The state-level efforts on earthquake risk reduction must, of course, be complemented by national-level efforts. Some of these indicative activities are presented in Box 1.

Box 1: Indicative national-level activities to support local efforts at earthquake risk reduction

- Encourage country-wide seismic instrumentation and data sharing among the stakeholders for developing area-specific Earthquake Early Warning (EEW) system;
- Improve techniques for evaluating and rehabilitating existing buildings in the seismic-prone areas of the country;
- Incorporate salient findings from seismic microzonation studies being conducted across the country by Ministry of Earth Science;
- Encourage performance-based seismic design of structures;
- Increase consideration of socio-economic issues related to implementation of seismic hazard mitigation measures;
- Develop a national post-earthquake information management system;
- Develop advanced earthquake risk mitigation technologies and practices;
- Develop guidelines for earthquake-resilient lifeline components and systems;

- Develop and conduct earthquake scenarios for effective earthquake risk reduction and response and recovery planning;
- Facilitate improved earthquake mitigation at state and local levels;
- Create low-cost shake-shock table to demonstrate the efficacy of earthquake resistance measures in buildings;
- Make all public utilities like water supply systems, communication networks, electricity lines, etc., earthquake-proof. Creating alternative arrangements to reduce damages to infrastructure facilities;
- Construct earthquake-resistant community buildings and buildings (used to gather large groups during or after an earthquake) like schools, Dharamshalas, hospitals, prayer halls, etc., especially in seismic zones of moderate to higher intensities;
- Support R&D in various aspects of disaster mitigation, preparedness and prevention and post-disaster management; and
- Evolve educational curricula in architecture and engineering institutions and technical training in polytechnics and schools to include disaster-related topics.

Second, focus on building capacities at the local level—for both disaster response and mitigation. The District Disaster Management Plans developed by the state are very good plans covering a whole range of issues and practical guidance primarily related to disaster response. There is an opportunity to take these even closer to the communities. In doing so, the emphasis on disaster mitigation should be as much as disaster response. While a heavy rainfall event may trigger landslides, it is the underlying factors of environmental degradation, inappropriate management of slopes that determine its severity. All these factors need to be addressed at the local level. Future revisions of district disaster management plans should enhance their focus on mitigation and in each district try to go local.

Third, India has adopted the Sendai Framework for Disaster Risk Reduction. This means that the country is not only committed to implementing activities but also to reducing losses—loss of lives and loss to the economy and infrastructure. In order to measure progress and to ensure that disaster risk reduction activities are leading to specific results, it would be important to set up a system to systematically collect data and establish baselines. This is painstaking work that must be initiated in right earnest at the state level. This would be a cornerstone of all future DRR efforts.

CONCLUSIONS

Like all earthquake-prone areas of the country, safety of built environment is of paramount importance in Manipur. The state's built environment is changing steadily. Traditional form of building is giving way to more modern methods of construction. Evidence from the recent earthquake suggests that this transition needs to be supported by a more effective regulatory environment, capacity buildings of various actors involved in shaping the new built environment and large-scale public education and awareness generation.

Varying Disasters—Problems and Challenges for Development in Manipur

Jason Shimray

INTRODUCTION

Manipur is a state in the south-east corner of the North Eastern Region of the Indian mainland which is stretched between 23.80° to 25.68° north latitude and 93.03° to 94.78° east longitude and covers a total geographical area of 22,327 sq. km, which comprises 8.52% of the total area of the north-east and 0.67% of Indian landmass. Of the total area, about nine-tenths constitute the hills surrounding the remaining valley area. On the north, lie the Naga Hills and the Lushai Hills of Mizoram, and the Cachar Hills of Assam demarcate the west. Manipur has an international border with Myanmar on the east and south that makes it especially sensitive. It shares a 352 km long international boundary with Myanmar that includes the Chin Hills to the south. Manipur was upgraded from the status of a Union Territory and was made a full-fledged state on 21 January 1972.

Manipur is inhabited by three major ethnic groups—the Meiteis in the valley, the Nagas and the Kuki-Chin tribes in the hills. People are predominantly Mongoloid and speak Tibeto-Burmese languages. The

---

J. Shimray (✉)
Manipur State Disaster Management Authority, Imphal, India

© The Author(s) 2018 95
A. Singh et al. (eds.), *Development and Disaster Management*,
https://doi.org/10.1007/978-981-10-8485-0_6

Meitei language which is the mother tongue of the Meitei people is the state language.

Manipur is situated in Seismic Zone V, which is the most earthquake-prone zone in the country (Seismic Zones India 2001). It keeps on experiencing minor tremors off and on. The seismologists, on the basis of past pattern, have predicted that a major earthquake is likely in the North Eastern Region of India (Tiwari 2002). Earthquakes of low-to-moderate intensity are recorded here regularly. The State of Manipur has experienced a number of strong earthquakes, the biggest in recent times being the one in the year 1988, a tremor of magnitude M7.2. Most earthquakes in western Manipur are shallow. But some, especially those recorded in the eastern parts and along and across the Myanmar border, have greater depths. Areas in central Manipur are especially vulnerable to damage during earthquakes as they lay in the Imphal Valley, the lowest point of which is the Loktak Lake, an acknowledged heritage site. The valley floor here undergoes strong movement from even far off tremors as its soft soil amplifies and projects the wave motions (Fig. 1).

According to GSHAP data, the State of Manipur falls in a region of high to very high seismic hazard. As per the 2002 Bureau of Indian Standards (BIS) map, this state also falls in Zone V. Historically, parts of Manipur have experienced seismic activity greater than M6.0–7.0. Approximate locations of selected towns and basic political state boundaries are displayed.

In Manipur, while all the districts are vulnerable to earthquake, the damage is most likely to occur in Imphal town, which has multistoreyed buildings built without adopting any earthquake-resistant technology (Fig. 2 and Table 1).

Manipur has suffered earthquakes of varying intensity several times in a year, but of scale 6.0 (TS) and above; also, the state suffers not less than at least two or more every year. This holds the state back in developmental investments as a huge amount diversion occurs in repairs and retrofitting throughout every year. Since 1990s till today, the intensity is increasing to 6.7 (TS) and above 7.0 (TS). The problem of damages and losses is increased due to the geographical location of Manipur. Geographically, the terrain is an elevated plain surrounded from all sides by hills of recent formation. Manipur Valley (Imphal Valley) located in the central part of the state is a flat, rounded valley. It is divided into four districts: Imphal East, Imphal West, Bishnupur and Thoubal. The total geographical area of the valley is 1900 sq. km. that falls within the parallels $24°16'N$ to $25°2'N$ and meridians $93°41'E$ to $94°9'E$.

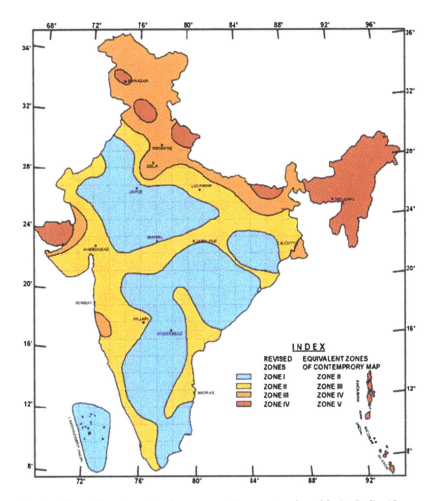

Fig. 1 Zone Mapping: Northeastern region most vulnerable in India (*Source* http://delhi.gov.in/DoIT/DOIT_DM/seismap.gif)

Flood is a primary and annual natural hazard in the area especially during the monsoon season damaging the crops and properties of the people. Flash flood occurs almost every year during rainy season due to poor drainage condition. The primary causes of flood in Manipur Valley are heavy run-off and less infiltration in degraded watersheds in the upper reaches of the rivers.

Manipur Valley is criss-crossed by the major rivers some of which are: Imphal, Iril, Thoubal, Sekmai, Wangjing, Khuga, Chakpi and Nambul

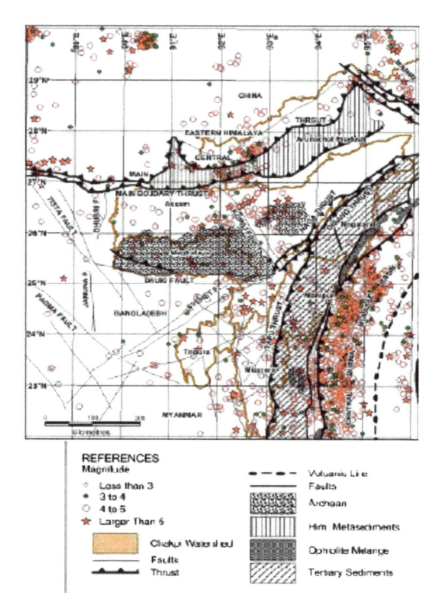

Fig. 2 Seismo-tectonic map of Manipur

Table 1 Major earthquake recorded in Manipur since 1926

Date	Magnitude	Remarks
18 August 1926	Earthquake 6.0 (TS)	East of Pallel, Manipur (Indo-Myanmar Border region), 23:58:48 UTC, 24.50N, 94.50E
15 March 1927	Earthquake 6.5 (TS)	East of Thaungd, Sagaing Division (Indo-Myanmar Border region), 16:50:32 UTC, 24.50N, 95.00E
20 May 1927	Earthquake 6.0 (TS)	Near Kangpat, Manipur (Indo-Myanmar Border region), 10:51 UTC, 24.50, 94.50E
11 July 1930	Earthquake 6.0 (TS)	North of Imphal, Manipur, 07:06:34 UTC, 25.00, 93.80E
22 September 1930	Earthquake 6.0 (TS)	Near Karong, Manipur, 14:19:14 UTC, 25.30N, 93.80E
2 June 1934	Earthquake 6.5 (TS)	East of Ukhrul, Manipur (Indo-Myanmar Border region), 05:04:27 UTC, 25.10N, 94.70
23 April 1935	Earthquake 6.0 (TS)	East of Ukhrul, Manipur (Indo-Myanmar Border region), 16:45:41 UTC, 25.10N, 94.70
9 September 1937	Earthquake 6.0 (TS)	East of Ukhrul, Manipur (Indo-Myanmar Border region), 23:37:27 UTC, 24.90N, 94.70E
6 May 1938	Earthquake 6.5 (TS)	East of Ukhrul, Manipur (Indo-Myanmar Border region), 03:40:57 UTC, 24.90N, 94.70E
21 March 1937	Earthquake 6.0 (TS)	SW of Kohima (Manipur–Nagaland Border region), 16:12:02 UTC, 25.50N, 94.00E
27 May 1939	Earthquake 6.7 (TS)	South of Pallel, Manipur, 03:45:37 UTC, 24.30N, 94.10E
11 May 1940	Earthquake 6.0 (TS)	Near Imphal, Manipur, 21:00:19 UTC, 24.90N, 94.10E
8 March 1947	Earthquake 6.0 (TS)	East of Imphal (Indo-Myanmar Border region), 14:33:05 UTC, 24.90N, 94.70E
30 April 1952	Earthquake 6.0 (TS)	SE of Kohima (Manipur–Nagaland Border region), 01:49 UTC, 25.500N, 94.500E
7 November 1952	Earthquake 6.0 (TS)	SW of Kohima (Manipur–Nagaland Border region), 04:33:57 UTC, 25.500N, 94.000E
1 July 1957	Earthquake 7.25 (TS)	Near Moirang, Southern Manipur, 19:30:22 UTC, 24.400N, 93.800E
30 September 1983	Earthquake Mb 6.0 (BKK)	East of Ukhrul, Manipur (Indo-Myanmar Border region), 10:39:27.0 UTC, 25.0393N, 94.6695E, 60.3 kms depth
5 March 1984	Earthquake Mb 6.2 (HFS)	East of Palel, Manipur (Indo-Myanmar Border region), 21:26:42.0 UTC, 24.5160N, 94.6204E, 67.50 kms depth

(continued)

Table 1 (continued)

Date	Magnitude	Remarks
6 May 1984	Earthquake Mb 6.0 (HFS)	NE of Aizwal (Manipur–Mizoram Border region), 15:19:11.0 UTC, 24.2152N, 93.5256E, 31.60 kms depth
18 May 1987	Earthquake Ms 6.2 (PEK)	Near Karong, Northern Manipur, 01:53:51.0 UTC, 25.2287N, 94.2076E, 52.80 kms depth
6 August 1988	Earthquake Mw 7.2 (HRV)	East of Imphal (Indo-Myanmar Border region), 00:36:24.6 UTC, 25.149N, 95.127E, 91 kms depth. Three people were killed in this earthquake. Tremors were felt over much of eastern and north-eastern India, Bangladesh, Bhutan, Eastern Nepal and Myanmar. Felt as far as Kolkata and Patna. Some damage was also reported from Homalin in northern Myanmar
15 April 1992	Earthquake Mb 6.3 (HFS)	NW of Mawlaik, Chin Division (Indo-Myanmar Border region), 01:32:11.0 UTC, 24.2680N, 94.9275E, 130.90 kms depth
18 September 2005	Earthquake Mb = 5.7	A moderate earthquake at Myanmar—Manipur border, 24.653N, 94.807E, D = 82 kms, OT = 07:26:00 at 12:56 IST causing isolated minor damage to property in some parts of Manipur. The earthquake was felt at many places in Northeast India and Bangladesh as well as in tall buildings in Northern Thailand
4 September 2009	Earthquake	Myanmar-Manipur border, Mw 5.924.381N, 94.712 E, D = 97.6 kms, OT = 19:51:03 UTC. A moderate earthquake struck the Myanmar-Manipur border, at 01:21IST. It was felt widely in Northeast India and in Bangladesh
4 January 2016	Earthquake Mb 6.7	A massive earthquake measuring Mw 6.7 in Richter scale with epicentre at Tamenglong District, Manipur has struck the state claiming altogether 10 lives and injuring 120 persons besides causing extensive damage to buildings to both public and private
13 April 2016	Earthquake Mb 6.8	A massive earthquake measuring 6.8 in the richter scale has struck the Indo-Myanmar border region with epicentre at Mawlaik in Myanmar. The earthquake has claimed 1 live beside causing further damage to public and private buildings

Fig. 3 Photographs of the recent floods and landslides in Manipur during July–August 2015

which either drain directly into lakes or indirectly connect (through lakes) with Imphal river which further down is known as Manipur river (Fig. 3 and Table 2).

Landslides are among the major hydro-geological hazards that affect large parts of the country. Most of the North-Eastern Regions are bristling with landslides of bewildering variety. North-Eastern Region, because of its continued evolution, fragile geological formation and structures, is highly prone to mass movement causing landslides. Since landslides are mostly triggered by events of heavy rainfall and seismicity, which could be followed by flood in the plains, the local populace fills the impact of this location caused by landslides.

As it is a hilly state, landslides and mudslides are quite common. In 2004, severe landslides affected Senapati District. Even at present, mudslides due to construction of Jiribam-Tupul Railway line have affected many families in Tamenglong District. Landslides are one of the natural hazards that affect at least 15% of land area of our country exceeding 0.49 million sq. km.

Forest Fires

Forest fires are common and frequent in the forest areas all over the state. The incidence of forest fire is more in the forest areas adjoining the valley. Villagers set fire to forests to get flush of new grass for their cattle

Table 2 Major floods recorded in Manipur since 1916

Date, month and year	Remarks
30 September–8 October 1916	The flood affected major areas lying on the eastern side of the Imphal River in the Imphal East district, i.e. Wangkhei, Khurai, Kongba, Porompat, Bamon Leikai and Soibam Leikai. Intensity of the flood was severe
June 1929	Flood in the Valley lasted for three days. The flood was of low magnitude
June 1941	Some of the areas lying on the western side of the Imphal river affected particularly Yaiskul area, due to breach of Imphal River embankment at Moirangkhom. The flood was of moderate magnitude
September 1952	Intensity of the flood was moderate
October 1953	Flood of moderate magnitude occurred in the Valley
1965	Flood of moderate magnitude occurred in the Valley
June–July 1966 and October 1966	Flood occurred in the Valley. Some of the areas such as Hiyanglam, Sugnu, Arong, Nongmaikhong, Wangoo, Tanjung were inundated from June to October. Breach of embankment took place at 60 places. The intensity of the flood was severe
July–August 1989	Flood occurred in Manipur Valley at its devastating worst. Altogether 361 localities were inundated. Breached of embankment took place at 40 places. 7 lakhs of people were affected and 97,500 hectares of paddy fields were damaged. Altogether 49,069 houses were damaged and 41,000 domestic animals were affected due to this flood. The magnitude of the flood was severe
14 October 1992	Due to the incessant rainfall in the upper catchment area of the major rivers of Manipur Valley, water level of all the rivers rose rapidly. The daily precipitation in the form of rainfall on 14, 15 and 16 October 1992 was very high and heavy discharge occurred in the rivers and caused breached, overtopping and piping at some of the places. Serious breaches took place at 4 different places. The flood was of moderate magnitude
September 1997	Flood occurred in Manipur Valley. All the rivers flowing through Manipur Valley were rising rapidly from 25 September 1997. Breaches of embankments took place at four different places of Nambul River, two places of Wangjing River, one place of Merakhong River, two places of Imphal River, two places of Thongjaorok River, one place of Khujairok River and one place of Khabi River. Due to the flood, damage caused to houses rose up to 4965 numbers. The flood was of high magnitude
July–August 1998	Flood occurred in the Valley in July 1998 affecting some areas of Iroisemba. In August, breach of river embankment took place at one place of Wangjing River, as a result inundating the areas of Lamding Nashikhong, Lamding Laishram Leikai and some adjoining areas. Magnitude of the flood was low

(continued)

Table 2 (continued)

Date, month and year	Remarks
September 1999	There was incessant rainfall from 24 August to 3 September 1999. The flood mainly affected the southern parts of the Valley. Not less than 7300 houses and 15,300 hectares of paddy fields are affected. The flood was of moderate magnitude
September 2000	Flood occurred in Manipur Valley. Breaches of river embankment take place at 30 different places. Not less than 2400 houses and 7800 hectares of paddy field were affected. Breaches of river embankment take place at 11 places of Thoubal River, 6 places of Wangjing River, 2 places of Arong River, 2 places of Sekmai River and 3 places of Manipur River. The flood was of moderate magnitude
June–July 2001	Flood of low magnitude occurred in some parts of Manipur Valley. On 7 June, breach of embankment of Nambul River took place at Nambul, Kongkham: inundating Kongkham, Sabal Leikai, Maibam and Naorem. On 1 July, Nambul River overflowed, inundating Uripok and Khwairamban Bazar. On 3 July, Chandranadi River, a tributary of Nambul River, overflows on the southern side, inundating cultivated lands of Chajing, Haoreibi and Karam
August 2002	Severe flood was occurred in Manipur valley. Breach of embankment took place at 59 places. Due to incessant rain in the catchments, all the rivers flowing in and around Imphal, Thoubal and Bishnupur districts were rising from 11 August 2002. On 13 August 2002, the water levels in all major rivers/streams in Manipur valley were rising alarmingly crossing the R.F.L on the same day. The water levels of the major rivers were so high on the above day that even the deckings of the bridges on the rivers were badly submerged under water. The flood mainly occurred in the south-eastern parts of Manipur valley. About 10,000 houses and 20,000 hectares of paddy fields were affected
July–August 2015	As a result of the heavy incessant rain beginning from 28 July 2015, the State of Manipur had experienced the worst flood and landsliding the recent past. The four valley districts were affected by the flood out of which Thoubal and Bishnupur are the worst-affected districts. All the major river had overflown causing havoc and washing away connecting bridges, breaching of embankments, cutting off many villages from the mainland. Further, the hill districts particularly Chandel and Ukhrul districts were severely affected by the landslide. As reported, the flood and landslide have claimed 16 lives, 78,925 houses have been damaged either partially or fully, 39488.98 hectare of agriculture land and 5833.98 hectare of pisciculture farms besides causing damage to public and private properties in the affected areas. Essential services like electricity and telephones were disrupted, large number of roads, culverts, canals and dams suffered extensive damage

and for collection of firewood. Regeneration (natural as well as artificial) is completely wiped out and wildlife including rare plants is severely damaged. The hill forests get burnt every year due to wildfire spreading from the burning of Jhum. Most of the fires occur in unclassed forests wherein the Forest Department has very little or no control and maintain the records. The extent of area affected by forest fire is estimated to be about 2000 sq. km annually.

HAILSTORM

These are very common throughout the state and cause severe damage to crops and houses. In March and April 2010, severe hailstorm affected the state, resulting in deaths of two children, apart from severe damage to buildings. Hailstorms cause heavy damage to crops and vegetation. Secondary hazards like snapping of electric poles due to uprooting of trees and disruption of communication links are also attributed to hailstorms.

The powerful storm left a trail of destruction. The heavy storm accompanied by hail and rain caused extensive damage to crops and houses, rendering hundreds of people homeless in 9 (nine) districts of Manipur and also brought down numerous trees and some electric poles along the Imphal–Dimapur highway and nearby villages. Parts of Imphal West and adjoining Senapati districts were the worst-hit. Many trees fell or were uprooted along the state and national highways, blocking traffic and causing long hours of snarls. At several places, tin-roofed houses, shops and other facilities were flattened by the thunderstorm. Some people who were asleep were injured by the flying corrugated iron sheets and crumbling portions of houses. The hill districts were the worst affected.

Due to the heavy thunderstorm, the State of Manipur experienced the worst disaster in the recent past. The 4 (four) valley districts were affected by thunderstorm out of which Imphal West, Bishnupur and Thoubal are the worst-affected districts. Further, the hill districts particularly Chandel, Ukhrul and part of Senapati District were severely affected by the thunderstorm.

The thunderstorm has caused extensive damage to public and private properties in the affected areas besides injuring altogether 20 persons. Essential services like electricity and telephones were disrupted; large number of dwelling houses, schools, buildings both private and government and community assets like community halls and waiting sheds were damaged.

The State Government along with the District Administration promptly responded to mitigate the disaster situation and provided

Fig. 4 Photographs of the recent hailstorm/ thunderstorm during the month of April 2016

food, shelter, medicines and other essential items to the affected people. According to the reports received from Deputy Commissioner of the Districts concerned, 20 persons got injured; 24,713 houses were fully or partially damaged; and 41 numbers of community assets, 2 government buildings, 20 school buildings, 1 PHSC and 1832.50 hectares of agricultural land were damaged (Fig. 4).

Pest Attack

A very peculiar phenomenon takes place in the Northeast India where bamboo grows extensively. These plants flower once in 50 years. When the seeds develop, rodent population, which eats these seeds, proliferates. These rodents then attack the rice fields destroying all the crops leading to famine (Bagla 2001). This phenomenon started in Manipur in 2006 and caused severe crop failure in Churachandpur, Tamenglong and Chandel Districts, forcing the State Government to take special measures to provide food grains to the public.

Drought

For some years now, the state has been facing the prospect of drought due to diminishing rains, which are a direct result of loss of forest cover, due to excessive felling of trees. In the context of Manipur, causes of drought are natural and human activities, both. Several types of weather changes have also altered the normal rainfall pattern in the area leading to a proneness to drought. In last few years, it has been observed that human activities like soil erosion and deforestation have also caused drought. In 2009, all 9 districts were affected by the drought. The State

106   J. SHIMRAY

Government of Manipur declared drought in respect of all nine districts on 25 June 2009. The deficit in rainfall up to the end of July 2009 was 47%. Rainfall from 1 to 12 August was 51.8 mm as against 87.4 mm during the corresponding period in 2008 a shortfall of 40.7%. Total 1.02 lakh hectare. area remained unsown against a total kharif crop area of 2.34 lakh hectare.

Epidemics

As Manipur valley is overcrowded, occurrence of water and airborne epidemics cannot be ruled out. In 2007, almost 150,000 poultry birds had to be culled and safely disposed of after the H5N1 strain of the Avian Influenza Virus was detected in some dead birds (Parsai 2007). In 2009, one case of Swine Flu has been reported from Manipur. In 2010, six people reportedly died following outbreak of Japanese Encephalitis in Manipur.

Besides the above-mentioned natural disasters, the people of Manipur are also vulnerable to various man-made disasters such as insurgency, ethnic clash viz. Meitei–Meitei–Pangal Clash (1993), Naga–Kuki Clash (1993), Kuki–Paitei Clash (1997–1998) and innumerable bandhs and blockades.

INSURGENT CONFLICT AMONG ARMED GROUPS

Today, Manipur is one of the worst-affected states in the north-east where at least 12 insurgent outfits are active at present. A report of the State Home Department in May 2005 indicated that 'as many as 12,650 cadres of different insurgent outfits with 8830 weapons are actively operating in the state'. According to government sources, the strength of those concentrated in the valley districts is assessed at around 1500 cadres for the Revolutionary People's Front (RPF) and its army wing, the PLA; 2500 cadres for the UNLF and its army wing Manipur People's Army (MPA); 500 cadres for the PREPAK and its army wing Red Army; while Kanglei Yawol Kanna Lup (KYKL) and its Yawol Lanmi army are assessed as having a strength of 600 cadres. The Kangleipak Communist Party (KCP)'s strength is assessed at 100 cadres.

Due to the problem of militancy, the investments meant for infrastructural development have been diverted into countering the growing unemployment in the state. There has been increase in educated unemployed youth in the state, and they get grabbed into the militant outfits.

The cases of extortion are also increasing. Militants have resorted to extorting from almost all places including places of worship, educational institutes, health centres and commercial establishments. This has led to closure of quite a few establishments in the state.

BANDH AND BLOCKADE

Bandhs, general strikes and blockades are a regular feature in Manipur. Bandhs are generally called by sociopolitical organizations. Besides, other organizations having linkages with insurgent groups are also known to give bandh calls. Bandh calls are generally made to protest against the inaction or indifference of government authorities towards a particular public (or a section of public) grievance. The bandh is supported by volunteers who, more often than not, are educated yet unemployed youth. This often leads to mindless violence, casualties and damage to public and private properties. Ironically, these protests, which are meant to be peaceful, often turn violent and have far-reaching consequences in the people's day-to-day lives and in the government's discharge of its duty. During a bandh, buses are set ablaze causing irreparable loss to the economically fund-starved State Government; vehicular traffic is sparse; commercial activities come to a standstill; and government and private offices are shut down. Every activity related to education such as attendance in schools, private tuition, coaching is automatically affected. Not only this, food item dealers, water, electricity, medical stores and hospitals all become inaccessible.

DISASTER MANAGEMENT IN MANIPUR

The Department of Relief and Disaster Management, Government of Manipur, was set up in the year 2006 in pursuance of the provision of the National Disaster Management Act 2005 in order to respond effectively to various disasters/calamities since the state is situated in Seismic Zone V, which is very much prone to earthquakes and other hazards due to its fragile ecology.

Under the Disaster Management Act 2005, the State Disaster Management Authority (SDMA) under the Chairmanship of the Chief Minister, Manipur; State Executive Committee (SEC) under the Chairmanship of the Chief Secretary; and District Disaster Management Authority (DDMA) under the Chairmanship of the Deputy Commissioner of the District have been constituted.

The Department has set up Emergency Operation Centre (EOC) at the state level and district level for emergency communications from district to state and from state to the Central Government. Further, the Department has notified/constituted the State Disaster Response Force (SDRF) with 35 numbers of personnel drawn from various units of the Police Department under section 45 of the Disaster Management Act 2005 as a specialist response team to any disasters in the state. The above SDRF personnel have been trained in disaster management particularly on search and rescue, first aid, etc., at National Civil Defence College, Nagpur. The Manipur Fire and Emergency Services have also been declared as the first responder to any disaster in the state.

The Department relates to capacity building mainly on sensitization, awareness programme, etc., of various stakeholders on disaster management. The various awareness programmes on disaster management are being conducted at the Disaster Management Institute (DMI), Babupara, at the state level and at the district and grassroots level too.

TRAINING AND CAPACITY BUILDING PROGRAMMES CONDUCTED DURING 2015–2016

During the period under report, various trainings have been conducted on School Safety and Preparation of School Disaster Management Plans, Building Codes and Designs and Basic Disaster Management Concepts to various stakeholders, viz. teachers, engineers, masons, police officials, VDF, NSS, NCC, NYKS officials and community volunteers and elected members of the local bodies as given below:

The National School Safety Programme (NSSP) under Other Disaster Management Programme (ODMP) is a centrally sponsored scheme with 100% funding from Government of India to be implemented by National Disaster Management Authority (NDMA) in partnership with Ministry of Human Resource Development (MHRD), Government of India, in 43 selected districts in the country with a view to provide safe learning environment. In the State of Manipur, NSSP is implemented in Imphal East and Chandel Districts covering altogether 200 schools each during 2011–2012 and 2012–2013 with a total fund allocation of Rs. 17.43 million. Under the programme, altogether 400 schools (200 schools in Imphal East and 200 schools in Chandel District) have been covered with training on non-structural mitigation measures such as

School DM plan preparation and mock drills. Both activities have been conducted in all 400 schools and altogether 800 teachers (400 each in Imphal East and Chandel district) have been trained on school safety programme. Retrofitting of two schools, one school in each district, had also been taken up. The NSSP scheme is almost fully implemented in Manipur excepting one component on 'Sensitization of Teachers' in respect of Imphal East District which is being taken up at present.

The Directorate of Civil Defence, Manipur, was set-up just after the Chinese Aggression in 1962 and subsequently placed under the Department of Relief and Disaster Management, Government of Manipur, in 2006. It has so far trained 2000 (approx.) volunteers of NCC, home guards, local club members and students of schools and colleges, DLOs [expand], police personnel including MCS/MPS [expand both] officers during the period under report under State Plan Fund and capacity building of 13FC grant. Under the Centrally sponsored scheme of the Ministry of Home Affairs on strengthening of Civil Defence set-up and mainstreaming of Civil Defence in disaster risk reduction programme, all 9 (nine) districts of Manipur have been identified as most vulnerable districts (MVDs) and a sum of Rs. 92.32 lakh per MVD is authorized to every district for various components during the entire regime/duration of the scheme. The 1st instalment of Grant in Aid released during the financial year 2014–2015 by the MHA for Government of Manipur is 179.68 lakh at Rs. 19.96 lakh per district. The amount has been deposited with the DCs/EDs/DRDA [expand] of each district infrastructure building namely setting up of Emergency Operation Centre (EOC) in their respective districts.

Major Shortcomings

Rugged and fragile geophysical structure, very high peaks, high angle of slopes, complex geology, variable climatic conditions, active tectonic processes, unplanned settlement, increasing population, weak economic condition and low literacy rate have made Manipur vulnerable to various types of natural disasters. Apart from these, the lack of coordination among agencies related to disaster management, no clear-cut job description of those agencies, resource constraint, the lack of technical manpower, diverse ethnic groups, insurgency, absence of modern technology, lack of public awareness, very remote, rural and difficult geophysical situation in the state and absence of modern technology are other factors that come in the way of coping with natural and man-made disasters in Manipur.

In addition to the above-mentioned facts, the following are the deficiencies in the Disaster Management set-up in the state:

1. Rigid bureaucratization, lack of structured involvement of community-based organizations and non-governmental organizations;
2. Lack of implementation of the existing standard operating procedures, communication disconnect and lack of coordination;
3. Deficient early warning system and inadequate resources for mass evacuation;
4. Low budgetary allotment in the absence of recent disasters;
5. Minimal involvement of community; and
6. Lack of large-scale research and flawed decision-making.

Conclusion

The intricate cultural and ethnic mosaic, which the North-Eastern Region represents, with over 200 ethnic groups with their own language and sociocultural identity resulting in ethnic conflict coupled with factors such as geographical location, difficult terrain and poor road connectivity, poor infrastructure, problem of law and order due to insurgency, pose a variety of challenges for ensuring a responsive administration and governance.

The region is backward in terms of quality and supply of public services such as education, health care, drinking water and sanitation, housing and building a network of villages and district roads and state highways which will require a significant commitment of additional resources by both the State Government and the Central Government for improving these basic utilities.

The standard of living of the people of the region as measured by per capita Gross Domestic Product (GSDP) has lagged significantly behind the rest of the country. The region lags behind the rest of the country not only in terms of the GSDP but several other development indicators as well. People do not have access to basic services in adequate measure. The standard development indicators such as road length, access to health care and power consumption in the region are below the national average. Therefore, the entire North-Eastern Region particularly Manipur State may be described as vulnerable to various hazards and disasters.

The quantum of good governance in the region should be understood in a broader perspective of larger people participation and cooperation in administrating its functions particularly when we talk of a sustainable disaster management. In the given context of Manipur, a clear and coherent administrative and policy-decisions are required to be taken by the governments for evolving a mechanism where all the stakeholders, civil society, NGOs and different communities can participate and collectively respond to any disaster situation at different levels particularly at the grass root level. Past experiences have shown that the State Administration or District Administration alone could not single-handedly manage an emergency situation arising out of disaster without the cooperation and participation of the general public. Therefore, disaster management should not be seen only in terms of imposition and execution of the provisions of the Disaster Management Act 2005 but rather, it should be understood as continuous planning, organizing and integration of both materials and resources in order to enhance the coping capacity of a society/community to any disaster situation. It should also aimed at changing the mindset of the people in developing a collective preventive culture to any disaster situation which will ultimately facilitate in achieving a resilient community from all sorts of hazards and disasters in the state in the near future.

REFERENCES

http://www.huffingtonpost.com/nehginpao-kipgen/intricacies-of-kuki-and-naga_b_2531115.html?ir=India.

Parsai, Gargi. (2007). Bird flu: Tripura, Mizoram tighten vigil, The Hindu, August 5th, 2007, 07:58 PM, http://www.hindu.com/2007/08/06/stor...0657900100.htm. Accessed 20 July 2017.

Seismo tectonic Atlas of India and its environs, Geological Survey of India: Government of India, 2001.

State Disaster Management Plan (2015–2016), SDMA, Manipur.

Tiwari, R. P. (2002). Status of seismicity in the northeast India and earthquake disaster mitigation. *ENVIS Bulletin 10* (1), 11–21.

Manipur's Tryst with Disaster Management Act, 2005

Nivedita P. Haran

THE CHALLENGE OF MAKING DMA 2005 ENFORCEABLE

The DM Act, 2005 has definitely led to some major structural changes in the management of disasters in most of the states in India. In the State of Manipur, the Disaster Management Department is now a separate administrative entity, the State Disaster Management Authority (SDMA) has been set up and the DM Policy framed at the state and the district levels. Now, at a time when the DM Act has completed its 10th anniversary, an assessment of its impact in the field and at the grass-root level has become necessary. There could not be a more opportune moment to do so.

The effectiveness and the success of any Act lie in the extent to which its implementation brings about improvement in the life-quality of the common people, especially the EWS and the vulnerable. In the State of Manipur, assistance is provided to those affected by a disaster, from the State Disaster Response Fund. However, it is observed that the administration remains relief-oriented with little effort towards mitigation, resilience-building and prevention. Administrators are also difficult to

---

N. P. Haran (✉)
Disaster Research Programme (DRP), Jawaharlal Nehru University,
New Delhi, Delhi, India

© The Author(s) 2018
A. Singh et al. (eds.), *Development and Disaster Management*,
https://doi.org/10.1007/978-981-10-8485-0_7

113

reach especially for those residing in remote areas or places with poor road connectivity. Where efforts are visible, it is oftentimes half-hearted and half-baked: projects are found to be not always need-based with little stakeholder participation. For a state like Manipur that lies on the foothills of the Himalayan range, at the cusp of some major geotectonic fault-lines and where all the districts are placed in earthquake vulnerability category 5, the necessity for a robust, energetic and effective disaster-preparedness can never be overrated nor overstated.

Due to its distant location and poor internal accessibility, some of the good practices carried out in a few of the other states do not easily come within the radar of the administrative set-up here. This paper attempts to identify a few of these practices and recommends the same to the state administration. The north-east has some good centres for research and learning. Yet, technology researched in these Centres does not reach the policy-makers. Its comparatively higher literacy rate, its women power, the wide network of NGOs and its demographic dividend and 'strong club-culture', that is IT- savvy, are positives that need to be utilized gainfully to create a win-win situation.

The paper brings out the different ways in which the benefits of the DM Act and Rules can reach the people, how the funds can be utilized more effectively, ensuring better transparency and accountability and some novel methods of making the citizens disaster-resilient. Manipur can show the way to the other north-eastern states on how to make a state truly risk-reduced and disaster-resilient that engages with the people at the grass-roots level. The paper also emphasizes the need for the states to learn from each other. The central agencies should have done that, but now the Disaster Research Programme (DRP) is aptly placed to fill this gap.

DM ACT: A TEN-YEAR ASSESSMENT IN MANIPUR

There could not have been a better time than now to assess the impact of the Disaster Management Act based on the perceived and measurable results, as the Act completes 10 years of its promulgation. The Act could not have come at a more opportune moment: the nation faced natural disasters with disconcerting regularity, the administration remained woefully relief-focused and the southern coastal states were just recovering from the impact of one of the most destructive disasters of all time, the Indian Ocean Tsunami that hit South and South–east Asia in December

2005. The Act was drafted to ensure that every state sets up its SDMA, declares its State Disaster Management Policy and drafts the State Disaster Management Plan and the District Plans. To bring about better administrative coordination, the states were also required to set up a separate Disaster Management Department.

India is a sub-continent in itself. From the coastal belt in the south and the hinterlands within to the fringes of the Saharan desert to the west and the Gangetic floodplains and the Himalayan ranges up north, the latter considered to be one of the youngest mountain formations geomorphologically that house some of the highest peaks and that are still growing! The terrestrial instability and the presence of dual tectonic plates make this region geologically unstable and prone to major earthquakes. The south-west coast forms the gateway to the annual monsoon whose unpredictability even in the presence of satellite imageries, digital rain gauges and radars makes floods an annual feature. Climatically, geologically and geographically, India as a country embraces the entire gamut of climatic conditions that vary wildly from one end to the other. Notwithstanding its size, India, with its second largest population in the world, has high population density and has one of the fastest rates of urban growth. With the rapidly increasing urban centres, there is a severe need for urban planning and socio-economic infrastructure. That is why the emphasis laid by the government through the promulgation of the Disaster Management Act could not have come at a more opportune moment.

For a country as large as India, the role of the Early Warning System (EWS) is especially important primarily due to its socio-economic conditions, the infrastructure status and its physiography. Over the years, the basic rain gauges and seismographs have been replaced by digital ones. Indian Meteorological Department (IMD) now has geostationary satellites, and it has a system to monitor rainfall on a continuous basis during the monsoon season using the data from its dedicated weather stations and to release three-hourly 'now-casts'. Yet, agriculture continues to depend on the vagaries of nature, fishermen's livelihood stumbles from one storm surge or cyclone to another and economists base their analysis and growth projections on whether monsoon would be normal or sub-normal. For this reason, the latest technological innovations need to be incorporated into the functioning of the IMD and more importantly, the data and knowledge base made more accurate, analysed real-time, utilized and shared for more effective management of disasters.

CONGREGATION OF POLICY MAKERS AT MANIPUR

The Workshop on Earthquakes and Landslides held in Manipur provided an apt occasion to assess the level of preparedness in the state. The state had the DM framework in place through the Policy and Plans. It has been holding training programmes for officials, NGOs and PRIs on a regular basis. But these activities were more on paper. The real test lay in ascertaining the extent to which the activities of the state had impacted the people on the ground. A good occasion to test this out arose when the teams attending the Workshop visited the remote villages and interacted with the local leaders, elected or otherwise, and other inhabitants to understand what changes had taken place in their lives vis-à-vis disaster mitigation and response efforts. The state now has a Disaster Management Department with its own administrative secretary. The SDMA is headed by the Chief Minister and meets occasionally. The state faces the impact of floods every year, sometimes more than once, and to provide relief to those affected, the funds are released from the Calamity Relief Fund, now renamed State Disaster Response Fund. Relief is provided primarily for loss of life, house repair and reconstruction, for loss of agricultural crops and cattle and for other items as listed in the guidelines. Funds are also utilized for the repair of damages to roads and bridges, and for the repair of electricity network and water supply lines. Apart from the above post-disaster response mechanisms, action is taken for tackling public health needs to prevent the outbreak of water-borne diseases and epidemics, vector-borne diseases and managing solid waste.

The SDMA, chaired by the CM, has 5 more ministers as members. Manipur is located in the earthquake high-vulnerability zone and experiences an earthquake of moderate intensity every few months. The Civil Defence Institute provides training not only for the members of the Civil Defence but also for the Home Guards and officials from related departments, viz. Revenue, Home, Health and Irrigation. The CD Institute has created a team of about 40 Master Trainers who now provide training at the district level. In addition, specialized training is obtained through the north-east Police Academy. The Fire and Rescue Services is normally the first responder in any disaster event and over the years the department has been strengthened. The emphasis laid on Civil Defence is a big advantage as Manipur with its hilly remote areas requires trained personnel on the ground. Only half of the districts have elected PRIs; the balance are governed by the Village Authority Act. Strengthening of

PRIs and district administration is, nevertheless, a priority whose importance especially in the context of management of disasters can never be undermined. The geographical remoteness of the state and inadequate development is a stark reality that cannot be ignored. Out of a total of about 240 villages, less than 20% have road linkage, less than 10% have pipe water supply and about half the villages remain non-electrified. Under these circumstances, the government has to work doubly hard to ensure better road connectivity, good communication and availability of equipment. In 2015, the entire village of Jamual was swept away by heavy rains and landslide. Reaching search and rescue teams to that area itself was a challenge.

However, the commitment shown by the DM Department and the SDMA staff in Manipur is heart-warming. Their keenness to widen their knowledge base was evident from their presence at the Workshop organized by the DRP team of JNU in April 2016. A number of initiatives are in the offing. The earthquake of 4 January 2016 that was recorded as 6.2 on the Richter scale proved to be a wake-up call. With better awareness among the people and a changed mindset, they demand disaster-resilient houses. Amended Building Bye-laws making disaster-resilient houses mandatory has been piloted in one district and could soon be made applicable throughout the state. Similarly, Jhum cultivation that has been a practice in the area for centuries is becoming unpopular as the people realize the adverse impact of Jhum on climate change. Manipur has had a very strong NGO base and the government is in the process of drafting a set of guidelines that would streamline the involvement of all non-state actors in implementing mitigative and preventive measures, in the capacity building especially through schools and colleges through the NSS scheme. The involvement of the Rayburn College at Lamka, Churachandpur district, was an eye-opener. While recounting his experience post-January 2016 earthquake, the Principal of the college became emotional as he was a witness to houses collapsing, bridges developing cracks, roads split open and lives lost. Reverend Khenpi told us how those rendered homeless and even those afraid to stay indoors gathered in the courtyard facing the chapel within the college premises. They were provided with immediate assistance in the form of food and clothing and plastic sheets to hang in place of collapsed roofs. Since the mobile towers were affected, connectivity snapped. The affected families needed to reach out to their relatives outside Churachandpur to keep them informed of their welfare. The assistance from the district authorities was slow in coming. In such situations, immediacy of reaction is crucial. Rev Khenpi further recounts how he addressed the gathering during mass in the chapel, telling them to take care of their family's safety and above all to stay calm. He laments that he has no expertise nor training on how to manage the situation post-disaster and if only he had the expertise he would have been able to guide the people with greater confidence. This is a typical case of the ability of a non-state actor to act as 'the voice' of the government especially in times of crisis and to assist the governmental institutions in distribution of relief, in guiding them on the do's and don'ts and basically to assuage frayed nerves and generate hope and confidence. Undoubtedly, no one can do it better than the Principal of the local college, one of the most respected persons in the community. He has the stature and the wisdom; the people look up to him; and therefore, they will listen to him.

With its amazingly active network of non-state actors, Manipur has to learn to make good use of these agencies which it has been unable to do until now. The local NGOs, youth clubs, NSS teams in colleges and NCC in schools, residents' associations and even women's groups are all potential volunteers who can do wonders in delivery of relief in the wake of a disaster. What is even more crucial is to use these agencies in generating awareness, capacity building, training and in raising civic sense amongst the people. As an illustration, Manipur has a multitude of rivers, streams and rivulets. It is highly essential to maintain the cleanliness of the rivers by not allowing the dumping of solid waste on river banks and by preventing the release of sewage into rivers. This requires the involvement of all stakeholders which include the residents' welfare associations, the youth organizations and the chamber of commerce and industry. The SDMA can identify such agencies in every district, every panchayat, engage with them, provide them the necessary input and training so that they can function as the eyes and ears and the delivery mechanism in a disaster.

Lessons to Be Learnt

Among the states in the north-east region, the Government of Manipur has made a sincere effort at disaster management that is impressive. However, there is a need to build on the efforts made to make the structure more robust and interactive. The training given to the public servants is of little use if these officials do not get the opportunity to use their expertise. The trained personnel should regularly be engaged in sharing their knowledge with other staff members which would also strengthen the resource levels at the training institute. The training programme can now bring within its fold the non-state actors too, viz. the NGOs, the private/corporate sector, PSUs, the PRIs.

The Fire and Rescue Services, the Village Office and the local Police are the first responders in any disaster. These can be strengthened by providing them with basic equipment such as torchlight, floodlight, pick-axe, rope, smartphone (with GPS) and basic supply such as blankets, tarpaulin sheets and emergency lamps. The constant refrain that came from the residents is the absence of officials available in the field. Being able to approach a District Collector or a SDO and express one's grievance one to one solves a big part of the problem: first because psychologically a petitioner would feel assuaged if he can reach the concerned official directly with no go-betweens; second, an efficient and alert Collector can redress a grievance, in many cases, instantly at least wherever he is the decision-making authority; and finally, the decision in such cases is much more just and reliable as the decision-maker gets a first-hand knowledge of the situation on the ground, thus reducing the chances of graft at the mid-level and lower level. It is not known whether any statistics-based study by an independent agency has been carried out, but it can be stated with conviction that the extent of leakage from the Relief Fund is estimated to be between 50 and 80%, through distribution of food to the affected, engaging tanker lorries to supply drinking water, purchasing equipment in the name of disaster, taking cuts from cash distributed for house damage and agriculture loss and many more. It is undeniable that disaster is the time for illegal money-making. In this regard, a Task Force set up in every village/hamlet (depending on the size) can act as the ground-level responder in case of any disaster.

What is a Task Force? A Task Force as envisaged here consists of local residents who have or can be trained to have the basic expertise to assist those affected. That includes local teachers, nurses and doctors, public officials, members from the Civil Defence and Home Guard, youth trained in first aid, swimming, diving. It may be a group of 10–12 members who should be given basic training in disaster response so that they are capable of responding when there is an emergency. Such a team would be very useful for the Patwari (Village Officer), Tahsildar and the District Collector in identifying community shelters, in search and rescue operations, in providing first aid, convincing families living in danger zones to relocate in advance thus saving lives and in distribution of relief material.

RECOMMENDATIONS

The following recommendations are made based on the Manipur Workshop and the lessons that can be learnt from the good practices of some other states:

1. The inordinate time overrun in implementing a project has been the bane of the state. With time overrun, there is cost overrun, not to mention the loss in opportunity cost. To reach assistance after a disaster to the affected people, the place needs basic infrastructure which at present it lacks. The line departments need to be professionalized and also made accountable. Any project should have a time and cost schedule and if it is not completed by the assigned date and at the estimated cost, the officials responsible have to be held accountable. In short, improving the core competencies of the officials lies at the base of the disaster management work. Illustratively, the Khuga Dam conceived and designed in 1990 was completed in 2014 with a 1500 times cost increase. For such cases, responsibility should be fixed and loss to the public exchequer recovered.

2. Every construction, be it a road or a bridge, a market or a dwelling unit, should be safe for its inhabitants. Technology is available and it would be shame not to use it to protect life and property. The Ema market in Imphal town was constructed and inaugurated with fanfare. But it developed visible cracks post-January tremors and is now standing like a ghost structure. The women traders have gone back to squatting on the pavement. It is recommended that any structure constructed with public money should be constructed using disaster-resilient technology. The technology to make structures resilient to earthquakes is available in the country, so why not make use of it?

3. The designing of any public infrastructure, be it a road, a school or a hospital, a dam or a market-place should be done in a professional manner and after taking input from all stakeholders, especially the direct beneficiaries. It is an accepted truism that nobody understands as much about a project as a user. The social impact assessment carried out prior to land acquisition can to some extent take care of this requirement, but not entirely. The Ema Market would have been many times more useful if a couple of restrooms

and parking area were added to its design; the Khuga Dam would have had a different façade if the downstream farmers and residents of Lamka had been consulted. In this regard, traditional and inherited knowledge need to be systematically documented. It has been proved time and again that in the context of disaster management, traditional knowledge, be it statistical information or time-tested methods of resilience-building are invaluable. In states like Manipur, where the traditional culture still prevails, such knowledge-base needs to be conserved lest these get lost.

4. The need for setting up panchayat-level Task Force cannot be reiterated. As explained earlier, the Task Force would act as the bridge between the public officials and the citizens. It would help bring together the local expertise placing it at the service of the people. It would also help encourage a sense of voluntarism amongst the people. This would be a refined form of the Civil Defence. The Task Force was experimented in selected panchayats in the State of Kerala and with visible success. However, they need to be institutionalized in the absence of which it would remain a one-off concept and would slowly die off. Manipur could become the pioneer in this regard and could take the idea forward. The interest shown by the University of Manipur is commendable, and the Task Force concept can be taken forward by the disaster management group from the University. In the beginning, it can engage the NSS volunteers which can later be extended to other students.

5. The changes in the techno-legal framework is unavoidable to instil mitigation and preventive measures into the administrative structure. There is a need to make necessary amendments in the relevant statutes, primarily the Building Bye-laws, the Pollution Control Rules, the Land Conservancy Rules and many more. Manipur has shown the way by making use of disaster-resilient technology in certain districts mandatorily. This can be taken up as a good practice for replication elsewhere. Similarly, Kerala has done good work in inventorizing its public lands and river banks to protect them in order to prevent encroachments. This again is worth replicating.

6. Manipur also needs to upgrade its technical capability by setting up digital seismographs, rain gauges and river-level monitoring devices. It is high time that technology came to the aid of the common man. Also, the IMD data need to be accessed freely by all

concerned agencies. The Disaster Management Department needs to arrange access to the data without each agency having to pay as happens at present.

7. Finally, following the example of states like Kerala and Bihar, the State of Manipur should set up its own State Disaster Mitigation Fund without further delay. Manipur has a strong culture of non-governmental activities. The work done by these agencies need to be better coordinated. This is possible if the Fund is set up. Along with the Fund, a standing body that meets at least once a quarter, chaired by the Chief Secretary or Additional Chief Secretary rank officer with representatives of non-state actors can be of much help in coordinating their activities.

In conclusion, it can be safely stated that Manipur has the potential to become a forerunner in resilience building, capacity building and engaging the non-state actors. It is well -poised to become one of the first states to make disaster management a field-based activity. The Department of Disaster Management can take up the suggestions made above with due seriousness, and with the active support of the Manipur University, the colleges and the NGOs make them into a reality.

Reviving 'Public Policy' and Triggering 'Good Governance': A Step Towards Sustainability

Pankaj Choudhury

INTRODUCTION

The world today is witnessing a rapid change. Rising techno-centricity and commodity-fetishism is gradually leading to a dominance of free-market forces over the state, leading to a strategic retreat of state from its necessary functions. However, as we know, public policy plays a major role in protecting and promoting socio-economic well-being of a nation. The movements of the major macroeconomic variables like output, employment and price are largely determined and influenced by the type of public policies prepared and implemented. Well-designed public policies and actions can influence the development process in an astonishingly positive manner. The Directive Principles of State Policy in the Indian Constitution uphold the guidelines for creating a just socio-economic order. An act is required to protect environment and permit sound economic development. India is the largest pluralistic society in the whole of South Asia comprising of hundreds of social and ethnic

---

P. Choudhury (✉)
Jawaharlal Nehru University, New Delhi, India

© The Author(s) 2018   123
A. Singh et al. (eds.), *Development and Disaster Management*,
https://doi.org/10.1007/978-981-10-8485-0_8

groups with distinctive and near distinctive identities. India like many other multinationals and multiethnic states is experiencing a revival of the ethnic consciousness, which had become dormant under the overwhelming influence of the much larger national identity projected by the freedom movement. The ethnic factor which previously occupied a marginalized space in Indian politics has now become embedded in its core. The trend emphasizing the right of a community to maintain its distinct social and cultural identity has become concern. What follows is an overview of ethno-nationalism in different parts of India and of ethnic diversities in Indian politics in the post-independence phase (Ghosh Lipi 2003).

OBJECTIVES

The present paper seeks to address the significance of transparency and good governance in relation to the state disaster reduction programmes. The prime need of today is to balance between the state and the market and to set up a proper regulatory framework to monitor corporate bodies and bureaucratic institutions in delivering services for disaster management programmes. The paper also makes an attempt to address some of the critical issues of the dwindling nature of state, the alarming need for proper public policies and good governance. The study is purely based on the literature reflecting different dimensions of governance and disaster reduction programmes in India. It also captures some issues of natural hazards and hydro-electric project's in the State of Arunachal Pradesh.

INDIAN DEMOCRACY AND DIVERSITY

India's complex ethnic cleavages seem to defy classification as a single ethnic structure. The country is home to more than 1600 language groups and six major religions. Followers of the Hindu religion are further divided by hierarchical caste system. This makes it difficult to mobilize most Indians on a single cleavage even though the appeal of caste and religion parties has been on the rise. British colonial rule had contributed to the shaping of cultural identity sometimes in divisive ways. While the sun did finally set on the British Empire in India, it was partition into two independent nations of India and Pakistan (which was subsequently also partition to create Bangladesh). Unlike many nation states which are premised on the claim of unique nation, unique language, culture or race, the founding idea of India was an idea of the nation state as intrinsically diverse and plural. The transience of even national identity as it was born out of interactions between historical experience and potentially endless future constructions and reconstructions. Therefore, the Indian Constitution privileged the Concept of Universal Citizenship perceived as a critical dimension of the project of nation-building. Instead of identical strategy of accommodation towards different cultural communities, the state in independent India has devised different institutional mechanisms for giving recognition to their interest (Jayal 2006).

HISTORICAL AND SOCIOCULTURAL ROOTS IN INDIA

India's federal democracy functioning in the specific context of its considerable size and diversity has evolved in response to the challenges arising from increased political awareness manifested periodically in electoral mobilization but extending also to other forms of political participation. The demands for recognition of territorial and group identities are in conflict with a highly pluralistic society and the existing structures of governance. The need to accommodate sociocultural diversities within a single political unit has generally been a major motivating factor in the preferences for federal forms of government. The complexities of granting political recognition to territorial-based groups is enhanced with a radicalized sensitization of diverse communities for their right to exist and continue as distinctive entities. The founding fathers wisely rejected the rigidity imparted by classical theories to inherently flexible resilient and adaptable federal idea. The dialectical distinctive national identity provides the vital force in federal nation-building. The multiple and concentric identities are an integral part of a federal polity and assume greater significance through the mediation of political parties and electoral process. Federal governance is characterized by institutional arrangements for solving problems generated by pluralism on the basis of *consultation bargaining* and *mutual consent.* The strong centre ideology was premised on the lack of confidence in institutions embedded in the local context. This *dichotomy between governance and polity was unsustainable* to the extent that it was inconsistent with the premises of Republican democracy. The distribution of legislative and executive powers did not produce neat mutually exclusive compartments because of innumerable overlaps in practices. Even if we have a cursory glance at the list of Central Government departments, it would reveal that the involvement of Union Government is simultaneously required. This leads to growing irrelevance of compartmentalized demarcations in practice and the increased marginalization. The lack of a coherent policy in tackling ethno-regional assertions of identity results into prolonged wars of attritions, which cost dearly in so many ways. The expectations that such movements will eventually lose momentum or that their leaders will tire and settle are all based on the undesirable per se and hence can only be conceded under irresistible political pressures of the right kind (Mukarji and Arora 1992).

NORTH-EAST DEVELOPMENT ISSUES

All the states of the north-east are basically agrarian and industrially backward. As these states are land-locked and the peculiar topographical features do not allow them to easily expand the market within, as well as between the states without heavy investment on roads and communication facilities, many modern industries cannot be set up for want of a viable scale of production. The north-east states should try to improve *infrastructure of the economy, change land and labour policy and maintain law and order.* Poverty and income inequality will increase if the region fails to involve itself value-adding production processes. Economic development of an area depends upon the quality and quantity of resources, level of technology adoption and the size of the markets. Development is measured in terms of its various indicators, which are built upon the basis of data of desired quality and quantity. Beside fertile land and abundant water resources, the region is endowed with variety of natural resources awaiting its proper utilization. The sources of water in the region are abundant. A large perennial river system, high rainfall, rich underground aquifers, tanks, ponds, lakes and derelict water combine to contribute to the vastness of this resource. Surplus water resources must be converted to growth opportunities, rather than to a source of destruction. Population pressure reduces the carrying capacity of natural resources, upon which the livelihood of people depends. Abundant water and other natural resources are the potential opportunities, but the perennial scarcity of financial resources has been a threat in several ways. To harness abundant resources, one needs to harness scarce financial resources appropriately in order to streamline the economic management of the resources. Financial resources are necessary to propel the regional economy to greater heights. The 73rd and the 74th Constitutional Amendments by the Central Government throw a ray of hope in this direction. In order to strengthen the entitlement power of people, measures to accelerate the process of development of the region are indispensable. In the flood-prone areas, infrastructure for farm sector needs to be expanded and strengthened on a priority basis through public sector investment.

COMMUNITY PARTICIPATION

A Community can be described as a group of people living in a geographical proximity of one another, within a definable habitation unit like a village or a colony. The quality of life appertaining to, or being, held by all in common; joint, or common ownership, tenure, liability, etc. Community is a body of persons living together and practising, more or less, community of good. (Singh et al. 2000)

The community is important because it is typically seen as a locus of knowledge. It's a site of regulation and management. There is a tendency to question the right of a specific community to determine how the state should act. The debates about the merits of community involvement in conservation come out of one or both of two reasons: (1) a sense of dissatisfaction with the conservation status of the areas being managed by governments; and (2) a sense of dissatisfaction with the manner in which the resources are being used and allocated, especially regarding the access being provided to the people, including the local communities, protect their areas against degradation or destruction authorized by the government. The debate is about the rights of local communities to use the resources that they have been traditionally using (and conserving) and that are part of their natural surrounds. Just as sovereign nations today lay claim on all the resources within their boundaries so would sovereign communities start laying claims on all their proximate resources (Singh et al. 2000). The status of biodiversity around the world is a cause for concern. The Global Environmental Facility has been supporting biodiversity conservation programmes in many countries and is helping the countries to develop Biodiversity Status Reports, Strategies and Action Plans. Strategy for conservation is what has become known as the participatory strategy of conservation. There is a belief among many individuals and institutions, both within and outside governments and international agencies, that the best, perhaps only, way of effectively conserving in situ biodiversity is by handling over the task and related controls to local communities. There is also a tendency to gloss over the differences between communities, ecosystems, socio-economic conditions and indeed, the objectives of conservation. *First*, in many situations unique biodiversity resources might be lost for good if a community-based approach is attempted inappropriately. *Second*, these few failures might pave the way for a turnaround in the currently popular support for community management, resulting in a universal re-establishment of centralized government control, which has already proved itself to be disastrous (Singh et al. 2000).

RESILIENT SOCIETY AND RISK ANALYSIS

We are seriously at risk from events for which we are 'un-prepared' not because we fail to remember what happened before but because we will encounter newly emerging risk that differs completely from what we faced before. It's also called systemic risk in a modern post-industrial society where a single physical disaster can trigger a spread of secondary and tertiary effects on other social systems or organizations resulting in the collapse of entire system supporting our economy as well as our social welfare (OECD Report 2003). Other than systemic disaster risk, we face structural facilities. Flood control projects where the priority has been on early completion with economy efficiency have degraded in the long run river environment by reducing biodiversity shrinking; the habitats of aquatic flora and fauna were degrading the water environment and changing the water soil cycle. Rapid urbanization in the former floodplain has weakened disaster preparedness on the part of local residents due to the decline of traditional local communities.

- What is the nature of disaster events that can occur?
- How likely is a particular event?
- What are the consequences?

RISK ANALYSIS FLOW CHART

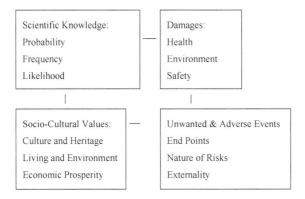

Risk = Hazard * Vulnerability
Risk = Hazard (Probability) * Loss damage/Preparedness (Resilience)
A natural disaster is generally defined as a serious disruption of the functioning of a community or a society due to the occurrence of an abnormal or infrequent hazard. Risk pervades all human activity (Stefan Hochrainer 2006).

A CASE STUDY OF ARUNACHAL PRADESH

Natural hazards are severe, and extreme weather and climate events occur naturally. Natural hazards can happen at any point of time, day and night thus are unexpected and unmanaged events. Natural hazards in Arunachal Pradesh (hereafter A.P.) include earthquake, landslide, cloudburst, flash flood and forest fire. Earthquakes, landslides and the cloudburst are the most destructive, in terms of loss of life and destruction of the property and environmental degradation. The geology of A.P. is very complex as it exhibits three different mountain systems of different origins in *juxtaposition*. These are: (1) the Himalayan ranges, (2) the Mishmi hill ranges, and (3) the Naga–Patkai–Arakan ranges. The Himalayan ranges in A.P. form a continuation of that from Darjeeling, Sikkim and Bhutan in its western part and continue up to the eastern part in Upper and East Siang districts and partly in Dibang Valley and Lohit district. The Mishmi hill ranges which form a part of the Shan-Malaysia Plate, but against the Himalayan ranges along the Tuting-Tidding Suture Zone, are mostly present in the Dibang Valley and Lohit districts. The Naga-Patkai-Arakan ranges that are against the Himalayan and Mishmi hill ranges are present in Changlang and Tirap districts. These are represented by the Schuppen belt of the upper tertiary sequence. Three major tectonic features of the regional scale are present in Arunachal Himalaya, viz. Himalayan Frontal Fault (HFF), Main Boundary Thrust (MBT) and Main Central Thrust (MCT). It is a well-known fact that A.P. falls under the High Seismic Zone V. It is because of its geographic position and geodynamically active three mountain systems. Movements in these mountains cause large number of earthquakes periodically along the thrust/faults planes, and quite a large part of the region is being affected by frequent landslides. This seismic activity in A.P. Himalaya is not uncommon and rather is a regular natural phenomenon, which, however, cannot be predicted in terms of time, magnitude and place. Unplanned developmental activities in A.P. particularly the concrete construction (like hydro-electric projects hereafter HEPs) for their modern lifestyle are increasing without considering the vulnerability to the earthquakes. The natural hazards particularly the earthquake and landslide hazards thus need attention for comprehensive plan for preparedness and mitigation for sustainable reduction in disaster risk in hazard-prone areas through an integrated approach with active participation of the scientific community and society. The need of the day is to educate the general public regarding the danger of the earthquake hazards and their mitigation. It is also important to follow strict enforcement of proper building codes. People themselves should realize to adopt the suitable building codes for their own safety (Singh Trilochan 2016).

The Government of Arunachal Pradesh has started a proactive comprehensive and sustained approach to disaster management to reduce the effects of and overall socio-economic development of the state by forming Arunachal Pradesh Disaster Management Authority (hereafter APDMA) a part of National Disaster Management Act-2005. It's coordinating and monitoring the implementation of rehabilitation activity between APDMA district administration, local authorities, Non-Governmental Organization (hereafter NGO) and local community. It also runs capacity building programmes to strengthen NGOs and local communities, and other stakeholders to become more resilient to cope with disasters. Disaster management programmes are run in three phases:

- Pre-disaster phase;
- Disaster impact phase; and
- Post-disaster phase.

The Government of Arunachal Pradesh (hereafter GoAP) has realized need-based consideration and not extraneous factors for relocation of people at the sight of HEPs. Relocation efforts will include:

- Gaining consent of affected population;
- Land acquisition;
- Urban/rural land use planning;
- Customizing relocation packages;
- Obtaining legal clearances for relocation;
- Getting necessary authorization for rehabilitation; and
- Livelihood, rehabilitation measures for relocated communities wherever necessary.

Tawang Valley Case of A.P.: Based on the available imagery, evidences supported by adequate ground truthing paraglacial deposits are present in the Tawang district above the elevation of 3500 m a.s.l. Tawang district is at an elevation of 5000 m and above. Since no glacier retreat data for Eastern Himalayas in Tawang are available but it's presumed that glaciers were at least 3700 m a.s.l. before 100 years. So no HEPs should be constructed above 3200 m. Also, project-specific strict environmental safeguards mitigation measures should be undertaken for projects above 2500 elevations. Natural habitats must be protected. Advanced and appropriate machineries are used during construction. Disaster Risk Plan:

- Dam breakage preventive measures surveillance and evacuation plan;
- Regulation of barrage water discharge;
- Seismic disaster management; and
- Glacial lake outburst flood monitoring early warning systems mitigation and preparedness.

132   P. CHOUDHURY

A Critic on HEPs in Tawang Valley of A.P.: HEPs are flawed as benefits are overstated and costs understated and adverse environmental impacts. It has also social impacts on riverine communities in the form of cultural identity due to relocation. Local institution should have more involvement to assess impacts. Democratic control is needed for riverine ecosystems and communities. Probing questions:

- What sort of development we are aiming?
- How to ensure equitable distribution of benefits and burdens opportunities and risks?

PROPOSED ACTION PLAN

Proposed Action Plan may be given as follows: (1) set-up of seismic observatories, (2) study of historical earthquakes (palaeoseismicity), (3) landslide studies by preparation of database monitoring landslides and high-risk areas and then suggest remedial measures, (4) set-up of global positioning system stations, and (5) public awareness by education on natural hazards mitigation. There are particular regions which are periodically exposed to the same hazards but many times other hazards are striking the regions unexpectedly. Communities that periodically face the wrath of nature have to be helped to cope up with these disasters and prepare well in advance so that losses are minimized. It is also important to know how to act in an emergency situation to avoid accidents arising from panic and ignorance. The natural hazards thus need attention for comprehensive plan for hazards assessment preparedness and mitigation for sustainable reduction in disaster risk in hazard-prone areas through an integrated approach with active participation of the community and society (Singh Trilochan 2016). Advanced planning is most important activity which aimed at providing basic directions for creating an environment for long-term protection. This may be given as follows: (1) identification of hazard-prone regions on the basis of historical and current knowledge as well as conceptual anticipation; (2) design of engineering specifications for various kind of structures particularly construction of more than two-storey building in urban areas; (3) assessment of the vulnerability and risk faced by existing structure and design for retrofitting, wherever necessary, special attention may be given to all essential buildings like hospitals; (4) design and operational readiness of the protocols for effective rescue, like relief measures, prevention of epidemics and emergency operation, of critical services; and (5) regular dissemination of information through carefully designed bulletins to evoke a constructive response and avoid panic. This strategy aims at meticulous scientific preparedness through development of low cost and locally supportable technologies. Efforts should be made for improvement of slope stability land reform classification, etc. Further research needs to be taken up in estimating and mapping the hazard intensities in the threatened areas. Educating the general public regarding the danger of the natural hazards and their mitigation is important. The natural hazards particularly the earthquake and landslide hazards need attention for comprehensive plan for preparedness. Adoption of suitable measures can minimize loss of life and property.

Suggestion and Conclusion

Globalization is a powerful force undermining environmental regulation. The notion of sustainability is about our obligation to the future. A practical approach model.

Risk	Community
Law	Resilience

We should choose policies that will be appropriate over as a wide range of possible circumstances as we can perceive. Liability to error is the law of life. The future is not adequately represented in the market. In principle, government could serve as a trustee, as a representative of future interest. The theory of late modernity posits the demise of the grand narrative and inception of multiple social realities or world views. While few would argue that science has no role to play in solving the world's problem, science as a way of knowing has lost its lustre. No attempt is made to reconcile the differing world views. Sociologists are concerned with risk communication and concentrate on the transfer of information between experts and layman. Empirical evidences suggest that selective and well-designed public policies with good governance in north-east region can largely influence the development process in an astonishingly positive manner. Assimilation and accommodative policies can strengthen social integrity of north-eastern states of India in general and Arunachal Pradesh in particular. Lack of authentic knowledge and information about customary usages, regulating diverse spheres of life in discrete tribal situations, is the root of problems. Participatory strategy of conservation is highly needed to cope with the problems of natural hazards in Arunachal Pradesh. Need of hour is political resolution.

REFERENCES

Barik, S. K. (2015). *Perspective Plan for Development of Tawang River Basin.* Department of Botany, NEHU Shillong, IIT, Guwahati, WWF, Tezpur, Assam. Foundation for Revitalization of Local Health Traditions (FRLHT), Bangalore, June. Chapter 1, pp. 3–9.

Ghosh, Lipi. (2003). "Ethnicity and Issues of Identity Formation". In Aleaz, Bonita, Ghosh, Lipi, and Dutta, K. Achinitya (Eds.), *Ethnicity, Nations and Minorities: The South Asian Scenario.* Kolkata: Manak Publications, p. 85.

Jayal, Niraja Gopal. (2006). *Representing India: Ethnic Diversity and the Governance of Public Institutions.* New York: Palgrave Macmillan, pp. 2–3.

Mukarji, Nirmal, and Arora, Balveer. (1992). "The Basic Issues". In Mukarji, Nirmal, and Arora, Balveer (Eds.), *Federalism in India: Origins and Development.* New Delhi: Vikas Publishing, pp. 3–17.

Report of OECD (2003): Available at: www.oecd.org.

Singh, Shekhar, Vasumathi, Sankaran, Harsh, Mander, and Sejal Worh. (2000). *Strengthening Conservation Cultures: Local Communities and Biodiversity Conservation,* November. Paris: Man and the Biosphere Programme, UN Educational Scientific & Cultural Organization. See also: Bauman, Zygmut, *Community: Safety in an Insecure World.* Cambridge: Polity Press, 2003.

Singh, Trilochan. (2016). "Natural Hazards and Mitigation Measures with Special Reference to A.P.". *International Journal of Innovative in Science Engineering and Technology,* Vol. 5, Special Issue. 6, May, pp. 8–12. Presented at National Workshop on Disaster Management (29–30 August 2014) at NERIST, Nirjuli, 791109, A.P.

Stefan, Hochraine. (2006). *Macroeconomic Risk Management Against Natural Disaster: Analysis Focused on Governments in Developing Countries.* Deutscher Universitals Vertag: Geneva. Chapter 1. Introduction, pp. 1–10.

Role of Heavy Machinery in Disaster Management

Keshav Sud

DISASTER MANAGEMENT AND HEAVY MACHINERY

Technology has been underutilized in managing disasters, and in India, it has largely been a very disproportionate investment into inappropriate disaster technology which fails disaster management when required. Disasters in India and much of the South Asian subcontinent have been the work of samaritans and martyrs brought in from the Army or more recently from an army like battalion such as the National Disaster Relief Force (NDRF) or Central Industrial Security Force (CISF). Huge amount of funds are pumped in post-disaster relief and recovery despite the fact that much of this would have been saved if the right investment had been made into technology which is appropriately suited, located at

---

Dr. Keshav Sud has obtained M.Sc. and Ph.D. in Mechanical Engineering (University of Illinois, Chicago, USA). He has worked closely on several aspects of construction machines during his research in Caterpillar, and Volvo. He has closely studied the usage of these machines in disaster management in USA.

K. Sud (✉)
Amazon Robotics, Boston, MA, USA

© The Author(s) 2018
A. Singh et al. (eds.), *Development and Disaster Management*,
https://doi.org/10.1007/978-981-10-8485-0_9

places where the community is already trained to handle it and also available to vulnerable zones.

This paper focuses upon the role of heavy machines in the varieties of hill disasters which have prevented normalization of development in the north-eastern states of India. One can see a tremendous role of these machines in preparedness and in post-disaster rescue and recovery. A right technology which is appropriately timed can reduce and sometimes even prevent many major disasters from affecting human lives. Little has been done so far to link research on preparedness with the usage, design, operation and affordability of these heavy machines in disaster management despite the fact that many lives could have been saved and damage reduced if the disaster management agencies had equipment of their own and didn't have to depend on the city municipal corporation or a private developer.

UNDERSTANDING THE MARKET OF HEAVY MACHINES IN RESCUE OPERATIONS

Construction equipment is spotted working alongside rescue teams at disaster sites around the world. These machines symbolize human strength and commitment to bring life back to normal as soon as possible. Emergency vehicles and fleets of construction equipment are amongst the first that get mobilized to a disaster-stricken region to begin the rescue and recovery process, and they are the last ones to leave after the rebuilding process is completed and normalcy of life has been restored. During the rescue and recovery process, the heavy machinery operators have 3 main objectives, assist in active rescue operations, clean up roads to allow access for ambulances and supply vehicles and if required build a temporary shelter for the safety and medical needs of those impacted (see Fig. 1).

The rescue and recovery phase is predominantly led by governmental or defence-owned assets; however, the rebuilding process is often contracted through private companies. Heavy machinery also comes in various types; each type plays an important role, and hence, it is critical to have operators cross-trained across machines and also in disaster management.

The Indian Construction Equipment sector had an estimated market size of 2.4–2.6 billion USD in the year FY15 which was a small fraction of the global market at over 75 billion USD. However, it has been

Fig. 1 Major roles of heavy machinery in disaster management

growing at an average of 30% annually compared to the global growth of only 5%. India is one of the top 10 markets for construction equipment sales; the Indian market is catered by about 200 domestic manufacturers. Bharat Earthmovers Ltd. was the first company to begin manufacturing construction equipment in India in 1964 under licensing from LeTorneau Westinghouse, USA. Today all leading manufacturers such as JCB, Komatsu, Caterpillar, Hitachi, Terex, Volvo, Case, Ingersoll-Rand and John Deere are present in India either as joint ventures or have set up their own manufacturing facilities.

Lower cost of manufacturing and lesser regulations by the Indian Government also enable domestic machines to cost much less than their internationally sold counterpart. For example, CAT 320D2 excavator costs about $160,000 in India, whereas the North American version CAT 320D costs above $200,000.

For disaster management, typically a fleet of machines with different functionalities is deployed, and some common types are dumpers, loaders, excavators and bulldozers (Fig. 1). This study will focus on the most commonly used excavator machine as it makes up the majority amongst

Fig. 2 Landslide in village of Malin on 30 July 2014 in Pune, Maharashtra

all machines sold in India. JCB India is the leading manufacturer in excavators. Amongst excavators, the 20-ton excavator makes up almost 45% of all construction equipment sales in India and is by far the most popular construction machine for commercial application and disaster management. The total Indian market for the 20-ton excavator averaged about 5000 units per year for FY13, FY14 and FY15. About 75% of the total demand is from the government sector and 25% from the private sector, but 80% of machines from either sector were put to use in the coal industry. Mathematically, this leaves only 1000 20-ton excavator machines available for non-coal industry-related work, which would include disaster management. This number is significantly insufficient to provide support across the country.

Twenty-ton excavators (Fig. 2) are very effective in disaster management due to their size, ease of transportation and ease of operation. Typically, with 80 hrs of training an operator is able to operate this machine effectively.

Unfortunately, developing countries such as India still face some challenges with insufficient fleet of government-owned machines available for disaster relief, a lack of trained operators and lack of bilateral agreements with private construction contractors to provide support in relief activities.

DISASTERS AND PREPAREDNESS

Disasters can be categorized in sizes from small that impact a few hundred to global catastrophes that impact hundreds of thousand people. Any disaster can be broken down into 5 phases: prediction of an impending event, preparing for it and then recovery, rebuilding and resettling.

Human technologies have the well-known ability to adapt and learn. Current technology has come far enough to be able to predict some types of natural disasters and also significantly simplify and shorten the recovery and rebuilding process.

The most critical aspect of disaster management is shortening the recovery and rebuilding process and getting life back to normalcy. The effectiveness and efficiency of this process are significantly dependent on the use of heavy machinery.

DRR AS HIGHLY LOCALIZED AND CONTEXT-SPECIFIC

Figure 2 shows JCB and Volvo excavators supporting the rescue of the estimated 160 residents from 44 households buried under the mud as a landslide hit the small village of Malin in the early morning hours on 30 July 2014. Whatever happened in Malin for the first time is an everyday experience in varying degrees in the Churachandpur and Tamenglong regions of Manipur. Landslide was caused due to heavy rains that continued even after the event making rescue efforts difficult. Over 400 National Disaster Response Force soldiers carried out rescue operations with the help of 4 excavators, yet it took 8 days to remove the mud and search for all survivors. But for the availability of construction equipment in a private construction site nearby which was immediately diverted to assist in relief activities, many lives could not have been saved. This sends a strong signal to the local preparedness agencies that remain closed to the usage of these machines in a terrain vulnerable to landslides of this magnitude (Fig. 3).

Fig. 3 Landslide in village Phamla, Tawang district of Arunachal Pradesh on 21/22 April 2016 (*Source* NDMA site)

During the earthquake of 26 January 2001 in Gujarat and other neighbouring regions, about 7904 villages were affected in 182 talukas in 21 districts, and about 40% of 37.8 million population were in distress. The rescue and relief efforts were assisted by 543 bulldozers and excavator, and 2853 loader and dumpers.

In September 2014, the Jammu and Kashmir region was hit by torrential rainfall that caused widespread flooding and landslides that claimed 200 lives in the first few days alone. Relief operations began immediately and continued over the next few days, and by 16 of

September, the rescue efforts had been completed and over 226,000 people were rescued from different parts of Jammu and Kashmir. Ten thousand personnel from the Army Engineer Corps and Border Roads Organization, equipped with over 400 bulldozers, excavators and dumpers, were involved in this rescue and reconstruction effort.

In July 2015, large sections of villages in three subdivisions of Darjeeling were lost to the Teesta River. More than 50 people lost their lives and many went missing. The regions of Mirik, Lava, Sukhia, Mongpong and Sevoke Kalibari, Rohini and Gorubathan were the worst affected. Rescue teams were equipped with only 2 construction machines that were operated by the Sashastra Seema Bal and the Border Roads Organization, the teams also faced resource capacity[1] limitations in terms of operational management and upkeep of the machines. One of the excavators got buried and destroyed under the landslide. Along with the rescue operation, the entire process of rebounding life to normalcy by clearing roads to make way for hill services-supply network,[2] connecting regions through temporary bridges over mountain streams, lakes and rivers relied upon construction equipment and their trained operation.

A very large number of SSB personnel who were deployed in this life-threatening rescue operation could have been assisted with more construction equipment with in-house service and support.

In any disastrous event, failure to respond quickly and effectively worsens the loss of life and property. Research shows that probability of rescuing victims trapped under debris decreases 50% after the golden 24 hrs due to dehydration and lost endurance 2. In the 4 cases summarized above, the distribution of heavy machines for disaster management varies from the highest ratio of 1 machine per 40 people impacted in Malin to lowest ratio of 1 machine per 4500 people impacted in Gujarat, and even in case of Malin, the recovery efforts took 8 days to complete

---

[1] The Border Roads Organization took several days to clear just one flank. Heavy equipment has been brought from Gangtok, and in the absence of trained local operators, the terrain management issues had delayed the support further. As many as a total number of four heavy machines were grounded due to lack of trained operators and retrofitters.

[2] National highways are the lifeline to many important cities in the Himalayan region, impact to these highways cuts off access to these human habitats; for example, the severe damage to NH10 after the Bhotey Bhir and Rangpo disasters affected movement over the whole Pankhabari, Kurseong and Lopchu regions, this prevented food and medical support reaching the impacted regions. Rare and endangered animals that live in national and protected forest regions of these hills, also got trapped and isolated.

142   K. SUD

and availability of machines nearby was purely incidental. This highlights the need for more heavy machines in disaster relief.

In comparison, we examine disaster management of Hurricane Katrina on August 2005, which was the largest and 3rd strongest hurricane ever recorded to hit the US mainland. Hurricane Katrina impacted 90,000 sqr miles in 3 American states of Florida, Louisiana and Mississippi. It is estimated that over 15 million people were affected in different ways varying from having to evacuate their homes to indirect economy suffering, 1836 lives were lost and 705 people are still unaccounted, and caused 81 billion USD in property damages.

On 26 August 2005, 3 days prior to first landfall most states declared emergency and recommended voluntary evacuation; however, on August 29 when the hurricane hit, it was estimated that over 100,000 people still remain in the affected areas and would need to be rescued. Federal, state and private agencies including FEMA and American Red Cross rapidly mobilized troops and heavy machinery to the 3 states. Over 27,000 rescue workers, National Guard troops, active military and trained civilian first responders along with 32,000 construction machines worked tirelessly for 9 days to almost complete the rescue operation 3, 4. In this case, there was 1 machine available for every 3 people impacted; hence, this massive recovery effort spanning 3 American states was completed in merely 9 days. This response measures way beyond that seen during the disasters in India.

The above comparison highlights the contrast of disaster response between developed and developing countries in terms of availability of trained relief personnel and infrastructure deployment (DRI). Such a monumental response was made possible as disaster management is taken very seriously in America. Along with professionals, there is a significant investment in training of civilian first responders. Technical training programs begin in high schools to develop interest in the engineering and operation of machines, understanding job site safety and first aid. Since Katrina, the American Government has invested 60 billion USD in bolstering the infrastructural and operational capacity of FEMA.

MODERNIZATION OF DISASTER RELIEF INFRASTRUCTURE IN INDIA

Natural disasters affect the world equally; however, 90% of the disaster-related injuries and deaths are reported from developing countries, and if you look at the top 10 countries for disaster-related deaths from

1996 to 2015, India ranks at number 5. There is a clear correlation between the insufficiencies in DRI leading to higher numbers of disaster-related deaths. DRI is a critical metric characterizing the overall developmental state of any country. It is not surprising that countries ranking lower on global development indices have placed DRI very low in their budgetary priority. This can be noticed in the statistical data.

At the '2009 World Economic Forum Conference', the Global Agenda Council on Humanitarian Response predicted a decline in the assistance traditionally provided by the 'international alliance of aid donors, UN agencies and international NGOs'. So developing countries need to address the insufficiency in their DRI domestically and immediately.

Some improvements in the Indian infrastructure began back in 2002 with the establishment of the Disaster Relief Network (DRN) in conjunction with Hindustan Construction Company (HCC) and the Construction Federation of India (CFI). Since its establishment, DRN has supported many rescue and relief operations, and it has two primary roles—providing training for efficient response to emergency situations and allocation and mobilization of heavy machinery and other construction assets through collaboration with private engineering and construction companies.

For instance, a part of the HCC's ground staff is trained in disaster management. Should a disaster occur near where HCC is working they are able to redirect their machinery and manpower immediately to relief work.

Almost all of the top heavy equipment manufacturers are present in India offering low-cost solutions to the domestic market thus enabling the Indian Construction Equipment market to grow at an average of 30% annually, but 80% of these machines are working in the coal sector, so despite the growth the availability of machines for disaster management remains insufficient.

There is also an increasing gap in the training and availability of skilled operators as most of the private sales goes to small contractors that do not prioritize training of staff in safe machine operation. It is projected that by 2020 about 100,000 trained operators will be required for full-time jobs, plus the additional need for civilian operators to assist in disaster management.

So despite the ease of availability of machines, the efforts to improve the DRI are not keeping up with the rate of increase in population density and urbanization as there is lack in commitment and collaboration.

144   K. SUD

Unless there is a focus on rapidly strengthening India's DRI, the impact of future disasters is projected to get worse.

CONCLUSION

India ranks 5th in the number of disaster-related deaths from 1996 to 2015 and has begun efforts to strengthen its DRI with the establishment of DRN in 2002; however, the rate of strengthening is insufficient to support the spread of the country, increasing population density and rate of urbanization.

It can be concluded that heavy machinery is a crucial part of this DRI as it plays a crucial role in all aspects of disaster management. The efficiency and effectivity of a rescue and recovery operation are dependent on the availability of trained operators and machinery.

India is a growing market in heavy machinery manufactures with an abundance of machine availability with lower than their average global costs. It needs to leverage its growing market in instituting a robust and distributed national disaster relief infrastructure. Investment into the DRI ranks very low in its budgetary priority and needs to be significantly increased.

Heavy machinery engineering and operation skills need to be marketed in universities as a desirable skill as the current Indian Construction Equipment market is at 2.8 billion USD and growing at a rate of 30%. This offers opportunities for employment in the fields of engineering, manufacturing, servicing and operations all of which will also carry a soft but notable impact in bolstering DRI.

As pointed out earlier, manufacturers invest significant resources to make machines easier to operate and an operator can be trained to operate an excavator with just 80 hrs of training, so there should be training programs deployed to prepare first responders through public private partnership.

Bilateral partnerships with key private construction companies to gain their support in disaster situations are essential. This partnership needs to be somewhat incentivized, including disaster management training for ground staff and the ability to redirect the company's decentralized operator and machine network to respond to disastrous situations.

BIBLIOGRAPHY

Annual Disaster Statistical Review: The Numbers and Trends 2008 (2009, June). Brussels: CRED.

Arup International Development Website.

Bissell, R. A., L. Pinet, M. Nelson, and M. Levy. Evidence of the Effectiveness of Health Sector Preparedness in Disaster Response: The Example of Four Earthquakes. *Family & Community Health, 27*(3), 193–203, July 2004.

Climate Change Human Impact Report: The Anatomy of A Silent Crisis. (2009). Geneva: Global Humanitarian Forum.

Construction Contractors Involvement in Disaster Management Planning by Peter Stringfellow, Queensland University of Technology, Australia.

Ellen. "Hurricane Katrina: Humanitarian Obligations and Lessons Learned." University of Denver. Accessed 2 March 2014.

EM-DAT: The OFDA/CRED International Disaster Database. Université Catholique de Louvain, Brussels, Belgium. www.emdat.be.

Feredal Emergency Management Agency—https://training.fema.gov.

Global Humanitarian Report. (2009). UK: Development Initiatives.

The watchers news network—http://watchers.news.

Witt, Emlyn, Kapil Sharma, and Irene Lill. Mapping Construction Industry Roles to the Disaster Management Cycle. *Procedia Economics and Finance, 18*(2014), 103–110. https://www.sciencedirect.com/science/article/pii/S2212567114009198.

World Economic Forum. (2010). Engineering & Construction Disaster Resource Partnership—A New Private-Public Partnership Model for Disaster Response.

https://georgewbush-whitehouse.archives.gov/reports/katrina-lessons-learned/appendix-b.html.

http://homelandmag.com/indian-military-undertakes-herculean-jk-relief-effort.html.

https://miningandblasting.files.wordpress.com/2010/03/indian-mining-construction-equipment-industry.pdf.

www.eadrcc.org.

PART II

Vulnérabilité Studies

Co-Seismic Slip Observed from Continuous Measurements: From the 2016 Manipur Earthquake (Mw = 6.7)

Arun Kumar and L. Sunil Singh

INTRODUCTION

Earthquake is one of the major natural disasters and has become a threat to life and properties. The large, great and mega earthquakes may also induce landslides, floods, large on the ground. The undersea bed earthquake can also produce the tsunami hazards.

According to plate tectonics theory, the plate's movement along with a subduction and transform boundaries of lithosphere plates can cause the earthquake hazards due to build-up of strain and its accumulation. The release of strain along these plates' margins triggers the earthquakes.

As the earthquake took place early in the morning, the commercial areas like malls, schools, in the city were almost vacant, hence reducing the death counts. Significant damages in Imphal city are observed such as women market complex, interstate bus terminal, BSNL building, National sports complex.

A. Kumar (✉) · L. Sunil Singh
Department of Earth Sciences, Manipur University, Imphal, India

© The Author(s) 2018
A. Singh et al. (eds.), *Development and Disaster Management*,
https://doi.org/10.1007/978-981-10-8485-0_10

Manipur is included in the high seismic zone and a number of small-to-moderate magnitude earthquakes trigger frequently. The Mw 6.7 earthquake occurred on 4 January 2016 in Manipur is the largest event in last 60 years. Prior to this event, the region was visited by an earthquake of 7.3 in 1957. The local network of broadband stations, which are being operated at Manipur University, produced adequate data to analyse this event (Fig. 1).

The epicentre of the event falls at 24.86°N and 93.65°E near the Noney village of Tamenglong district of Manipur. The focal depth of the event has been estimated to be 50 km. The entire NE region of India falls under the Seismic Zone V which is considered to be most vulnerable to earthquake hazard. Figure 2 shows seismicity of NE India recorded for the period of 1964–2015. Manipur University has set up a continuously operating GPS permanent station at Imphal (IMPH) was installed in 2003, later on three more Hengkot (HENG), Sompi (SOMP) and Moreh (MORE) permanent GPS station were added in the region (g.1) with the collaboration with National Geophysical Research Institute (NGRI), Hyderabad. These stations are located in the IndoBurmese Arc (IBA) region. The epicenter of recent Mw 6.7 earthquake occurred on 4 January 2016 in Manipur is located near the Noney Village of Tamenglong district on the west of IBA. The 10 January 1869 Cachar-Manipur earthquake was the most severe earthquake in the available 2000 years of written historical records of Manipur (Singh 1965; Parratt 1999). The sediment-led valley region has remained the centre of inhabitation since historical times. The valley is located in the central part of IBA. Thus, the region could not have escaped from damages caused due to any larger earthquake than that of on 10 January 1869. Therefore, the recent shock (Mw 6.7) has caused more damages in the city of Imphal in comparison with the region surrounding the epicenter.

The area lying close to the epicentre observed a maximum slip of approximately 5 cm due to the main shock. This caused a large number of small fissures and cracks on the ground surface. Post-examining the damages in detail, it was found that the damaged buildings were built with poor construction practices. The event has given an idea about the possibilities of hazard scenarios in Imphal due to local and regional tectonics sources. It was also validating seism tectonics of Burmese Arc region.

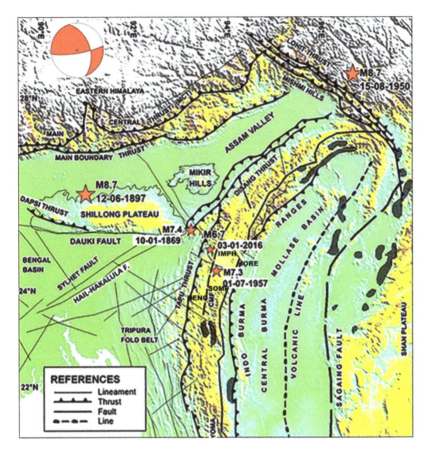

Fig. 1 General tectonics in the Indo-Burmese region and location of GPS permanent station

SPATIAL TECHNOLOGY

GPS spatial geodetic technology assisted in monitoring of the crustal movements. The GPS technology is equipped with high precision observation, fast operating efficiency, and small size equipment. GPS system ensures that at any instance on earth, the satellites can collect data of latitude, longitude and height of the points to provide positioning,

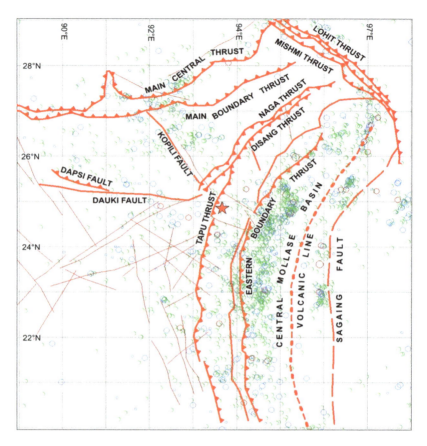

Fig. 2 Map showing plot of seismicity in Northeast India within a window of and 20–29°N latitude and 89–98°E longitude during the period 1964–2015

navigation and timing capabilities. GPS consists of three parts—constellation, ground monitoring part, and signal receivers. GPS positioning is of two types—absolute positioning (directly determining the coordinates of observations stations, where Earth's center is taken as the coordinate point), relative positioning (when GPS receivers were placed at the two ends of baselines).

The absolute positioning system is used to take distance by taking references as the known instantaneous satellite coordinates, in order to determine the user receiver antenna at the corresponding point. It is difficult to maintain strict synchronization with the help of the micro-time satellite bell and the receiver bell; this difference is called the 'pseudo-range'. The satellite bell error can be rectified by making use of navigation messages, but it is not possible to determine the clock error in advance. In order to eliminate the same, relative positioning is used.

The Indian Plate motion which is at the rate of 37 mm/yr (± 0.2 tolerances) has already been estimated towards NNE direction in reference to European plate. This has been found using GNSS receivers and a new global network which spreads across the geographical and azimuthal coverage, including all the plates surrounding India.

GPS ARRAY IN MANIPUR

In the IBA region, there are five permanent GPS towers for monitoring the crustal deformation at Imphal, Manipur. The analysis from these sources suggests that the site moves at the rate of about 36.3 mm/yr (± 0.5 tolerances) towards N55 degrees in the ITRF 2008 (Singh et al. 2014). The plate moves at the rate of 16.77 bmm/year towards N222 degrees (which is towards southwest), with respect to the Indian plates. The site is situated about 15 km east of Churachandpur Mao Fault (CMF), which is reported to accommodate a part of India—Sudan motion. The motion of this site usually stays unaffected due to the earthquakes in the nearby region; however, the 2004 Sumatra–Andaman earthquake had caused a co-seismic displacement of approximately 3.5 mm towards the south-west. This site has a linear motion which might be supported by some seasonal variation, and it does not show any evidence of slow earthquake or estimated slip on the CMF or along the plate boundary.

The earthquake on 4 January 2016 in Manipur with a magnitude of 6.7 had its epicentre 20 km west of the CMF at about 60 km depth. It occurred in a very sparsely populated area where most houses were built usually out of wooden frames. Hence, the damage caused due to the earthquake was minimal. However, it had some ill effect causing damages to building and loss of eight lives in the neighbouring Imphal valley. This earthquake and other historic events reveal the role of local site effect in Imphal valley.

GPS data processing and analysis of GPS data received from the permanent station of IBA are used to determine the quasi-static displacements that were caused due to this earthquake.

The GPS data received from the stations were converted into RINEX observation files and translation, editing and quality checking (TEQC) software was made use of for performing quality checks to the data. Post-quality checking, data with high cycle clips, multipath and of duration more than 18 hours were eliminated.

Figure 3 shows the locations of the GPS stations which are maintained at Manipur University. The data received from these sources are regularly processed at the Department of Earth Sciences, Manipur University,

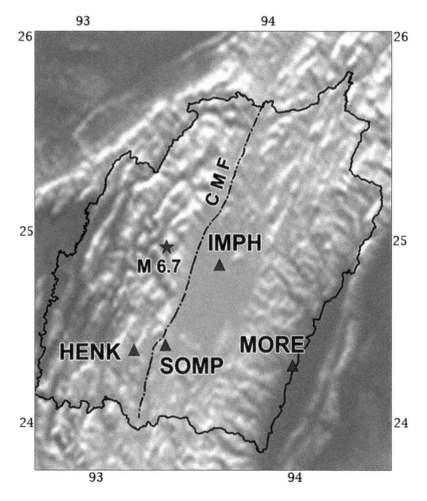

Fig. 3 GPS stations in Manipur

using software like GAMIT/GLOBK (Herring et al. 2010a, b). For analysis purposes, data are also accounted from International GPS service (IGS) sites, namely BAHR, HYDE, LHAS, IISC, KUNM, KIT3 and PLO2. The data from IGS are processed on a daily basis which produces loosely constrained station coordinates and satellite orbits. This data are further combined with loosely constrained solutions of nearby IGS station data available from Scrippe Orbital and Positioning Analysis Centre (SOPAC). Position estimates and velocity stabilization were achieved by making use of GLOBK software (Herring 2005).

Coulomb 3.3 software (Toda et al. 2005; Lin and Stein 2004) is used for calculating stresses changes at any depth caused by fault slip. For the interpretation of our analysis, we used permanent stations that are deeply anchored in bedrock and have proven to have predictable motions that can be adequately modelled as a constant velocity over several years. Out of four continuously operating GPS permanent stations in this region, only IMPH station has been operating for more than 10 years old. So at the time of the earthquake, the other new GPS stations had only been operating for about 1 year, and hence did not had well-determined site velocities. Time series of IMPH site is shown in Fig. 4. In the time

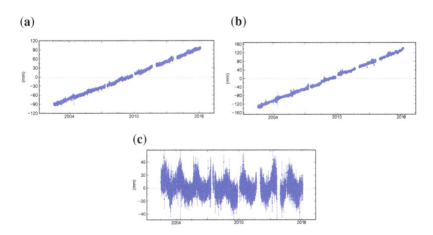

Fig. 4 Time series at IMPH in ITRF 2008. **a** North component (20.9 ± 0.5 mm/yr). **b** East component (29.7 ± 0.5 mm/yr). **c** Up component (0 ± 19 mm/yr)

Table 1 Average co-seismic displacement at permanent GPS stations using GPS data of December 2015 and January 2016

GPS station	Distance from epicentre (km)	North measured (mm)	East measured (mm)
IMPH	30	3±0.5	8±0.4
MORE	85	2±0.3	3±0.2
SOMP	50	5±0.3	7±0.3
HENG	55	3±0.5	3±0.4

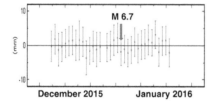 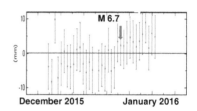

Fig. 5 Time series at IMPH showing coseismic offset due to 2016 Manipur earthquake

series, north component shows deformation 20.9±0.5 mm/yr, east component shows deformation 29.7±0.5 mm/yr and up components shows deformation 0±19 mm/yr. In ITRF 2008 reference frame, the IMPH site shows a velocity of 36.3 mm/year towards N55°. The estimate slope for east component is also given. Co-seismic displacement is observed very clearly from two nearby permanent GPS stations IMPH and SOMP (Table 1). At the time of the shock time series of IMPH, a co-seismic displacement is evidenced of about ~8.5 mm in ENE (Fig. 5). The offset in the North component is not obvious; however, it is quite distinct in the East component. In the North component, the offset is ~3 mm and in the East component ~8 mm. Similar co-seismic displacement is observed of about ~8.6 mm in NE is observed from the continuous operating GPS station SOMP (Fig. 6). In this station, the offset in the North component and East component is quite distinct. In the North component, the offset is ~5 mm and East component is ~7 mm. The time series of the observed displacement (Fig. 6) shows that the main shock is followed by a significant post-seismic signal occurring in the same direction as the co-seismic movement period lasting 10 days shows a rate of about 2 mm/day (during this period, the strong

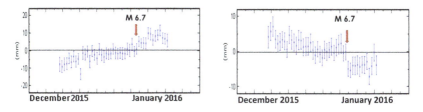

Fig. 6 Time series at SOMP showing coseismic offset due to 2016 Manipur earthquake

aftershock (Mw = 6.7) of January 4 is seen as a step of about 4 mm on the time series. A slight pre-seismic displacement (1–2 mm) is detectable during 2 weeks before the main shock. It is particularly evidenced as a change in the slope of the time series during the 30 days. The coulomb stress change values were compared again with the earthquake focus distributions which were obtained from USGS between the years 1950 and 2016. It is noticed that the earthquakes, which occurred near the modelled fault, are seen on high-stress region (red-coloured areas). The earthquake was coherent with the high-stress region at the North and South boundaries of the modelled fault (Fig. 7).

Conclusions

A transregional or a global seismic monitoring network should be established in order to overcome the problems faced in monitoring and forecasting earthquakes all around the world. GPS can be used to monitor crustal movements in the short-term and mid-term precursors before the earthquake and the instantaneous changes post-earthquakes. This may reveal the characteristics of the crustal movement caused by the earthquake. Few characteristics that are being researched around the world for this analysis include active faults in the displacement and changes, earth's rotation parameter of change, earth's interior structure of inversion.

The crustal deformation studies have proved to be precise in the present context to evaluate the co-seismic displacement before and during the earthquake. The daily analysis of GPS data from the permanent stations along with three newly established stations for a duration of one month prior and 4 months after the earthquake shows a co-seismic displacement of 3–5 mm in the North and 3–8 mm in the East.

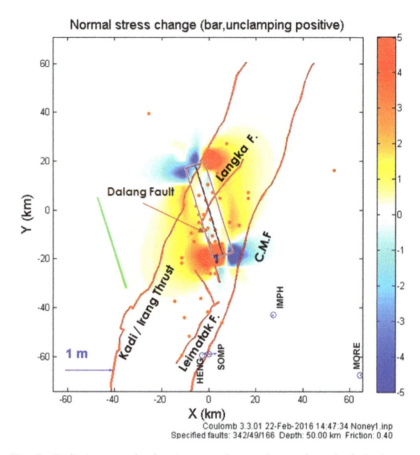

Fig. 7 Preliminary results showing normal stress change along the fault plane

During the time of shock, co-seismic displacement is evidenced of about ~8.5 to ~8.6 mm at the nearby two permanent stations IMPH and SOMP. A weak pre-seismic signal is emerging from the background noise during the month preceding the main shock and is evidenced by a change in the slope of the time series. The main shock is characterized by an abrupt change of the time series. A significant after-slip movement is seen for the month following the co-seismic rupture.

The estimated seismic movement in the region and the stress drop suggest that the hypocentre of the earthquake falls within the interpolate region of IBA. This region has an impact of past-earthquakes which propagated the existing strike-slip fault. The crustal deformation studies are precursory in nature, and continuous monitoring by GPS is one of the techniques which is helpful for earthquake hazards assessment.

Acknowledgements The financial assistance provided by Ministry of Earth Sciences and Ministry of Science & Technology Government of India to carry out the present investigations is thankfully acknowledged.

REFERENCES

Herring, T. A. (2005). *GLOBK: Global Kalman Filter VLBI and GPS Analysis Program Version 10.01*. Cambridge: Massachusetts Institute of Technology.

Herring, T. A., R. W. King, and S. C. McClusky. (2010a). *GAMIT Reference Manual: GPS Analysis at MIT, Release 10.4*. Cambridge: Massachusetts Institute of Technology.

Herring, T. A., R. W. King, and S. C. McClusky. (2010b). *GLOBK Global Kalman Filter VLBI and GPS Analysis Program, Release 10.4*. Cambridge: Massachusetts Institute of Technology.

Lin, J., and R. S. Stein. (2004). Stress Triggering in Thrust and Subduction Earthquakes and Stress Interaction Between the Southern San Andreas and Nearby Thrust and Strike-Slip Faults. *Journal of Geophysical Research* 109, B02303. https://doi.org/10.1029/2003jb002607.

Parratt, Saroj Nalini. (1999). *The Court Chronicle of the Kings of Manipur: 33–1763 CE, The Cheitharol Kumpapa, 1*. London: Routledge, Taylor and Francis Group, p. 305.

Singh, R. K. Jhaljit. (1965). *A Short History of Manipur*. Imphal, India: Manipur Sahitya Parishad, p. 365.

Singh, L. S., V. Gahalaut, and A. Kumar. (2014). Nine Years of GPS Measurements of Crustal Deformation at Imphal, Indo-Burmese Wedge. *Journal of the Geological Society of India* 83(5): 513–516.

Toda, S., R. S. Stein, Keith Richards Dinger, and Bozkurt, S. (2005). Forecasting the Evolution of Seismicity in Southern California: Animations Built on Earthquake Stress Transfer. *Journal of Geophysical Research* 110, B05S16. https://doi.org/10.1029/2004jb003415.

Post-Earthquake Geodisaster of 6.7 Mw Manipur Earthquake 2016

P. S. Ningthoujam, L. K. Lolee, C. S. Dubey,
Z. P. Gonmei, L. Thoithoi and T. Oinam

INTRODUCTION

In the wee morning hours of 4 January 2016, people in major parts of Northeast India, West Bengal, and Bangladesh were forced out of their beds. An earthquake of 6.7 Mw magnitude had struck the region. According to United State Geological Survey (USGS), the earthquake which hit at 4.35 a.m. (IST) on 4 January was located in an isolated remote rugged terrain of western Manipur with its epicentre located in a remote village near Noney in western Manipur. The region of Northeast India had been categorized by Geological Society of India (GSI) as Zone V in their Seismic Zoning Map of India (Bureau of Indian Standard). At least eleven people were killed, and over 200 others were injured and numerous buildings were damaged.

The study of post-earthquake effects is very important for any major earthquakes as it can prevent further future disaster caused by landslides, soil liquefaction, fire, and flood, etc. This study, therefore, focuses on the

---

P. S. Ningthoujam (✉) · L. K. Lolee · C. S. Dubey ·
Z. P. Gonmei · L. Thoithoi · T. Oinam
Department of Geology, University of Delhi, New Delhi, India

© The Author(s) 2018

A. Singh et al. (eds.), *Development and Disaster Management*,
https://doi.org/10.1007/978-981-10-8485-0_11

analysis of post-earthquake geological and geomorphic changes in parts of Imphal and Tamenglong districts of Manipur. Study of such features will help us to understand the causes of the earthquake, nature of the movement of crust and any possible future disaster. In this study, we will analyse cracks, landslides, subsidence-uplift of land, groundwater seepage, etc.

STUDY AREA

USGS reported that the epicentre of the 4 January 2016 earthquake of western Manipur lies at latitude 24.804°N and longitude 93.650°E, near Noney town with an uncertainty of 5.7 km. The earthquake depth was reported to be 55 km below the surface (USGS). Geologically, the study area lies within Disang, Barail and Surma Group of Tertiary period (Ningthoujam et al. 2012). Disang is soft, dark grey splintery shales and intercalation of shale siltstones with sandstones. Barail Group is subdivided into three formations: Laisong, Jenam and Renji, stratigraphically arranged and comprised mostly of massive to thickly bedded sandstones with intercalations of shale (GSI 2011; Ningthoujam et al. 2012; and Gonmei 2015). In the study area, Surma Group is exposed as Bhuban and Bokabil Formations, which are thick beds of fine-to-medium sandstones with intercalations of sandy claystone sandstones (GSI 2010; Gonmei 2015).

Structurally, the study area falls within the Schuppen belt of Indo-Myanmar Subduction Zone (IMSZ) in western Manipur (Ningthoujam et al. 2012, 2015). There are two major structural features truncating the region: first the N–S trending thrust-fold system as a result of the E–W convergence of Indian Plate and Sunda Plate (Brunnschweiler 1966; Nandy 1981); second the NNW–SSE trending strike-slip fault systems (Gonmei 2015), spawned as a result of the clockwise-dextral rotational movement of the region (Gahalaut and Gahalaut 2007).

GEODISASTER AND POST-EARTHQUAKE MANAGEMENT PLAN OF THE STUDY AREA

For every earthquake disaster, a natural phenomenon which happens without any warnings, there are various pre- and post-management plans mapped out by Government of India and Government of Manipur such as earthquake preparedness guide for home dwellers, earthquake safe constructions and buildings, retrofitting of existing buildings. However, these plans are viable only for urban areas which are well connected with the nearest locations with modern amenities and basic supplies. The study area where the epicentre is located lies in a remote terrain which is not well connected. Therefore, every possible measure for awareness of earthquake based on geology has to be given first.

Imphal Valley

Imphal valley is the state capital and has a population of more than 270,000 (Census of India 2011). As per USGS report, maximum portion of the valley has experienced the shaking with intensity more than 6.0 Mw. However, some portion of the north-western part of the valley was struck by more than 6.5 intensity. Lots of buildings (mostly concrete, RCC) were damaged in the event; however, there are no reports of any geological changes in the area.

Maranging

USGS in their preliminary information has reported the epicentre of the event in the remote hill of Maranging. In this area, tunnelling work of Indian railways is in progress. The information suggests that the epicentre was 600 m north-east to the upslope of Tunnel no. 26 P1. We have done detailed field work in the area to find out geological and geomorphological changes; however, any sign of variations were observed in the area. In this area bed, rocks are in intact position.

Kabui Khullen

Kabui Khullen lies 27 km north of the Noney town in Tamenglong district. This village is very remote and is connected with the nearest town Noney by a kuccha jeepable road—a road which is the lifeline for the villagers—that could only be used during the dry seasons. The village covers an area of 679 m × 50 m, and there are 58 houses in the village with a population of 385. The maximum amount of destruction was observed in this remote village. Many houses were destroyed; however luckily, there were no fatalities. The village lies at the crest of a ridge (spur) and the spur align at NW–SE direction. The spur is covered with 1–5 ft overburden, and bedrocks in the area are weathered. The rock layer beneath

164 P. S. NINGTHOUJAM ET AL.

the spur is comprised of intercalated layers of thinly bedded sandstone and shales. The bedding of the rocks is gently dipping towards NE with an angle of 4–10 degree. Many cracks were developed along this crest of the spur. The total length of some major cracks was found to be around 900 m. Maximum cracks were observed here in this remote village with 11 parallel cracks spaced by 2–4 ft apart from each other in the football ground of the village.

From the perspective of post-earthquake disaster, the village can be divided into two portions: southern and northern.

Southern Portion
There is deep gorge of 600 m with aslope angle of 45–60 degrees in the south of the village, and in the recent earthquake, cracks are developed 10–15 m away from the gorge in the southern portion of the village. Therefore, there are very high chances of collapse/sliding of the southern portion of the village in rainy season or in future earthquakes.

Northern Portion
While in the north of the northern portion of the village, there is a 200-m-deep slope dipping 25–30 degrees towards the north. There is some cultivated land in the northern slope area that indicates that the area is covered with soil. Therefore, even though cracks are developed in this portion, there are very low chances of a collapse of the region.

This study concludes that the southern portion of the Kabui Khullen village will be not safe in the rainy season and future earthquakes. Therefore, the villagers in the southern portion of the village should be advised

1. To evacuate their houses in times of heavy rainfall, earthquake events.
2. To take shelter in stable and safer northern portion of the village in time of extreme events.

Noney Area

Noney is a large village located in Nungba of Tamenglong district, Manipur, with a total of 635 families residing. The Noney village has a population of 3854 as per Population Census 2011. This village lies in ~4 km. North-west of the reported epicenter (USGS) of Maranging.

Politically, the village is divided into 4 parts, viz. Noney part I, II, III and IV. The locations of post-earthquake effects in Noney. In Noney part I, cracks were developed along the slope of the hill of Noney part I. A new spring has also emerged in the hill slope of Noney part II, Lungkhuijang.

Noney Part I
Detailed study of the geological and geomorphological conditions of the area was carried out in the area to know the causes and possible future disaster. The study makes us to know that the cracks are developed in the old land-slided zone that has a thick overburden cover of 10–20 ft. Villagers also reported that the area downslope of the crack is a sinking zone. Study of DEM of the area revealed that the slope of the area is 20–30 degrees towards east (N100), and cracks are aligned perpendicular to the slope (N20E). Therefore, it can be concluded that the cracks were originated due to earthquake-triggered sliding of some unstable portion of the overburden towards downslope. Gravity pulling further widens up the cracks.

In the field investigation, a subsidence of 6 inches was observed in the overburden of NE portion of the crack, and villagers have reported widening of cracks (up to 4–5 days). This may be due to gravity pulling, and in the future rainy season, when the overburden gets saturated, the condition may get worsen.

Therefore, we would like to advise for the construction of an RCC wall 20–30 ft downslope of the cracked area. It will stop the creeping of overburden in the slope and the area will be stabilized. Emission of smoke was reported from a major crack, and the villagers were alarmed thinking it was a kind of volcanic smoke. We have explained to them that it has nothing to do with any sort of volcano. Actual evidence suggests that it was water vapour rising out due to the difference in the air temperature between the soil layer and the air.

Noney Part II
As an effect of post-earthquake, seepage of water was reported in Noney part II (Lungkhuijang). The villagers were frightened by this unprecedented condition. Spot analysis of the event reveals that the water seepage is due to immergence of a new spring from a dried-up drainage at an elevation of 840 m. The earthquake has triggered a small landslide in the location, and 4 springs holes were exposed in the carved out portion.

Geological investigation of the area revealed that the area is covered with 5–8 ft-thick soil layer, and they are underlained by weathered bedrock dipping towards the west. Bedrock in the area comprises of weather Disang shales intercalated with thick clay layers. Weather Disang shales are good aquifer rock, while clay layers are impermeable to water. This condition developed from a pocket of perched ground water in the subsurface of the area. Effect of the recent earthquake has made soil cover above the perched water pocket to slided down and the water in the pocket flows down as spring. A spring is a point where water flows out of the ground. The flow of water may continue depending on the amount of water stored in the pocket. The spring water is safe to use, and it will be beneficial to villagers as well.

Conclusions

This study focuses on post-earthquake geodisaster study of 6.7 Mw Manipur earthquake 2016 to prevent from further future disaster triggered by the event. The most significant amount of geodisaster due to the event was observed in Kabui Khullen village where 11 parallel cracks spaced by 2–4 ft apart were developed. Since the village lies in the crest of a spur, there is a very high chance of massive landsliding that may destroy the southern portion of the village. The villagers were advised to evacuate their houses to take shelter in the stable and safer northern portion of the village if the landsliding continue due to future extreme events. An earthquake-triggered landslide was also observed along the slope of a hill in Noney part I village due to which cracks were generated in the scarp, and we have advised for the construction of an RCC wall 20–30 ft downslope of the cracked area to stop the creeping of overburden in the slope.

References

Brunnschweiler, R. O. (1966). On the Geology of the Indo-Burma Range. *Geological Society of Australia, 13*, 127–194.

Gahalaut, V. K., & Gahalaut, K. (2007). Burma Plate Motion. *Journal of Geophysical Research, 112*(B10), 402.

Geological Survey of India (GSI). (2010). Geological Map of Manipur, from https://www.gsi.gov.in/webcenter/portal/.

Geological Survey of India (GSI). (2011). *Geology and Mineral Resources of Manipur, Mizoram, Nagaland and Tripura, Miscellaneous.* Publication No. 30 Part IV, 1, no. (Part 2).

Gonmei, Z. P. (2015). Crustal Shortening and Tectonic Studies in Western Manipur, Indo-Myanmar Range. Unpublished M.Sc. Dissertation, University of Delhi.

Nandy, D. R. (1981). Tectonic Pattern in NE India. *Indian Journal of Earth Science, 7*(1), 103–107.

Ningthoujam, P. S., Dubey, C. S., Guillot, S., Fagion, A. S., & Shukla, D. P. (2012). Origin and Serpentinization of Ultramafic Rocks of Manipur Ophiolite Complex in the Indo-Myanmar Subduction Zone, Northeast India. *Journal of Asian Earth Sciences, 50*, 128–140.

Ningthoujam, P. S., Dubey, C. S., Lolee L. K., Shukla, D. P., Naorem, S. S., & Singh, S. K. (2015). Tectonic Studies and Crustal Shortening Across Easternmost Arunachal Himalaya. *Journal of Asian Earth Sciences, 111*, 339–349.

United State Geological Society (USGS). (2016). http://earthquake.usgs.gov/earthquakes/.

www.bis.org.in/other/quake.html IS 1893 (Part 1): 2002 Criteria for Earthquake Resistant Design of Structures.

www.census2011.co.in/census/city/184-imphal.html.

Landslide Zonation in Manipur Using Remote Sensing and GIS Technologies

R. K. Chingkhei

LANDSLIDES A NATURAL RECURRENCE OF MANIPUR HILLS

As we know that landslide is the movement of mass, rock, debris or earth down a slope,[1] as one of the major natural hazards, accounts each year for enormous property damage in terms of both direct and indirect costs. Landslide is a regular natural phenomenon in most of the northeastern states of India and Manipur is not an exception. Manipur being a hilly state, land-slides of various types including mudslides are quite common mostly during the rainy season (June–September). This phenomenon occurs mostly in the hilly regions of Senapati, Tamenglong, Ukhrul and Chandel districts. Again, since Manipur is part of Himalaya that is geologically young and geodynamically active to triggering large number of earthquakes and intensive soil erosion, is highly prone to landslide hazards. The risk and vulnerability of these districts has increased during the last few decades due to increase in population and their properties.

---

[1] Cruden (1991).

R. K. Chingkhei (✉)
Department of Forestry and Environmental Science,
Manipur University, Imphal, Manipur, India

© The Author(s) 2018
A. Singh et al. (eds.), *Development and Disaster Management*,
https://doi.org/10.1007/978-981-10-8485-0_12

The migration of the villagers towards these highways, to increase their income from the cash crops and other related activities, is also one of the triggering factors for landslide hazards due to unplanned and unscientific approach in their expansion.

Some of the major landslides that have affected the State of Manipur during the last few decades of worth mentioning are the Nung Dolan (Sivilon) Landslide of 22 August 2003 along the NH-37 that results in the highway blockade for about a month. Subsequently on 7 September 2010, 22 June 2012, July 2013 and July 2016 also, the incident recurred. The slide material is mostly rocks. Along the NH-2 mention may be made of simultaneous landslides at Mouzhu (3 km from Tadubi) and Tadubi on 21 May 2002. The slide at Mouzhu recurred on 5 August 2002. In 2004 severe landslides and mudflow affected Phikomei (Mao) and Gopibung and several minor landslides occurred simultaneously along NH-2. The Phikomei landslide occurred on 9 July 2004 that rejuvenated again on 11 and 14 July 2004 destroying at least 50 houses and more than 100 houses were badly affected. The Gopibung mudflow occurred on 10 July 2004 and destroyed 10 houses and many hectares of paddy fields. On 8 September 2007 landslide occurred at lower Sajouba, about 2.5 km from Tadubi towards Tadubi-Ukhrul road, and damaged 49 houses and many paddy fields. On the same date simultaneous landslide was also triggered at Tadubi. On 29 July 2010 landslide occurred near Maram bazaar blocking the highway for about three days. During April and July months of 2015 few landslides also occurred at Senapati district along the NH-2 at Maram Paren and near Senapati Old District Hospital area that hampered the transportation for few days. Yet in a very rare case of landslide during August 2015 along NH-2 at Joumol village located under Khengjoi Sub-Division in Chandel District 20 villagers died and 12 houses were swept away. During 2015 several landslide incidences were reported along NH-202 at different locations that hampered the normal life of the villagers for few days. It was also reported that the NH-202 has sunk at least three feet over a stretch of 50 m at Naga gate near Ukhrul district hospital, Hundung due to heavy rainfall.

It is quite clear by now that landslides, a regular natural event, affect the people of Manipur to a great extend and the government needs to develop serious action plans base on scientific inputs on how to mitigate the damages in the near future. It is with these few points in mind that the present study has been taken up to identify the vulnerable areas and map these areas as various hazard zones and ultimately to develop the management map to mitigate the hazard. To achieve these goals certain objectives has been set viz. (1) preparation of parametric thematic layers from various sources in GIS domain. These layers include lithology, geomorphology, lineament, fault, percent slope, slope morphology, slope aspect, drainage, landslide incidence, soil texture, soil depth, rock weathering, slope-dip relation, landuse/landcover; (2) detailed field to generate a soil & terrain information system; (3) generation and integration of GIS thematic layer, and (4) preparation of Landslide Hazards Zonation (LHZ) maps and Landslide Hazard Management (LHM) maps through the integration of various thematic maps using GIS software. The integration of GIS with Remote Sensing derived thematic maps may highly facilitate the assessment and estimation of Regional Landslide Hazards.[2] This integrated approach of Remote Sensing and GIS is highly useful in evaluation, management and monitoring of natural hazards such as landslides in a short period of time. Routine use of remote sensing data and its analysis of a hazards prone area also help to monitor the changes in surface feature.

The hazard zones in a LHZ map contains useful information regarding the varying amount of stability that are calculated based on the importance of the causative factors of instability. The LHZ maps can provide useful inputs to the planners thereby enhancing the implementation processes of various developmental schemes, especially in the hilly regions.[3] If we can identify areas with different landslide hazard zones then, it will be easier for the planners to develop appropriate mitigation programme suitable for that particular area. The LHM map on the other hand provides useful mitigation information based on the input parameters of a region that can be utilized by planners during programme formulation.

STUDY AREA

The National Highways under study are NH-39 (NH-2), NH-53 (NH-37) and NH-150 (NH-202) that connects Imphal-Mao; Imphal—Jiribam and Imphal—Ukhrul respectively in the State of Manipur (Fig. 1).

---

[2] Yuan et al. (1997).
[3] Anabalagan (1992).

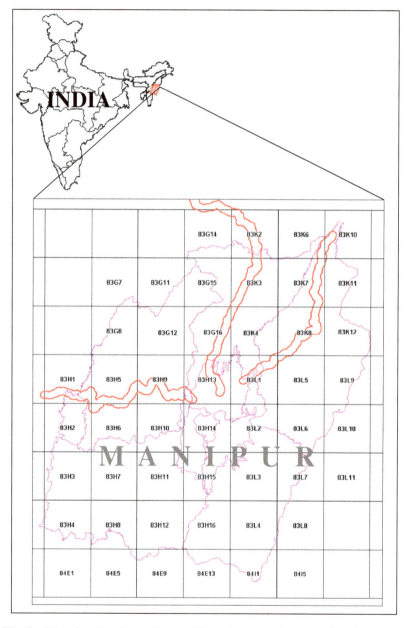

Fig. 1 Map showing the study area. The red polygon indicates the 2 km buffer area along the national highway where study has been conducted

Manipur lies between the latitudes 23.830°N to 25.680°N and longitudes 93.030°E to 94.780°E with annual average rainfall of 1467.5 mm. The altitude ranges from 40 m at Jiribam to 2994 m at Mt. Iso Peak near Mao Songsong. Geologically the state is of recent origin. It is a part of the Trans-Himalayan geological formation from the sea of Tethys in the Archaean period about one billion years ago and is closely linked with the evolution of Neogene Surma basin, Inner Palaeogene fold belt and Ophiolite suture zone associated with Late Mesozoic-Tertiary sediments. The soil of Manipur can be broadly divided into two viz. the red ferrogenous soil in the hill area and the alluvium in the valley with normal pH value ranging from 5.4 to 6.8. In general, the soil contains small rock fragments, sand and sandy clay and depicts various forms. Comparatively the top soils on the steep slopes are very thin and are prone to high erosion resulting into formation of sheets and gullies and barren rock slopes whereas in the plain areas, especially flood plains and deltas, the soil is of considerable thickness. During the last few decades the villagers have migrated towards the National Highways with a view to improve their economy by doing some appropriate roadside business based on their agricultural/cash cropping.

METHODOLOGY

All the thematic layers corresponding to the causative factors responsible for the occurrence of landslides in the region are prepared from the Remote Sensing data and 1:50,000 topographic maps of Survey of India (SOI) supported by the field data in GIS domain. The systematic combination of these GIS themes, based on their assigned relative importance weightages, generates the zonal hazard map.[4, 5, 6, 7] The detailed methodology adopted in the present study is shown in the flow chart (Fig. 2).

[4] Sarkar et al. (2004).
[5] Kanungo et al. (2006).
[6] John Mathew et al. (2007).
[7] Gupta et al. (1999).

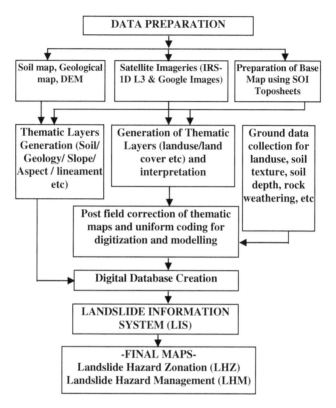

Fig. 2 Flow chart of the steps involved in the present study

Preparation of Base Maps

The base maps of the study area are prepared using 1:50,000 SOI toposheet. The base features like major roads, other roads, rivers and streams, and settlements, anthropogenic factors, were incorporated in the base maps.

Preparation of Input Layers

Input Thematic layers were prepared in GIS domain using ARCGIS and ERDAS. These thematic layers include Anthropology, Baseline, Basepoly, Basepoint, Bound, Drainage, Fault, Geology, Geomorphology,

LANDSLIDE ZONATION IN MANIPUR USING REMOTE SENSING ...   175

Table 1   Parameters used in the preparation of LHZ and LHM maps

S. No.	Parameters	Categories	GIS input	Feature type
1	Anthropology	(1) Road/slope cutting (2) Mining (3) Toe removal/erosion/cutting (4) No activity	anthro	Polygon
2	Drainage		drainage	Line
3	Fault	(1) Minor fault (0–500 m) (2) Major fault (500–1500 m) (3) Mega fault (>1500 m) (4) Thrust	fault	Line
4	Geomorphology	(1) Low dissected denudo structural hill (2) Moderately dissected denudo structural hill (3) Highly dissected denudo structural hill (4) Alluvial fans (5) River terraces (6) Flood plain (7) Valley (8) Toe removal/erosion/cutting (by river) (9) Mass wasting deposit (10) Scree	geom	Polygon
5	Geology	(1) Disang (2) Barail (3) Surma (4) Tipam	geo	Polygon
6	Landuse	(1) Dense vegetation (2) Medium vegetation (3) Degraded vegetation (4) Barren land (rocky) (5) Barren land (non-rocky, waste land) (6) Agricultural land (7) Water body	landuse	Polygon
7	Landslide	(1) Active (2) Old	lslide	Polygon
8	Lineament	(1) Minor lineament (0–500 m) (2) Major lineament (500–1500 m) (3) Mega lineament (>1500 m)	lineam	Line

(continued)

176   R. K. CHINGKHEI

Table 1   (continued)

S. No.	Parameters	Categories	GIS input	Feature type
9	Lithology	(1) Shale (2) Sandstone (3) Sandstone with shale/siltstone (4) Unconsolidated sediment/ material	litho	Polygon
10	Slope aspect	(1) North facing (2) South facing (3) NE facing (4) NW facing (5) SE facing (6) SW facing (7) East facing (8) West facing (9) Flat	asp	Polygon
11	Slope dip	(1) Dip facets parallel to slope (2) Dip facets opposite to slope	sldip	Polygon
12	Slope morphology	(1) concave (2) convex (3) Straight (4) Break in slope	slmorph	Polygon
13	Slope	(1) 0–15 (2) 15–25 (3) 25–30 (4) 30–35 (5) 35–40 (6) 40–45 (7) 45–60 (8) >60	slope	Polygon
14	Soil depth	(1) <50 cm (2) >50 cm	soildep	Polygon
15	Soil texture	(1) Rock outcrop (2) Coarse (3) Medium (4) Fine	soiltext	Polygon
16	Rock weathering	(1) Low (2) Moderate (3) High (4) Very high (5) Very low to nil	weath	Polygon

Landslide, Landuse, Lineament, Lithology, Slope Aspect, Slope Dip, Slope Morphology, Slope, Soil Depth, Soil Texture, Rock weathering. The sources of these layers are the SOI topographic maps (1:50,000 scale), Remote Sensing data (IRS 1D LISS-3 satellite data), climate and earthquake data (India Meteorological Department), soil maps (National Bureau of Soil Survey and Landuse Planning), geological maps (Geological Survey of India). Besides, the field data, published literatures and other related data are also used during the study. A brief description of the input layers that are used during the generation of LHZ and LHM maps is given in Table 1.

FIELD DATA COLLECTION

Using the information from the basemap a detailed field investigation is carried out in the study area to check and validate the thematic maps in the field and also to collect field data pertaining to landuse/landcover and landslide triggering factors. During the field visit, verification of the thematic maps such as different geomorphological features, landuse/landcover categories, landslide types, slope forms, major and minor fault and lineaments are checked. Detailed field data about type of landuse/landcover, structural discontinuities along which the failure is taking place, hydrological and slope conditions, anthropogenic activity, mechanism of failure, existing stability measures and vulnerability are collected. The soil data are also collected during the field. Various field photographs are also taken for post field data rectifications. This field information is then used to update the thematic maps and necessary modified are done accordingly.

FINDINGS AND SUGGESTIONS

Prediction of potential landslide areas has been very difficult because of the complexity of the factors involved and the relationship to each other which is wide ranging. Selection of factors and preparation of corresponding thematic data layers are crucial components of any model for landslide hazard zonation and management mapping.[8] In order to generate the landslide hazard zonation, map a model has been developed

[8] Chingkhei et al. (2013).

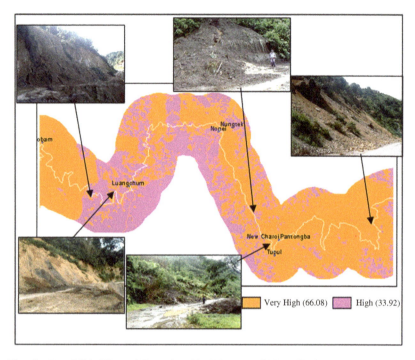

Fig. 3 Landslide Hazard Zonation (LHZ) map of 83 H/9 along NH-37

in a GIS environment. Data in the form of parametric thematic layers as given in Table 1 are put into GIS and using Landslide Information System (LIS) the LHZ and LHM maps are generated. In this paper, only the representative example of LHZ and LHM maps of the Highways are shown (Figs. 3, 4, 5, 6, 7, and 8) to reduce the space and congestion. The LIS software categorizes six hazard zones and each zone is assigned a colour where lower wavelength colours represents the low hazard zones while the high hazard zones by the higher wavelength colours in the maps. These categorization of landslide hazard zones is based on ranking and weightage assigned to each categories in different thematic layers prepared. The values obtained at the lower ends are classed as low and very low hazard zones represents relatively stable while those at higher ends severe, very high and high hat indicates relatively less stable. The middle values (moderate zone) indicate moderate instability of the area.

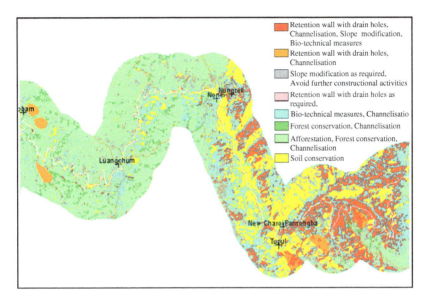

Fig. 4   Landslide Hazard Management (LHM) map of 83 H/9 along NH-37

The management maps show suggested management practices having various combinations (Figs. 4, 6, and 8). The management practice is solely dependent on the input parameters. Altogether, there are there are twenty-one suggested management practices that are already fed to the LIS (Table 2). Out of these only those management suggestions relevant to the area under study are shown on the LHM maps.

LHZ AND LHM MAPS ALONG NH-37

The LHZ and LHM maps along this road section have been prepared based on the area extent of 1:50,000 SOI toposheet. The LHZ and LHM maps of SOI toposheet 83-H/9 between Kotlen and Taobam has been selected as an example along this road as this section of road shows maximum landslide incidences. The LHZ shows two landslide hazard zones viz. very high and high. The area percentages of very high and high are 66.08 and 33.92% respectively (Fig. 3). The overall result suggests that the area is highly prone to landslide hazard. After the creation of the Landslide Hazard Zonation map validation were done by direct

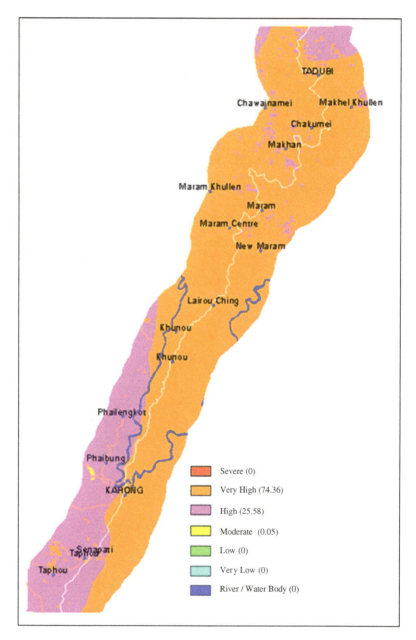

Fig. 5 Landslide Hazard Zonation (LHZ) map of 83-K/3 along NH-2

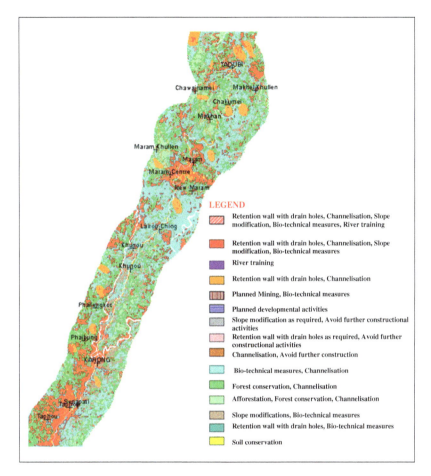

Fig. 6   Landslide Hazard Management (LHM) map of 83-K/3 along NH-2

field visit and it is found that the areas given in the zonation map which have very high zone have active landslides which confirms the accuracy of the LIS generated Landslide Hazard Map (Fig. 3).

The LHM map (Fig. 4) suggest construction of Retention wall with drain holes, Channelization, Slope modification, Bio-technical measures; Retention wall with drain holes, Channelization; Slope modification as required, avoid further constructional activities; Retention wall with drain holes as required, avoid further constructional activities;

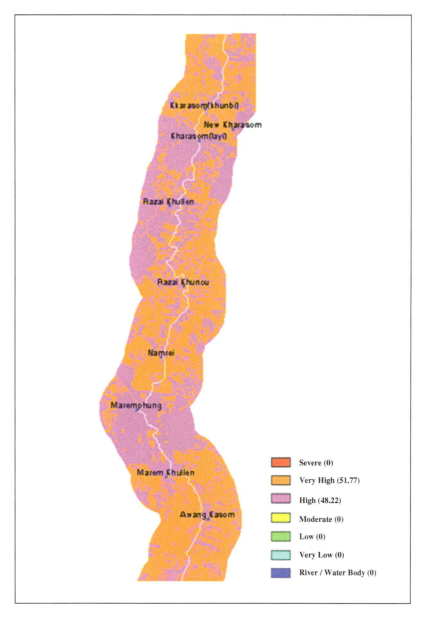

Fig. 7 Landslide Hazard Zonation (LHZ) map of 83-K/7 along NH-202

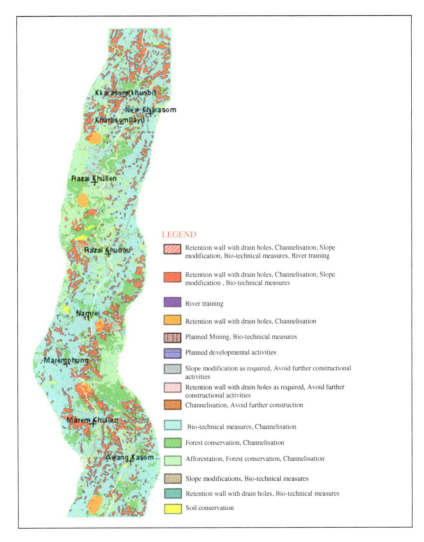

Fig. 8   Landslide Hazard Management (LHM) map of 83-K/7 along NH-202

184   R. K. CHINGKHEI

Table 2   Landslide management scheme in LIS

S. No.	Suggested management
1	Forest conservation and channelization
2	Afforestation, forest conservation and channelization
3	Slope modification and afforestation
4	Afforestation and channelization
5	Bio-technical measures
6	Slope modification, bio-technical measures, controlled grazing and grassland development
7	Bio-technical measures and controlled grazing
8	Grassland development and channelization
9	Slope modification, terrace cultivation and grassland development
10	Retention wall with drain holes and bio-technical measures
11	Channelization and bio-technical measures
12	Soil conservation
13	Planned development activity
14	Retention wall with drain holes, channelization, slope modification and bio-technical measures
15	Channelization and bio-technical measures
16	Retention wall with slope modification along the buffer zone (25 m each side)
17	Retention wall with drain holes and channel diversion, anchoring and bolting
18	River training
19	Slope modifications as required and avoid further constructional activity
20	Retention wall with drain holes as required and avoid further constructional activity
21	Channelization and avoid further construction

Bio-technical measures, Channelization; Forest conservation, channelization; Afforestation, forest conservation, channelization etc. These management practices are location specific and recommended as per the site conditions.

LHZ and LHM Maps Along NH-2

Like in NH-37 the LHZ and LHM maps along this road section have been prepared based on the area extent of 1:50,000 SOI toposheet. The LHZ and LHM maps of SOI toposheet 83-K/3 between Taphou and Tadubi has been selected as an example for this road. The LHZ shows three landslide hazard zones viz. very high, high and moderate. The area percentages of very high, high and moderate are 74.36, 25.58 and 0.05% respectively (Fig. 5). The overall result suggests that the area is highly prone to landslide hazard. The LHM map provides various management practices as show in the map (Fig. 6).

LHZ and LHM Maps Along NH-202

The LHZ and LHM maps of SOI toposheet 83-K/7 between Awang Kasom and Kharasom (khunbi) has been selected as an example for this road. The LHZ shows two landslide hazard zones viz. very high and high. The area percentages of very high is 51.77% and that of high is 48.22% (Fig. 7). Considering the total percentage of the two zones it may be said that the area is highly prone to landslide hazard. The overall result suggests that the area is highly prone to landslide hazard. The LHM map provides various management practices as show in the map (Fig. 8).

CONCLUSION

Information on various degree of landslide hazard exposure of an area is highly needed to assess the potentiality of landslide occurrence so that appropriate mitigation programme may be formulated. The conventional methods may provide good results but is time consuming but with the advancement in the field of remote sensing and GIS techniques time constrain and the multi-parametric analysis at a single run to produce reliable information is now possible. In the present study the integration of Remote sensing and GIS techniques (LIS) for the generation of LHZ and LHM has proven its efficiency in providing valuable information in less time. Moreover, since the data in the GIS platform can be updated anytime, subsequent analysis and monitoring can be done with ease at in future also. The landslides management map provides the remedial measures for future.

Acknowledgements The author is thankful to Prof Arun Kumar, Department of Earth Sciences, Manipur University, for providing all the necessary support in conducting the present study. The financial assistance given by DTRL, Ministry of Defence, New Delhi is also thankfully acknowledged.

References

Anabalagan, R. (1992). Landslide Hazard Evaluation and Zonation Mapping in Mountainous Terrain. *Engineering Geology, 32*, 269–277.

Chingkhei, R. K., Shiroyleima, A., Robert Singh, L., & Kumar, A. (2013). Landslide Hazard Zonation in NH-1A in Kashmir Himalaya, India. *International Journal of Geosciences, 4*(10), 1501–1508. ISSN: 2156-8367. https://doi.org/10.4236/ijg.2013.410147. http://www.scirp.org/journal/ijg.

Cruden, D. M. (1991). A Simple Definition of Landslide. *Bulletin of the International Association of Engineering Geology, 43*, 27–29.

Gupta, R. P., Saha, A. K., Arora, M. K., & Kumar, A. (1999). Landslide Hazard Zonation in a Part of Bhagirathi Valley, Garhwal Himalayas, Using Integrated Remote Sensing—GIS. *Himalayan Geology, 20*(2), 71–85.

John Mathew, V. K. J., & Rawat, G. S. (2007). Weights of Evidence Modelling for Landslide Hazard Zonation Mapping in Part of Bhagirathi Valley, Uttarakhand. *Current Science, 92*(5), 628–638.

Kanungo, D. P., Arora, M. K., Sarkar, S., & Gupta, R. P. (2006). Comparative Study of Conventional, ANN Black Box, Fuzzy and Combined Neural and Fuzzy Weighting Procedures for Landslide Susceptibility Zonation in Darjeeling Himalayas. *Engineering Geology, 85*, 347–366.

Sarkar, S., & Kanungo, D. P. (2004). An Integrated Approach for Landslide Susceptibility Mapping Using Remote Sensing and GIS. *Photogrammetric Engineering & Remote Sensing, 70*(5), 617–625. American Society for Photogrammetry and Remote Sensing.

Yuan, R. K. S., & Mohd, I. S. (1997). Integration of Remote Sensing and GIS Techniques for Landslide Applications. *18th Asian Conference on Remote Sensing (ACRS)*. Malaysia, Poster Session 3.

Building a Resilient Community Against Forest Fire Disasters in the Northeast India

P. K. Joshi, Anusheema Chakraborty and Roopam Shukla

People (tribes) should develop along the lines of their own genius and we should avoid imposing anything on them. We should try to encourage in every way their own traditional arts and culture.
—Pt. Jawaharlal Nehru (1958)

THE PROBLEM

Forest fire distribution, severity and consequences are closely linked to modern human activities. Fire events are major cause of change in forest structure and function, challenging the supply of ecosystem goods and services. Among the different floristic regions, the Northeast India suffers the maximum due to spread of forest fires from age-old practice of shifting cultivation, Jhum fields. This happens in addition to the human negligence while interacting with forest resources. Newer approaches of forest fire management are required to solve the current problems and potential future challenges. Approaches that can ensure monitoring,

---

P. K. Joshi (✉)
School of Environmental Sciences, Jawaharlal Nehru University,
New Delhi, India

A. Chakraborty · R. Shukla
Department of Natural Resources, TERI University,
New Delhi, India

© The Author(s) 2018   187
A. Singh et al. (eds.), *Development and Disaster Management*,
https://doi.org/10.1007/978-981-10-8485-0_13

188 P. K. JOSHI ET AL.

interpretation, synthesis, archiving and distribution of forest fire incidences to the different stakeholders for timely decision-making are of utmost importance. This would require subscription to an early warning system for forest fires through effective risk modelling. Therefore, we advocate the use of satellite remote sensing inputs and geographical information system (GIS) that can assist in the identification of fire-prone areas. With this background, the focus of this article is on the broader aspects of forest vulnerability to fire, current methods used in mitigating its impacts and the need for more participative forest fire management plans in Northeast India.

INTRODUCTION

India is one among the biodiversity hot spots of the world. This is due to its rich and endemic forest resources across the Himalayas and Western Ghats. As per the Indian State of Forest Report 2015, India has forest cover of 701,673 sq. km, comprising 21.31% of the total geographic area of the country. The major forest cover in India is located in the north-eastern states (171,964 sq. km, 65.59%), which is a quarter of the total forest cover in the country (Table 1). Undoubtedly, the forests of Northeast India are rich in biodiversity, making it a biodiversity hot spot, with a complex cultural history. Apart from the rich biodiversity of the north-eastern forests, they regulate hydrological flows and provide

Table 1 Forest cover in north-eastern states of India (*Source* India State of Forest Report 2015)

State(s)	GA	VDF	MDF	OF	TT	PER
Arunachal Pradesh	83,743	20,804	21,301	15,143	67,248	80.30
Assam	78,438	1441	11,268	14,914	27,623	35.22
Manipur	22,327	727	5925	10,342	16,994	76.11
Meghalaya	22,429	449	9584	7184	17,217	76.76
Mizoram	21,081	138	5858	12,752	18,748	88.93
Nagaland	16,579	1296	4685	6975	12,966	78.21
Sikkim	7096	500	2160	697	3357	47.31
Tripura	10,486	113	4609	3089	7811	74.49
Grand total	262,179	25,468	75,400	71,096	171,964	65.59

GA Geographical Area; *VDF* Very Dense Forest (canopy density ≥ 70%); *MDF* Medium Dense Forest (canopy density between 40 and 70%); *OF* Open Forest (canopy density between 10 and 40%); *TT* Total Forest Cover; *PER* Percentage of Geographical Area numerous ecosystem services to some of the world's most densely populated agricultural lands. In the recent decades, rates of deforestation and forest degradation have increased rapidly due to shortened swidden (shifting cultivation or slash-and-burn or *Jhum*) cultivation cycles increasing land-clearing pressures with less time for regeneration of forests.

The region is diverse in many senses, which include topography, social beliefs and practices, forest types and their distribution. The causes and drivers of forest degradation and deforestation in the region are common which include shifting cultivation, requirement of timber, field wood and fodder, mining and other requirements of socio-economic development. However, the forest management practices in the region vary geographically (Table 2), as management practices are based on topography and population density. Therefore, policies and management framework, keeping in mind the distinct social set-ups, can help in contextualizing the specific interventions.

Forest Fire in the North-Eastern States

Along with other causes of deforestation and forest degradation, the forest fire is a major cause of forest degradation, deforestation and other loss to forest ecosystems. The impacts of the fire on the forests and its services in the North-Eastern Region are diverse. Besides directly damaging the forest trees, the fire also adversely affects its regeneration capability, microclimate, soil erosion and wildlife among others. It also has wide-ranging adverse environmental, economic and social implications in immediate locality and contribution in the global environmental change. Increase in population has resulted in higher frequency and subsequent damage to the forests due to forest fire. Globally, forest fires are reported to occur in varying intensity due to various factors. As per the information compiled in Global Forest Resource Assessment 2010, on an average one percent of all forests are reported to be significantly affected by the fires each year. However, the areas affected by fires are larger as the actual information on fires is missing or under-reported from many regions.

Forest fires in India are generally ground fires. Fires affect about 35 million hectares of forest area per year. About 95% of the forest fires are induced specially to promote new flush of grasses, collection of minor (NTFP) forest produce like seeds and flowers or to prepare land

Table 2 Major forest management context in Northeast India (*Source* Barik et al. (2006). *Forest sector review of Northeast India*. Background Paper No. 12. Community Forestry International, Santa Barbara, CA, USA)

Forest contexts (forest/capita)	State	Dominant forest authority	Primary forest land use	Future strategies
Lowland plains (0.14–0.23)	–Assam –Tripura	–State forest department –Traditional institutions with little control –District councils	–Mixed forests for timber production –Monoculture plantations –Protected areas	–Joint Forest Management (JFM) (benefit sharing, national model) –Joint protection in Protected areas
Central and Eastern hills (0.54–2.32)	–Assam (hills) –Manipur –Meghalaya –Mizoram –Nagaland	–Communities –Traditional institutions with strong control and effective	Swidden/Jhumland pool –Non-Timber Forest Products (NTFP) and domestic and local markets –Sacred and watershed forests	–Community Forest Management (CFM) supported through JFM programmes –Indigenous institutions –Special models
Greater Himalaya (6.0)	–Arunachal Pradesh –Sikkim	–State forest department and communities –Traditional institutions with less control and partially effectively	–Formal silviculture –Traditional *Jhum* –Forest gathering systems –Formal and indigenous conservation	–Combination of JFM and CFM strategies depending on legal status and capacity

for cultivation (in case of North-Eastern Region its shifting cultivation). While statistical data on fire loss in India are very weak, but one of the reports estimates that annually the proportion of the forest areas prone to forest fire ranges from 33% to over 90% in different states and forest types. Forest Survey of India (FSI), a government office monitoring forest, reports that around 50% of the forest areas are prone to fire with different degree of frequencies and damages, and 6.17% of the forest areas is highly prone to severe fire. In the Northeast India, most of the times, the practice of *Jhum* is the leading cause of forest fire. The most heavily affected areas by *Jhum* are Meghalaya, Nagaland, Manipur, Mizoram, Tripura and Arunachal Pradesh. In Assam, only a few limited pockets practice this. One of the estimates reports that 4.35-million-hectare forest area in the North-Eastern Region is affected by fire because of *Jhum* practice. This is significantly increasing with the growth of population in the mountains that is resulting decrease in the land/person ratio. As a result, more forest areas and bigger patches are being subjected to *Jhum* and the revisit period to the same locations has reduced. For example, one of the estimates found that the revisit period has reduced from thirty years to two years; this ultimately affects the regeneration capability of the abandoned (fallow) *Jhum* areas.

Earlier, collection of such information on forest fire was challenging, as the most of information was dependent on the data collected and reports generated by district administration. However, in present times, management and mitigation of forest fires require much more efficient information system, almost real-time, on distribution of forests and forest fire incidences. With the advent of earth observation system and satellite data with wavelengths sensitive to forest fire, information collection and dissemination have become efficient and effective. Forest Information and Resource Management System (FIRMS) developed by the University Maryland provides near real-time active fire locations across the world. This uses MODIS (Moderate Resolution Imaging Spectro-radiometer, 1 km pixel) and VIIRS (Visible Infrared Imaging Radiometer Suite, 375 m pixel) active fire products for use in support of fire management (near real-time alert systems) and other science applications requiring improved forest fire maps. The operational version of FIRMS known as the Global Fire Information Management System (GFIMS) delivers information to ongoing monitoring and energy project to wide range of audience including United Nation (UN) Originations. Bhuvan, the Indian geoplatform of ISRO, provides the same information

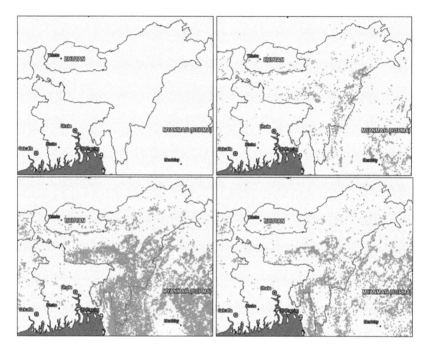

Fig. 1 Forest fire locations as per Fire Information for Resource Management System (FIRMS) over Northeast India (**a**) 01-07-2015 to 31-12-2015, (**b**) 01-01-2016 to 29-02-2016, (**c**) 01-03-2016 to 30-04-2016, (**d**) 01-05-2016 to 30-06-2016 (Adapted from Satendra and Kaushik, A. D. (2014): Forest Fire Disaster Management. National Institute of Disaster Management, Ministry of Home Affairs, New Delhi)

as the near real-time forest fire alerts. For example, Fig. 1 shows locations of forest fire Terra/Aqua MODIS reported on Fire Information for Resource Management System (FIRMS).[1] Being geospatial in nature, this can provide exact location, distribution and frequency of fire occurred. The data provided are archived and can be analysed to understand the pattern and behaviour of the forest fire along with the dominant seasons for a particular location. Remote sensing information can aid the conventional methods of forest management and in turn enhance

[1] https://firms.modaps.eosdis.nasa.gov/firemap/.

BUILDING A RESILIENT COMMUNITY AGAINST FOREST FIRE DISASTERS ... 193

Table 3 Forest fire alters for year 2014 and 2015 and the fire season in north-eastern states of India (*Source* India State of Forest Report 2015)

State(s)	Fire incidence		Fire season	
	2014	2015	General	Peak
Arunachal Pradesh	535	358	Jan–May	Feb–Mar
Assam	2536	1656	Jan–May	Feb–Mar
Manipur	1774	1286	Feb–Apr	Mar–May
Meghalaya	1123	1373	Jan–May	Feb–Mar
Mizoram	2189	2468	Feb–Apr	Feb–Apr
Nagaland	886	722	Feb–Apr	Feb–Apr
Sikkim	0	3	Jan–May	Mar–Apr
Tripura	1160	476	Feb–Apr	Mar–Apr

the effectiveness of forest fire management efforts. Table 3 represents a synthesized data on the forest fire alters for year 2014 and 2015 and the forest fire season in the north-eastern state of India.

Northeast India is the most affected site in India by forest fires. Occasionally, there are a few natural causes of forest fires (e.g. lightning) in the region. However, forest fire incidences are mostly related to human activities, such as (i) roadside clearing and burning of dried grasses (usually in February and March, the driest period of the year); (ii) burning of forest floors and dry grassland to destroy unwanted vegetation and facilitate growth of new shoots for grazing; (iii) burning of forest floors for collection of NTFPs (leaves, flowers and fruits/seeds); (iv) burning of forest floors and shrubs to improve visibility for hunting wild animals; (v) careless prescribed burning and fire line layout and construction; (vi) charcoal-making in the forests; (vii) cooking and camping by woodcutters and other forest users; and (viii) carelessness (dropping biri/cigarette butts) while passing through forest tracks. In the Northeast India, since the burning time coincides with the dry seasons when forest flora is full of fuel load, negligence by *Jhum* cultivators often leads to burning of adjacent forest areas. For example, *Jhum* is highly inflammable in bamboo flakes and locations rich in dry grasses and leaf litter.

In almost all the north-eastern states, the respective State Governments are taking forest fire prevention and control measures. For example, the Mizoram Government introduced the Mizoram (Prevention & Control of Fire in the Village Ram) Rules 1983 for effective prevention and control of forest fires. It has also set up state- (Chief Minister to be the head), district- (Deputy Commissioner) and village (Village Council President to be head)-level fire prevention and protection committees with identified rights, duties and functions. The *state*-level committee acts as the apex body for all the other committees and also interacts with the Central Government on the matter. The *district*-level committee has advisory, supportive and coordinating functions. The *village*-level committee is to mobilize volunteers for fire watching and firefighting in each village. The initiatives taken in Mizoram are the first and among very few examples of the forest fire management in the mountain regions across the world.[2]

States are also implementing Government of India's integrated forest protection scheme wherein forest fire management through a Joint Forest Management Committee (JFMC) is an integral component. The increasing incidences of *Jhum*-related forest fires have prompted local communities and JFMCs to take measures to protect forest in the limited vicinity. Table 4 provides state-wise causes and mitigation options of forest fires and the recommended prevention, preparedness and response activities.

The forest fire mitigation and management in the Northeast India needs to be carried out in a close collaboration with the local communities. Community-based forest fire prevention and management can be operationalized as a four-tier approach. First, the *Regulatory measures* in which, the community as a whole determines the phases, extent, degree and timing of *Jhum* burning. This should be determined based on the requirements and regeneration capability of the forest land under *Jhum*. The timing (preferably early in the morning or by early afternoon) of fire should be in such a manner so that the burning could be completed by evening. Second, the *Village Forest Fire Prevention Committee* (VFFPC) should nominate volunteers to be 'Fire Watchers' who remain active particularly during *Jhum* burning. They should be quick in communicating to larger group, identify and take measure for mitigation in case of wildfire and also understand any other measures to check and regular fire mechanisms, if needed. The third is *Community roles and actions* during a forest fire. This would include responding to information received from

---

[2]Darlong, V. T. *Traditional community-based fire management among the Mizo shifting cultivators of Mizoram in Northeast India*. FAO Corporate Documentary Repository.

Table 4 State-wise causes and mitigation option in case of forest fire (Adapted from Satendra and Kaushik, A. D. (2014). Forest Fire Disaster Management, National Institute of Disaster Management, Ministry of Home Affairs, New Delhi)

State	Causes	Mitigation options	Prevention, preparedness and response
Arunachal Pradesh	–Controlling weed and pest –Forest department burns degraded forest –Good grass growth –*Jhum* or shifting cultivation –Resin taping	–Community conserved area (CCA) –Village-level communities	–Alternative livelihood –Awareness programmes –Brigade for fire control –Committees empowered to fine the offenders –Community participation –Creation of fire lines, back counter fire, control burning –Environmental education
Assam	–Control burning by forest department –Escape of fire during road construction –Exhaust of vehicles –Good grass growth –*Jhum* or shifting cultivation	–do–	–Fire line clearance –Involvement of local communities –Involvement of village councils and local NGOs –Litter burning –Modernization of existing firefighting system
Manipur	–Hunting and Timber mafia –*Jhum* or shifting cultivation –Management of grassland –Occasional fires by lightning –Trespassers to forest	–Control measures through JFMC –Training and awareness generation programmes	–Appointment of fire watchers –Fire line clearance –Litter burning –Modernization of existing firefighting system

(continued)

Table 4 (continued)

State	Causes	Mitigation options	Prevention, preparedness and response
Meghalaya	−Collection of NTFP −Driving away wild animals −Hunting −*Jhum* or shifting cultivation −Dropping burning biri/cigarette butts	−do−	−Locals can help in locating and suppressing fire −Traditional methods to detect and suppress fire
Mizoram	−Burning farm residues −Clearance of land −Collection of NTFP −Good grass growth −*Jhum* or shifting cultivation −Dropping burning biri/cigarette butts	−Individual to entire village −The Mizoram (Prevention & Control of Fires in the Village Land) rules −Village council −Village Forest Fire Prevention Committee (VFFPC)	−Forest department making appropriate arrangements −People participation is being given more emphasis −Public awareness and training programmes
Nagaland	−Collection of NTFP −Cooking near the forest area −Drive away wild animals to save crops −Dropping burning biri/cigarette butts −Escape fires from the burning of farm residues −*Jhum* or shifting cultivation	−Village councils −Village Development Boards impose fines for deliberate forest fires and due to negligence	−Community helping, suppressing and detecting forest fire −Village Councils the local self-government bodies also cooperate

(continued)

Table 4 (continued)

State	Causes	Mitigation options	Prevention, preparedness and response
Sikkim	–Bonfire during winter –Hunting and illegal felling –Natural lightening on high hills –To grow good grass crop –Drive away wild animals	–Awareness generation programmes through JFMCs –Control burning in the vicinity of villagers and fire watchers during the season –Control rooms at the headquarters and each district reporting fire incidence –Restricting agriculture slash on dry and windy days	–Awareness among community –Constitution of forest fire protection committees –Develop institutional mechanism –Involve community in forest fire management –Traditional methods to control and prevent fire
Tripura	–Burning of farm residues –Collection of NTFP –Cooking near the forest area –Drive away wild animals –Dropping burning biri/cigarette butts –*Jhum* or shifting cultivation	–Control measures through JFMC –Training and awareness generation programmes	–Appointment of fire watchers –Modernization of existing firefighting system –Traditional methods like fire line clearance and litter burning

the 'Fire Watchers' and suppress the fire in case of incidence. They are also expected to contribute in preparation of fire lines in strategic locations and also arranging means and measures that would be necessary to regulate and mitigate forest fire. They should be empowered to conduct enquiry to determine the causes of fires and identify the culprits, if any, in case of wildfires. Finally, the *Individual family members* should also share the responsibilities on the day of *Jhum* burning. For example, the members notify village authorities, immediate neighbours and those having adjoining *Jhum* fields about the date and likely time of burning. When the fire is ignited in the field, the members should remain in the field until the burning is completed and stay watchful to the spread of forest fire in the vicinity of their fields. Other adult members of the family should store water for immediate use in the event of fire in the vicinity. Members should also remain watchful for wind-borne kindling falling on the thatch roofs.

Some of these common practices could work as preventive measure for the forest fire. For example, at the time of slashing the vegetation clearing corridor should be about 8–10 m wide between the forest patches. This acts as fire lines in the *Jhum* field itself, thereby preventing the spread of fire to adjoining forest areas. The villagers can be encouraged to maintain leafy trees and bushes, which could be used quickly for extinguishing fires. Advance vigilance of the locations around the *Jhum* field can ensure fire lines, assess vulnerable points and plan strategies for action during the event of a fire escape. And as a practice, determining and imposing penalties on individuals, or collectively on the entire community, for deliberately or accidentally causing forest fires.

The Central Government, represented by the Ministry of Environment, Forests, and Climate Change (MoEF&CC), provides funds to the State Government to undertake fire management programmes. The activities include controlled burning, preparation of firebreaks, removal of fire hazards along roads and around plantation areas and awareness/sensitization programmes for fire management through meetings, seminars, poster campaigns, print media and electronic media. The Central Government also provides funds for firefighting equipment, special clothing for firefighters, vehicles, wireless communication equipment and training. The State Governments need to ensure and promote community participation in fire management and make effective use of the funds by clearly defining the common objective, developing a collaborative management plan for forest management (Table 5).

BUILDING A RESILIENT COMMUNITY AGAINST FOREST FIRE DISASTERS ... 199

Table 5 Framework for promoting community participation in fire management

Steps	Details
Common objectives	A clear set of common objectives is necessary for forest fire management
Collaborative management plan	The objectives need to address all stakeholders of forest resources and with regard to forest fire management
Management of budget	The local authorities and communities should be informed about financial challenges
	The local organization (e.g. sub-district councils) needs to support in case of financial challenges
	The community's participation should also work towards alternative and innovative funding mechanisms
Fire management	All stakeholders should participate in the management of fire in the forest areas
	The villagers should share information on forest fires and their effects on NTFPs, as per the preference
Paradigm shift from protection and suppression to management	Credible research and timely dissemination of appropriate technologies to influence adoption of improved practices
	Research leading to improved fire management and community involvement in decision-making

SUMMARY

Northeast India is increasingly experiencing major challenges in the traditional *Jhum* practice and fire management efforts. The challenges in the forest fire management are on account rising population, demand of more land for agriculture and the changing needs and aspiration of the local people. The intensity and extent of *Jhum* lands have increased over the period. The frequency of the *Jhum* has decreased to 2 years resulting in loss of huge parts of forest as abandoned *Jhum* that should have been converted to secondary forest by regeneration. The younger generation is migrating to town, and thus, less of the people remain in villages with interest for traditional community-based fire management system. This has weakened the community-based fire management systems, resulting in wildfire due to ill-managed *Jhum* cultivation.

The modern society is in transition, between the local needs and the modern aspired world leading to changes in traditional farming practices. At present, in a typical village, only about half of the families are fully dependent on *Jhum*. In addition, the traditional practices of land use management have also eroded because of dependency on government-initiated employment programmes. The government programmes have higher degree of interventions in natural resources including forest. This has resulted in the changes in the perspectives and interest of local communities towards forests. Local people perceive that the government-appointed VFFPCs are responsible for fire control, mitigation and management. Thus, the shift from community-based to government-initiated programmes has led to erosion of the community spirit, particularly in forest fire prevention and management.

In the present times, unregulated selection of *Jhum* sites constrains community-based fire management. Traditionally, the communities used to select area for *Jhum* cultivation in limited patches or blocks. This would enable the community to easily watch over and control any spread of fire to adjoining forested areas. However, in the present times, *Jhum* cultivation plots are being selected wherever possible, in smaller sizes, and with high and dominant bamboo shoots (because of under regeneration). This is result of higher pressure on land due to increased population and reduced per capita land availability. Owing to such fragmented, small and dispersed *Jhum* patches, the forest fire risks have increased and it has amplified the management challenges. The presence of higher percentage of bamboo forests around the *Jhum* sites calls for extreme caution. In such a scenario, only active community participation both in regularization of *Jhum* and management (prevention and control) of forest fire is an effective solution. Rather than substituting traditional systems and bringing in newer interventions, the government should enhance the effectiveness of traditional community practices with the help of real-time technical inputs from remote sensing outputs to combat the challenges posed by periodically reoccurring forest fire.

The Khuga Dam—A Case Study

Langthianmung Vualzong, Aashita Dawer
and Nivedita P. Haran

INTRODUCTION

Water is one of the most indispensable natural assets. There can be no life without water; in fact, life began in water. Basic living organisms emerge from the water. Civilizations have sprung up along rivers, and they have also been wiped out by water-linked disasters. Water has the highest potential for danger that can cause a disaster. The maximum number of fatalities occurs due to water-related disasters. Water is the cause of disputes between nations as well as within countries. Every state lays emphasis on and prepares plans for the effective, equitable and technologically robust techniques for managing water. Debates have been going on regarding the nature of technical interventions most suitable

---

L. Vualzong
Centre for the Study of Law & Governance, Jawaharlal Nehru University, New Delhi, India

A. Dawer
O.P. Jindal Global University, Sonipat, India

N. P. Haran (✉)
Disaster Research Programme (DRP), Jawaharlal Nehru University, New Delhi, Delhi, India

© The Author(s) 2018
A. Singh et al. (eds.), *Development and Disaster Management*,
https://doi.org/10.1007/978-981-10-8485-0_14

in managing water, the optimal size of a dam, the best ways to retain the organic character of a river and similar other dilemmas.

Intervention of state in managing the water resources is thus considered to be of prime importance. Since water is a common property resource that is often prone to the problem of free riding, there is an expectation from the state for efficiently governing the resource. State, with its power of eminent domain, in order to protect and further public interest, takes on the responsibility of yielding equity and efficiency. India is endowed with abundant water resources which if exploited astutely could result in boosting its economic growth. For this purpose, the Government of India has constructed various dams to increase availability of water supply, control floods, facilitate irrigation, allow inland navigation, control sedimentation and generate hydroelectric power.

Ideally, dams are expected to bring about development in the surrounding areas, thereby increasing the overall economic benefit to the nation. This development generates additional livelihood options, strengthening the community's resilience in an event of disaster. Dams are one of the early inventions of mankind. Nay, even beavers are known to build simple dams, that are temporary but that can control the flow of excess water. Dams that are engineered structures need the right location, optimal design and correct execution. However, in certain cases the dams that were supposed to bring prosperity have instead resulted in the deterioration of economic and social well-being of the people of the region. Such is the story of the Khuga Dam in Manipur.

The Khuga Dam

The Khuga Multipurpose Project, locally known as Mata dam, located approximately 9.65 kms away from Lamka town in the southern part of Churachandpur District of Manipur was approved by the Planning Commission of India for an amount of Rs. 15 crores and conveyed under their letter No. 20 (19)/80-I&Cad dated 25 July 1980 (Fig. 1).

The actual construction of the Khuga Multipurpose Project was started in the year 1983–1984 but soon after, the first contract was terminated due to tardy progress. The objective of the dam was to generate hydelpower of 1.5 MW, provide irrigation facilities for 10,000 Ha, as well as provide drinking water for domestic consumption with supply components of 5 MGD to serve the people of Lamka, Churachandpur district and the people of Manipur in general. The catchment area covered an area of 312 sq. km. Expected hydropower to be generated was raised

Fig. 1 The Khuga/Mata Dam

to 7.5 MW in the following year. The foundation stone of the project was laid by Dr. Manmohan Singh, the then Deputy Chairman, Planning Commission, Government of India on 3 October 1986. However, the construction of the dam remained suspended for many years until it was resumed in the year 2002. Financial constraints including irregular fund flow, poor law and order situation in the state and ethnic clashes at the project site are cited as the reasons for the delay. After 28 years, UPA Chairperson, Smt. Sonia Gandhi inaugurated the dam on 12 November 2010. Construction of the hydropower part of the project had yet to begin. Within a few weeks of its inauguration, the shutter gates were closed as some major cracks had been noticed in the irrigation canals.

A close perusal of the Annual Administrative Report of the Flood Control and Irrigation Department,[1] Government of Manipur, for FY 2011–2012 gives an accurate indication of the wide chasm that exists between such reports and the situation on the ground. The report gives an idea of the large number of posts created for the execution of the Khuga Multipurpose Project: An Assistant Chief Engineer at the apex with a Superintending Engineer and 4 Executive Engineers on the irrigation side and a comparable number on the electrical side. Two of the EE's are exclusively for the two

---

[1]Annual Administrative Report from Irrigation and Flood Control Department, Manipur, March 2012.

204 L. VUALZONG ET AL.

canal divisions, i.e. to supervise the canal construction work. In addition, there is a post created at the SE level to ensure quality control. According to the same Annual Report, under Major and Medium Irrigation Works, during FY 2011–2012 Rs. 226 crores has been expended for irrigation and water supply. Interestingly, the report of 2011–2012 admits that there is a provision of 5 MGD of raw water to be supplied from Khuga Multipurpose Project to the State PHED. The report declares that the headway component of the Khuga project was completed and the same commissioned in November 2010. However, the report makes no mention of the targeted objectives which had been achieved by that date. It is also silent about the shutting down of the dam due to the danger posed by the cracks in the canal system even though the department would have been in the know of it. Para 3.1 titled 'Khuga Multipurpose Project' is worth reproducing here:

3.1 KHUGA MULTIPURPOSE PROJECT:

The Khuga Multipurpose Project was sanctioned by the Planning Commission in July, 1980 for an estimated cost of Rs 15 crore. The Hydro-power component of 3 × 500 KW was separately sanctioned in October, 1983 for an estimated cost of Rs 1,23 crore and with the latest revised cost of Rs 16.68 crore (2008). The project is earmarked and monitored by the centre. The revised estimated cost of the project is Rs 381.28 crore for irrigation component at price level 2009 has been cleared and sanctioned by the Technical Advisory Committee (TAC) of Ministry of Water Resources in its 100th meeting held on 26-10-2009. The Head work component has been completed and commissioned on 12th November, 2010 with a partial irrigation potential of 10,000 Ha. The whole irrigation component 1/c canal system is targeted for completion during 2013.

The latest revised cost of Rs 434.65 crores has been approved by State PIB on 21st October 2011 and the same has been submitted to the CWC, New Delhi for obtaining Investment Clearance of Planning Commission, Government of India.

The outlay for 2011-12 is Rs 5.50 crore. The anticipated expenditure during 2011-12 is Rs 5.50 crore. The progressive expenditure up to March, 2011 is Rs 395.55 crore. The budget provision of the project for the next year, 2012-13 is Rs 5.50 crore.[2]

[2] Ibid.

Objective of the Study: The Khuga Multipurpose Project could not be successfully completed even after 32 years. The paper attempts to investigate a number of pertinent questions arising from the delay in the project, which are as follows:

- Who are the officials responsible for the non-completion? Can accountability be fixed?
- Was there a defect in the dam design? Or, was it a case of poor implementation?
- With a cost overrun of over a thousand times, who is responsible for the loss?
- What is the estimated loss in opportunity cost and what are its consequences?
- Is Khuga Dam a case of a unique failure or are there many more such cases?

In order to have an in-depth understanding of the project, the following methodology was employed: all available data and documents on the project were obtained which unfortunately were woefully little as the government claimed since many decades had passed, much of the data had either been destroyed or were misplaced! The project was discussed with the concerned department/s and with the local residents whose land had been acquired and who were potential beneficiaries of the project. The authors also scoured online to obtain news reports and any other documents on the Khuga Dam. The data so obtained were analysed and the results were presented below.

COST-BENEFIT ANALYSIS

The projected cost at the time of inception was Rs. 15 crores. However, over the years the estimated cost escalated manifold to Rs. 300.77 crores which further increased to Rs. 335.15 crores in 2009 and Rs. 380.98 crores and finally is reported to have been revised to Rs. 434.65 crores by 3 October 2011.[3] Due to sheer ignorance, apathy and oblivion on the part of the state, the 33 plus year old Khuga Dam multipurpose project has seen a 30-fold increase in project cost since its inception.

---

[3]Failed dams in Manipur? Any lessons learnt retrieved from https://cramanipur.wordpress.com/2014/08/18/failed-dams-in-manipur-any-lessons-learnt/.

According to the report of Irrigation and Flood Control Department, Manipur of 2008, the benefit-cost ratio was calculated to be at 1.08.[4] According to the report, it is found that that the benefit-cost ratio has been calculated by incorporating the benefit occurring only from increase in agriculture output due to irrigation. The benefit of electricity generation has not been taken into account. This is rather an ambiguous method to calculate the ratio as the first and foremost objective laid out at the onset of the project was to generate hydropower. In terms of actual benefits accrued, not a single unit of electricity has been generated since the completion of the dam. The assessment done by SANDARP in 2014 revealed that the operational feasibility of the power component was never studied.[5] Moreover, the irrigation requirement of the hilly region of Churachandpur is negligible due to the slash and burn (Jhum) cultivation prevalent in this region. Thus, the irrigation benefit from the dam is zero. Thus, the actual benefit-cost ratio would be negative.

NEGATIVE IMPACT OF KHUGA DAM

For the construction of the dam, the state exercised the power of eminent domain to acquire land from the people in the villages rendering them displaced. In 1992, total number of villages to be affected was assessed to be 16 with a population of 579. The compensation to be paid was Rs. 25,500 per family. However, in 2005, a joint spot survey was conducted by the District Level Committee (DLC) headed by the DC which recognized the total number of villages submerged to be 21 with 706 houses. The number of people affected was 4100.[6] Despite the promise of meagre sum of compensation, most of the people, especially non-patta[7] holders remain uncompensated till date. The livelihood of the internally displaced has changed drastically. A comparative primary study of six displaced villages with about 350 households of nearly

[4] Report from Irrigation and Flood Control Department, Manipur, October 2008.

[5] Present Tensed, Future Expensive: Large Irrigation Projects in Northeast India retrieved from https://sandrp.wordpress.com/2014/05/28/present-tensed-future-expensive-large-irrigation-projects-in-northeast-india/.

[6] Letter to the Commissioner, IFCD from A. Ibocha Singh, Deputy Commissioner/ Collector dated 5 July 2005.

[7] A patta is a legal document issued by the government in the name of the actual owner of a particular plot of land.

THE KHUGA DAM—A CASE STUDY   207

Table 1 Pre- and post-occupation of people

Types of occupation	% in 2005 [before]	% in 2009 [after]
Wet paddy cultivation	80	a
Jhuming	6.1	44.5
Collection of forest products (wood and charcoal)	6.4	50.5
Farming (livestock)	4.5	3.3
Petty business	3	1.7

[a]Paddy fields submerged under water

2000 population for the years 2005 and 2009 was done by S. Thangboi Zou. People of Churachandpur are primarily engaged in cultivation and collection of charcoal and firewood (84.6%) followed by secondary activities (9.1%) and tertiary activities (6.3%). However, since the construction of the dam and their displacement, their occupational structure has changed (Table 1).

According to a study by S. Thangboi Zou (2011), it is observed that before displacement, most of the population was engaged in paddy cultivation. However, when their fields submerged they shifted to jhuming, collection of wood and charcoal. The overall income reduced as is depicted in Table 2. The condition of houses worsened by 70%, electricity by 100%, employment by 94.5% and drinking water by 73%.

The authors of this case study visited the site of the Khuga Dam in April 2016. This was a team consisting of academics, research scholars, government officials and administrators from JNU, Manipur Central University, NDMA, NIDM, SDMA and other agencies attending the workshop organized jointly by JNU and the University of Manipur at Imphal. The team spent a day in the area visiting the dam site, gathering data and interacting with the stakeholders and other residents. A detailed questionnaire was sent to the Secretary, DMD, Imphal

Table 2 Income after displacement (ibid.)

Percentage (%)	Income less than Rs. 20,000	Income Rs. 20,000–40,000	Income more than Rs. 40,000
Before displacement	26.0	58.0	16.0
After displacement	44.8	46.3	9.9

seeking some further data. Since no response was received, a researcher was sent to Imphal once again in June–July, 2016. He visited the government departments, viz. Departments of Disaster Management, Irrigation and Power. He also visited Lamka and the site of Khuga Dam, interacted with the evictees and whose land had been acquired for the project and some journalists who had been investigating into the case for considerable period of time. The primary reasons for failure of the project can be identified as below:

- Poor design of the dam
- Lack of engagement with the direct stakeholders, i.e. those who were losing land, the farmers whose land falls along the irrigation canals, etc.
- Poor and tardy execution which can only be the result of poor supervision over contractor, inefficient contractor and lack of any serious corrective measure
- Absence of concern for taking up the hydelpower project despite its importance
- Inability of the executive to float the project as a base for politico-ethnic unity in the region
- The backward profile of the region and illiterate or semi-literate population unable to raise questions or demand accountability.

DAM DESIGNING AND EXTENT OF FINANCIAL LOSS

A lot has been discussed over the last few decades about the need for dams, their efficacy and efficiency and their impact on the environment. The designing of a dam is based on certain basic scientific principles that are clearly laid down. These pertain to the location of the reservoir, the length and direction of the canals and in case of a multi-purpose dam, the location of the hydelpower station. Based on the design, the project is then implemented. In the case of the Khuga Dam, it appears there were basic flaws in its design. This was compounded by inferior implementation, poor material, bad supervision and corrupt practices at different levels. Thus, a simple dam could not be constructed over 32 years and when only the irrigation part of it was said to be completed with the power generation part nowhere near completion, serious defects were found in the canal walls. The faulty dam could now pose serious threat to the lives of the local population as in case of a dam break over

Fig. 2 Cracks in the dam

20 settlements downstream would get washed away. The potential for a disaster is real and needs to be addressed.

Yet, the irony lies in the fact that the Flood and Irrigation Department of Manipur State or the Planning Commission has refused to admit the major flaws in the project, the loss it has caused to the public exchequer and the need to fix accountability. As indicated in the Annual Administrative Report of 2011–2012, it seems to be business as usual. In an open democracy, this is totally unacceptable. The slew of senior staff ranging from Assistant Chief Engineer to Executive Engineer whose posts were created exclusively to supervise this project, failed to detect any of the defects in construction. The government owes it to the land evictees to mete out suitable penalties on those who through act of omission or commission, failed to perform (Fig. 2).

As the CB analysis above indicates, there was 30-fold increase in the project cost. This is understandable given the lapse of over 3 decades. But if after 3 decades the dam had become a reality, at least an attempt could have been made to justify the time and cost overruns. When the dam proves to be a complete non-starter, there is no acceptable explanation for that. *First*, those whose land was acquired by the government have a strong reason to feel aggrieved. Records show that even though acquisition took place over 30 years ago, the compensation is yet to be paid to some evictees, the reason being some of those found to be living on the land or cultivating it did not possess valid patta or land ownership documents. Their case is still in court. Further, there is a strong dispute about the quantum of compensation paid. Interaction with one of the evictees revealed that the compensation amount was fixed at Rs. 25,000 per family. The Land Acquisition Act 1896 laid down the procedure and methodology for determining the compensation amount that included the land value, solatium and interest. In the absence of any records or data being made available by the government in this regard, it can only be presumed that the law was followed although the evictees do not seem to believe so. *Second*, there seems to be a complete lack of trust between the land acquisition authority and the evictees. Land acquisition is a difficult procedure that hits hard the landowner who loses his/her land. Hence, the process has to be addressed with understanding and empathy, which appears to have been lacking. It is necessary to take the evictees as partners in development rather than as adversaries.

Third, during the 32 year-period when the ill-fated dam was under construction, the local residents inform that cultivation of these lands came to a standstill. The crops cultivated include paddy, vegetables and other minor produce. With these lands being out of the domain of cultivability, the farmers were forced to seek out alternate lands. Vast majority of farmers in this region resort to subsistence farming. Ironically, the Khuga Dam was meant to stop 'the scourge of Jhum farming'. Instead, it pushed more farmers into Jhum. What is equally serious, more farmers gave up paddy cultivation, the staple crop in this region, and became dependent on collection on forest usufructs. Ironically, the Khuga Dam project that by providing assured irrigation was expected to encourage paddy cultivation in the area became a precursor for less area cultivated under paddy, higher incidence of Jhum cultivation and greater dependence on collection produce. With no organized set-up for sale of forest produce, the residents fell into the clutches of middlemen and moneylenders.

Fourth, with the construction of the dam it was expected that Lamka town would grow as an urban centre. With assured irrigation, more farmers would go into farming. Lamka town would provide better marketing avenues and better returns. It was also expected that pisciculture could come up in a big way along the banks of the reservoir that could take care of the fresh fish requirements of the state capital, Imphal. At present, fish supply to Imphal comes even from as far away as Kolkata.

The hydelpower generated would take care of the power needs of the entire district and would also encourage the sprouting of start-ups in trade and manufacturing sector. The reservoir and the picturesque surroundings of the hills and the flowing river were expected to provide an attraction for tourists bringing with it the ancillary business of cafes and restaurants, small-size hotels and curio shops. Alas, none of these could take place and they all remained a pipe dream. An entire generation passed by living under the delusion of 'a developed Lamka'. They suffered the hardship of an unsettled life, dust, grime and noise pollution, some were forced to relocate and at the end of it there was nothing but disappointment.

Loss in Opportunity Cost and Its Consequences

This brings us to the issue of loss in opportunity. The amount spent on the dam construction was, of course, the obvious loss from the public exchequer. The loss by way of the unsown crops, the small trades that could not be set up and most tragically, the generation that grew up on the hope that the dam would help them get employment, that loss cannot even be assessed. It would run into hundreds of crores. Moreover, Manipur has been a disturbed region. It is also a region that is vulnerable to natural disasters. One of the crucial ways to address the issue of local community-based and ethnic disputes is to ensure development of the area. With good social and economic infrastructure, the factional disputes get dissolved. All kids can go to school, the population has proper housing with water supply and power, there are adequate avenues for gainful employment and there are hospitals and hotels. The coming up of the dam project raised expectancy amongst the people, of jobs, of better infrastructure and resulting peace in the area. But their hopes were belied. The disappointment and dejection in the eyes of the locals when they talk of the dam is starkly visible. Further, far from being able to discourage and discontinue Jhum cultivation, Jhum has continued with greater vigour. The saving of forests has not taken place. The hills around the dam reservoir are proof of this; they stand completely denuded with only scrub and wild weeds growing.

Standing on the banks of the reservoir, as one takes stock of the surroundings, one can visualize how picturesque the area would have been with its virgin forests, pristine river and the fabulous flora and fauna. It was no less than a paradise on earth. If only the lives of the inhabitants could be improved with some basic infrastructure: a few roads, disaster-resilient electrified houses for all, potable water in their taps and water to irrigate their fields. That's what 'the babus promised us' says old grandpa with unseeing eyes dull with cataract, who lost a major part of his land and now is too old to go into the forest to collect firewood and is left with no land to till. There is no record to show any action taken against the engineers who designed or supervised the implementation of the project, the civil servants who failed to fix responsibility for the cost and time overrun and the below par work done, or the contractor who provided below-standard work. The details were sought from the government of any case filed before the court or the arbitrator against the contractor for inferior construction work, but no details were made available. Obviously, no accountability has been fixed on the elected representatives who won elections based on the promise of a dam. Nor are there any records to prove the bribes alleged to have been paid by the contractors, subcontractors and petty traders to the political leaders and bureaucrats, although the villagers and activists openly talk about it. There is no method to measure the impact of incomplete and poorly executed Khuga Dam on the lives of people. Apart from the lost opportunity, they now live under the constant fear of a possible dam break that can wash away all the downstream villages in one massive deluge.

CONCLUSION: IS KHUGA DAM A UNIQUE CASE?

To conclude, we need to answer the following questions: Is a case like the Khuga Dam unique? Are there many instances of such projects taken up with great fanfare, but that are failed projects in our country? The answer to the second question is unfortunately in the affirmative. In the irrigation sector alone, there are instances of project failures and the case of Khuga Dam is not unique. The system of central auditing is obviously inefficient. Audit reports would raise a few queries that would also be responded to by the implementing agencies but to no effect. Exemplary penalties on those responsible are non-existing, and the loss that occurs to the public exchequer remains unrequited. Allegations of graft within the irrigation department are widespread. The poor landowner who gave up his land for a pittance of compensation, the people of the area who underwent hardship and whose hopes were ultimately dashed and the unfortunate tax payer remain the ultimate sufferer (Fig. 3).

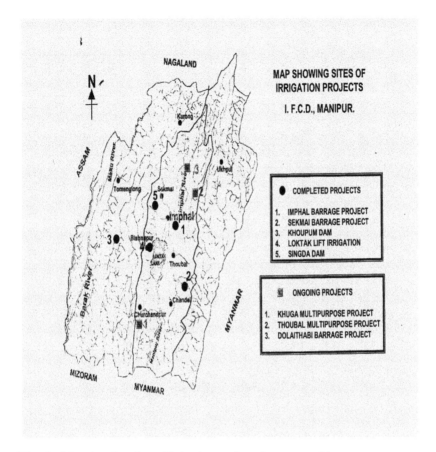

Fig. 3 Map showing sites of irrigation projects (supra note 1)

In the end, it is important to highlight the issues that require urgent attention:

1. The danger posed by the unsafe Khuga Dam needs to be addressed on an urgent basis. Simultaneously, similar other dams all over the country that have suffered huge time and cost overrun and are poorly executed, need to be identified and similar action be taken. The Government of India needs to take the lead to emphasize the enormity of the problem and to draw the attention of the State Governments towards it.

2. The Government of India (MHA, Niti Aayog and NDMA) needs to take up selected cases of 'failed dams', fix responsibility on the engineers, public servants, contractors or political executive and mete out strong exemplary punishment that could act as a deterrent to others.
3. There is a need for timely auditing and corrective action without waiting for decades so that more public funds are not thrown into bad projects. The role of the statutory audit must be strengthened and audit recommendations not treated in a routine manner.
4. The Khuga Dam Case Study has a lot to teach the administrators and researchers, policy makers and elected representatives. Such case studies should be made part of the course material used at the training institutions and institutes of higher learning.
5. The role played by activists in taking up such cases of failed projects, and to analyse and highlight them, is commendable. Such activists may be involved in generating greater awareness among the people, be it on the provisions of the Right to Fair Compensation and Transparency in Land Acquisition, Rehabilitation and Resettlement Act, 2013, or the responsibilities of public servants or the need for higher probity in public life.

References

Government of Manipur. (2008). *Report 2008*. Irrigation and Flood Control Department, Manipur.

Government of Manipur. (2012). *Annual Administrative Report 2011–2012*. Irrigation and Flood Control Department, Manipur.

Saikia, Parag Jyoti. (2014). *Present Tensed, Future Expensive: Large Irrigation Projects in Northeast India*. Retrieved from https://sandrp.wordpress.com/2014/05/28/present-tensed-future-expensivelarge-irrigation-projects-in-northeast-india/. South Asia Network on Dams, Rivers and People (SANDRP).

Yumnam, Jiten. (2014). *Failed Dams in Manipur? Any Lessons Learnt*. Retrieved from https://cramanipur.wordpress.com/2014/08/18/failed-dams-in-manipur-any-lessons-learnt/. Centre for Research and Advocacy Manipur (CRAM).

Zou, S. Thangboi. (2011). "Disrupted and Dislocated Livelihoods: Impacts of Khuga Dam on the Livelihoods of the Displaced People in Churachandpur, Manipur". *Journal of North East India Studies, 1*(1), July–December 2011, pp. 36–60.

Dams and Disasters in the Northeast India: A Collateral 'Ethical' Relationship

Prasenjit Biswas

INTRODUCTION

The lessons of Uttarakhand should not be lost on the north-east. For one, most of the region's hills and rivers are located on the same Himalayan faults, and to make matters worse, the soil is primarily made of sediments. The Centre's plan to construct a mega dam on every river that flows into the region, by way of huge monetary investments involving private parties, awaits a similar fate.

The proposed Tipaimukh project is proposed on the confluence of the Barak and Tuivai rivers in Manipur's south-western area, and there will be an earthen/rock-filled 178-metre high dam. In case of any breach, it has the possibility of not only inundating certain parts of Manipur's districts like Senapati, Tamenglong, Churachandpur and Jiribam, but also the whole of the Barak Valley of southern Assam. Silchar town will be under 70 feet of water, as estimated. Originally conceived to protect the lower regions of the Barak Valley and Sylhet district of Bangladesh from floods, the project is estimated to cost Rs. 8000 crore.

P. Biswas (✉)
North-Eastern Hill University, Shillong, India

© The Author(s) 2018   215
A. Singh et al. (eds.), *Development and Disaster Management*,
https://doi.org/10.1007/978-981-10-8485-0_15

The Union Ministry of Environment and Forests gave clearance to the Tipaimukh dam project located in Manipur's Churachandpur district many times, but every time it had to be withdrawn due to environmental vulnerabilities. This entailed clearing of about nine million trees across 311 sq. kms of pristine forest and is, at best, a man-made disaster waiting to happen. The justifying argument on generation of power and consequent industrial development exposes the humiliation of the facts emanating from the Lower Subansiri dam project in Arunachal Pradesh. Out of 2000 MW that the Subansiri project is supposed to generate, Assam will get less than 200 MW, that, too, at a high price. Similarly, from an installed capacity of 1500 MW at Tipaimukh, it is not clear yet how much the region will get because it has to sell power to Bangladesh and Myanmar as well as part of quid pro quo deals.

More significantly, the dam will totally alter the flow of water in the river basin on a short- and long-term basis. By containing the flow during the high season and by augmenting it in the lean season, the dam will endanger the hydrology of the river, affecting its entire downstream morphology. Experts believe that it will destroy biodiversity, as the Barak River is home to many rare water species along with the overground flora that will be permanently lost.

On the lower riparian Bangladesh side, damming the river will negatively impact the flow of water in the entire Surma–Kushiyara navigation channel by turning it into a deteriorated sediment deposit. Needless to say that starting from Tipaimukh to downstream Bangladesh via the Barak Valley, the entire agricultural and fishing zones will be reduced to desertification in no time. This will cause climatic changes and transform itself from a sensitive biodiversity zone to an area of frequent landslides, flood and other disasters. So far the Union Ministry of Environment and Forests has not been transparent in conducting mandatory public hearings, or it did not have a downstream environment impact assessment beyond 10 kms. What has been gathered from unofficial expert reports on the possible impact of the Tipaimukh dam is less than optimistic. Indeed, there is no field survey on the area of submergence, destruction of human and agricultural habitat and, hence, apprehensions galore. Even those who are funding the project are not very clear, as negotiations are on between several private parties and various government agencies. There is a complete lack of accountability in sensitive matters of protecting fragile ecology and river basin, crucial factors in maintaining a climactic balance in the flow area of the river.

Compounding such an unresolved possibility of disaster, there is a plan to construct mega or medium dams on Mizoram's major rivers. The Tuirial River flows from north to south to finally meet the Barak at Tipaimukh. A dam is also proposed on the Tuirial which flows from the north-western side of Mizoram. And there will be dams on other major rivers such as the Tuivai and Kaladan. There is a plan to generate 2500 MW from all the 13 rivers of Mizoram. Needless to say that all these proposed projects hardly meet the requirements of informed public consent with prior environmental impact assessment. Yet, the governments are in a great hurry to begin such projects without any guarantee that the projected quantity of power will be owned and distributed by the state concerned and their people. Building dams on all these rivers in southern Assam and Mizoram, with several shared underground hydrological strata in common, will only cause the weakening and depletion of underground water reserves, the most basic for the sustenance of life forms. Such unwarranted consequences of dam-led development in a geologically and ecologically fragile region need to be taken into consideration in the aftermath of what happened in Uttarakhand.

Conceptual Understanding of Possible Disasters Over the Barak River

The point of origin of the river Barak lies on the southern slope of Barail range in the northern part of Manipur's border with Nagaland where it is known as Kirong. From there on, it moves in a westerly-southerly direction crossing Senapati, Tamenglong and Churachandpur districts of Manipur and meet its tributary Tuivai to move to a place called Tipaimukh, where it meets its two other tributaries Tuirial and Tuivai, both flowing from Mizoram. This span between southern Barail and Tipaimukh covers a total stretch of 56 kms surrounded with thick forest of about 300 sq. kms with rare and rich biodiversity.

It is well known that the river Barak in its course, being part of larger Meghna–Brahmaputra–Ganges Basin, creates both underground and overground hydroscape that sustains livelihood and food chain of about 500 million of people starting from south-west of Manipur to downstream northern Bangladesh. The intrinsic connection between hydroscape and lived space of communities establishes a conceptual domain beyond a straightjacketed understanding of such so-called tropical rivers in terms of need for government intervention for food supply and support in the case of disasters turning it into an abstract notion of hydrological space (Linton 2008). As opposed to such a trivial understanding of hydroscape in abstract, the river Barak is looked upon as 'river of life' with flowing memory, myth, language, music and such cultural signifiers of community life which turns it into a heritage site (Arora and Kipgen 2012). Indeed, Hmars, the largest indigenous tribe living in and around Tipaimukh, in their cultural and historical memory locate a sacred place called *Rounglevaisuo*, a little downstream from Tipaimukh as a place of confluence of many ancestral tribes, called *Unau-Supuis* in their long journey from Irrawaddy Valley of present Myanmar, and they locate a place of dead souls called *Thiledam* (located upstream the river Barak above Tipaimukh). Apart from Hmars, the Zeliangrong people believe that the five most important places where King Jadonang's magic sword was hidden lie on the Ahu waterfall (Pamei 2001). Apart from this description of sacred landscape in the upper course of the river, the middle course that lies in Barak Valley of southern Assam is described as *Borobakro* in Kalikapurāna. The river Barak considered as lifeline of Barak Valley acquires an ingenuous geographical and intellectual context in terms of its linguistic and cultural ecology. The river in its deltaic downstream at northern part of Sylhet district of Bangladesh constitutes its own 'nationalized environment' and lower riparian rights in terms of flourishing agriculture and pisciculture. A translocal and transnational river such as Barak has always generated a discursive conflict between upper riparian narrative of multipurpose dam and lower riparian narrative of violation of riparian rights at the transnational level. Apart from such a mutually undercutting discursive conceptualization of the river Barak, it has also developed an indigenous narrative of 'stream of life' that connects both the upstream and the downstream in a South Asiatic food chain of wet rice and fish culture. The presence of a large number of aquatic Gangetic dolphin, fish species such as various kinds of catfish, gunnel, Hamilton and hilsa in the river and rare species of hoolock gibbon, phayre's leaf monkey and other such animal species surviving in the upstream thick forest turns Barak into a territory of biodiversity hotspot that has to be conserved as per India's commitment to conserve Gangetic dolphins. Overall, the run and cascade habitats of the rare and endangered fish species within river Barak and its tributaries create an inter-species habitat of aquatic animals and fishes that has an ecological dimension.

One needs to understand how this ecological dimension is affected by various technological and instrumental interventions. It is noted by Mazumder et al. (2014) that:

> Due to the construction of a sluice gate at the confluence of the Dhaleswari river, the visits of dolphins to this tributary ceased (...). In the Katakhal river also, the visits of dolphins have declined since the HPCL was established in 1970. Discharge of effluents from the paper mill–which has an unbearably pungent smell–into the south bank of the Barak river forces the dolphins to prefer the north bank while migrating upstream, thereby missing out the confluence. The confluences of the tributaries, Chiri, Jiri, Ghagra, Sonai, Badrinala and Jatinga, were previously reported to have small populations of the dolphins, sometimes throughout the year(...); however, they no longer congregate at these sites in winter, mainly due to depletion of prey fish species (...). Sadarghat Dhar,[1] Niyairgram Dhar, Lalmati Dhar and Narayan Dhar are among the reported wintering grounds of the dolphins, with all the necessary physical features (...). However, from 2012 onwards, no dolphin has been sighted by local people or fishermen in the winter months, apparently because of the depletion of the prey fish base. Lalmati Dhar and Narayan Dhar are among the most disturbed places in terms of fishing practices, while Niyairgram Dhar and Sadarghat Dhar are comparatively less disturbed although prey species are still limited. Thus, the Barak river has fewer prey fishes overall to support a resident dolphin population in the winter months.

These led the authors to conclude that unless habitat and movement spaces for Gangetic dolphins declared as the national aquatic animal are maintained with a zeal, it will be quite an ironical thing to contribute to their endangerment by using technologies for all kinds of purposes. Especially, they contend that construction of mega dam at Tipaimukh will remain as the single most factor for devastation of dolphin population which is already endangered in the Barak Basin. This possible devastation throws up a picture of an interconnected pattern of disaster: of an ongoing process of gradual drying up of the south bank tributaries of Barak river led to migration of dolphins upstream, who are still to relocate themselves in specific sites as they move upward on the north bank. The same pattern could be observed in the case of endangered

---

[1] The word 'Dhar' in local language would mean where river has taken a sharp turn and changed it direction of the stream.

220   P. BISWAS

species of fishes which, because of geomorphic structures of the river, have now to alter their path of migration from south to north on the upstream, which traditionally is supposed to be places only for breeding of the fish-seed. This author's personal observation with a team of nature lovers has found dead fishes in large quantity floating downstream near Sibapurikhal-Phulertal area, which are partly poached through poisoning plants and partly due to inability to find habitats downstream. This further presents what arose in the Krishna Water Dispute Tribunal as an absence of real-time data about hydrological cycle in any river in India, as there are sharply differing spatio-temporal scales: geomorphologic scale of run-off and dependable flow and the limited engineering scale of streamflow (D'Souza 2016). The two scales conflict with each other as one tries to come to terms with Barak's 'meander bend migration' in which more than one 'migration cycle' exists in a single meander (Laskar and Phukon 2012). The patterns of channel migration, bank erosion, sediment deposition in alluvial beds and unstable bedrock formations present multiple migration cycles across the entire stretch of 900 kms until it meets Meghna in Bangladesh. The resultant effect of too many misfit stream and oxbow lakes are created everywhere that are seen as large water bodies springing out of the channel-changing migratory flow of the river Barak. We can gauge this channel-changing behaviour from the following representative picture of the river channel migrating zones:

Figure 1 is explained by Laskar and Phukon (2012) in the following way,

> From the overlay of six temporal datasets (1818–2003), CMZs are delineated through identification of the active migration zone, active floodplains and palaeo floodplains. Active floodplains represent alluvialnplain adjacent to the active channel with recurrent flood inundation and are dotted with cut-off meanders (locallyknown as anua), whereas the palaeo floodplains are at a slightly higher topographic level and are characterized by overlapping sets of meander scars which are well discernible from the satellite imagery. It is observed that the CMZ is restricted where the river passes through bedrocks and becomes wide in the alluvial reaches. Four segments with well-developed CMZs are – Banskandi to Kashipur, Ramnagar to Masimpur, Barjatrapur to Badarpur and downstream of Srigauri.[2]

---

[2] Ibid., 82.

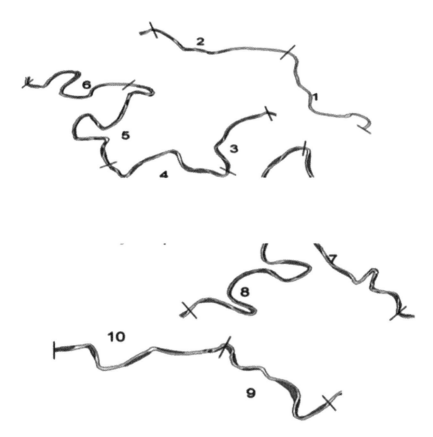

Fig. 1   Channel Migrating Zones (CMZ) of the river Barak

These meandering movements and migration cycles and pattern of changing alluvial and bedrock channels as part of the river basin turn it into an unpredictable zone of seismic activity, which is experienced in a major way during two major earthquakes in 1897 and 1984. In 1897 earthquake, the river abruptly changed its meander at Kashipur junction and formed an easterly-southerly ring by eroding the eastern part of Silchar town in a massive way. In 1985 earthquake, northern tributary of Barak, namely Rukni or Sonai, changed its course by almost 180 degrees and created a huge river-rift valley. These kinds of changes induced by seismic faults caused damage to lives and property, for which there is no appropriate accounting and understanding. Historically, there are records of huge damages to river banks and creation of land fissures that arose in 1869 earthquake in Barak Valley (Olympa and Kumar 2015). According to a NHC Background Paper (2006), geomorphologically weak river beds of Assam are described in a report in this way,

> The gradient of an alluvial river in a state of long-term equilibrium is mainly a function of water discharge (especially the bank full discharge that fills the channel to the tops of the banks), of input of bed sediment, and of the predominant grain size of riverbed sediments. Increasing flow tends to flatten the gradient, whereas increasing sediment input and coarser bed sediment tend to steepen it. The flattening of gradient through Assam suggests that effects of increasing water flow or reducing grain size (or both) override those of increasing sediment input.

This continuous flattening of the gradient in the case of river Barak, which has undergone multiple meanders, has also experienced rising river bed due to contributions of prominent grain size sediments from its tributaries. Flattening of gradients and channel shifting results into embankments hit by flash floods and inundation of densely populated floodplains. The river Barak and its basin are located in an area of high seismic volatility, as stated by Khandaker Mosharraf Hossain (2016),

> The area is surrounded by regions of high seismicity, which include the Himalayan Arc and Shillong Plateau in the north, Burmese Arc, Arakan Yoma anticlinorium in the east and complex Naga-Disang-Haflong thrust Zones in the northeast. The major Dauki Fault system along with numerous subsurface active faults and a flexure zone – called Hinge Zone – lie in the vicinity of the dam site. Manipur-Mizoram states are part of India Myanmar Hill Range formed by the interaction of the Indian, Eurasian and Myanmar Plates. These weak regions are believed to provide the necessary zones for movements in the North-East India Region, which experienced many major earthquakes during last 150 years and has been affected by small earthquakes occasionally.

Khandekar further argued that depending on the possibility of return of 1897 type earthquake due to sudden tectonic activity of Dawki fault and Disang-Haflong thrust, it is quite probable that a dam reservoir can crack even at a small intensity. The river Barak, lifeline of the Barak-Surma Valley extending into north of Sylhet district of Bangladesh, stands at the risk of Tsunami like rush of water in the towns like Silchar,

Sunamganj and Moulvibazar areas. Khandekar's position is supported by well-documented data about earthquakes in the North-Eastern Region (Kayal 1998). The Zone V features of instability and geomorphic vulnerability turn any weighty and high dam-like concrete structure into a potentially hazardous structure. The issue could be further understood by connecting it with catastrophic natural events such as Tsunami. Local people of Barak Valley observed a sudden rise of water level during winter months of 2004 in all the tanks, wells, lakes and other such water reservoirs. Also, there were cases of sudden land depression as water underneath alluvial soil forms an underground reservoir in coincidence with the full-to-the-brim state of Barak river damaging the built environment around it.

This above rough and ready picture of the river Barak and its ecological and geological conditions make it clear that any construction of a dam or barrage has the possibility of changing its flow, which would further affect the channel-changing behaviour in its fragile rock-bed. Alteration of water flow during peak and lean seasons in the distributaries of Surma and Kushiyara can result in drying up of many wetlands, oxbow lakes, underground aquifers and farmland situated in the northern Sylhet. Reduction of farmland size due to non-availability of water attributable to drainage congestion and sudden floods is directly connected to the hydrological and seismic uncertainties that are experienced in the course of the river Barak.

The watercourse convention 1997, to which both India and Bangladesh are signatories, provides elaborate guidelines for the purpose of non-navigational uses of international watercourse and its conservation in state-to-state negotiations. Article 5(1) is stating 'adequate protection of watercourse' and Article 6(1) is stating that conservation, protection and economy of the use of water to be taken into account in case a river is to be utilized by any one party. The spirit of the convention is that no one party will violate other's rights to access to equitable water. The co-riparian countries need to be guided by no-harm principle. As Bangladesh has pointed out, major harm of Tipaimukh dam would be multiple: hydrological, biodiversity disasters, loss of livelihood of both indigenous and non-indigenous people on both sides, potential desertification of north-east of Bangladesh and many other such deleterious impacts that cannot be fully grasped. A greater catastrophic event of release of saline water on the soil and sudden gush of water leading to large-scale inundation in towns like Silchar and Sylhet have caused a major concern.

Another very significant concern about river Barak is release of saline water due to oil exploration in south-west hills of Manipur covering 4000 sq. kms. The oil well spills of saline water, frequently leaked chemicals and gaseous particles, sewage and other environmentally dangerous residues will flow down the 30 oil wells at the hilly region from where the river Barak originates. These spillovers will flow through small streams to the main course of the river Barak. Due to monsoon flooding of the river, such oil spills will then flow into downstream Bangladesh causing not only loss aquatic biodiversity but by completely turning river water unfit for any human purpose of agriculture and other uses. This is how Immanuel Varte of 'Ensure sustainable resource management and justice' expressed his concern,

> It is known all over the world that oil spills during testing, actual drilling, accidents and bursting of pipelines has led to severe contamination of water bodies and rivers, Barak River, Irang River, Makru River and tributaries of Barak River will be the first to be impacted by such petroleum exploration and drilling related pollutions and contaminations.[3]

There is a further aggravated concern that pollutants will spread in the entire Barak–Meghna river system and will have long-term effects downstream. Already, contaminants from oil wells in Tamenglong district run by Jubilant Energy private company are flowing down the river without much of treatment, according to some local residents. The initial environmental impact assessment (EIA) prepared by the government only talked about endangerment of 'the existence of endemic species of Manipur and the rich flora, flora and biodiversity of Manipur. The report also undermines the fact that the entire Tamenglong, Churachandpur and Jiribam is an eco-sensitive zone where survival sources of wildlife and communities in the region go beyond national parks and sanctuaries' (Yumnam 2016). The pattern is same everywhere: innocent villagers at Tamenglong, Senapati and Ukhrul are often wrongly persuaded to give NOCs for oil exploration. Places where villagers resisted this, such as Parbung, Jiribam and Nungba, there is no respect shown by oil companies to the opinion of local people.

---

[3]Cf., https://www.causes.com/posts/888339 that carries a statement by Immnuel Varte, an indigenous civil society group leader from Manipur.

What is equally deleterious is possible impact of declaration of National Waterway No. 6 on the Lakhipur-Bhanga stretch of 121 kms on the river Barak. Needless to say, there has not been EIA based on free, informed, prior consent of people, apart from violation of lower riparian rights of Bangladesh, as an international river like Barak cannot be used for navigational purposes without adhering to UN convention 1997. In the name of development and reduction of transport costs, the deleterious impact of inland waterways on rivers recognized world over now is given a short shrift in such a declaration. It is to be noted that across Europe, rivers affected by navigational activities are recuperated,[4] while in India, such concerns are thrown into winds for petty invest- ments and commercial gains. As navigation is considered as a permanent danger to the life of a river due to hydromorphological alterations and hazardous substance leakage and its transportation,[5] what is even more germane is utilization of these waterways to the Manipur border for car- rying out oil exploration and dam building heavy machinery and other raw materials.

One can now see how the river Barak is facing multiple threats from various environmentally disastrous human activities such as dam build- ing, oil exploration, deforestation and poaching on river fauna com- pounded by geohydrological and seismic vulnerabilities that largely affect Assam–Manipur border, Barak Valley of Assam and downstream north- east Sylhet district of Bangladesh.

ETHICAL AND GEOPOLITICAL IMPLICATIONS

India's role as a bigger nation state in South Asia undergoes a percep- tion threat, if lower riparian countries are affected by activities described above. Ethically speaking, India has to not only ensure well-being of lives and livelihood on a vast stretch of seismically active riverine zone of South Asia, but it also has to take the lead in reduction of disaster risk from river-related activities. This will call for performance of due ethical and legal duties bestowed on India by international treaties, covenants and conventions. Further, India has to respect its own domestic laws that call for conservation of forest, biodiversity and protection of other

---

[4]An example of Rhine river recuperation is available at http://www.rivernet.org/rhin/ rhineriver.htm. Accessed 17 August 2016.

[5]See, https://www.icpdr.org/main/issues/navigation. Accessed 13 April 2016.

natural resources in tribal belts and blocks covered under various schedules of the Indian Constitution.

Ethically, India has to evolve certain 'best practices' that create conditions of social and political harmony by respecting the world views of the tribal communities of North-Eastern Region. This would require a sensitive understanding of the nature of inter-relationship between nature and culture, which is available in the form of indigenous knowledge about mountains, species-life and rivers among many of the tribes of the region. In the context of river Barak, such local knowledge is still a fresh pool of knowledge that validates much of the scientific knowledge. Oil exploration and dam building activities should not lead to destruction of sacred ecological spaces that lead to loss of indigenous and local knowledge.

From an ethical perspective, the question is, *ought economic activities undermine indigenous knowledge*? If such an undermining is permitted by using the instrumental power of laws, it not only undermines the sacred and generational storehouse of knowledge, but it results into a significant loss of community resilience in the vent of disasters. As India, as a responsible state, is engaged in building disaster resilience, it has to find moral and ethical foundations of people's own resilience, which is based on their ethical and cultural norms. This provides us the much needed fulcrum to argue about priority of these indigenous knowledge bases over technological and engineering manipulation of nature, which India needs to harness as a committed nation to mitigation of disasters.

References

Arora, Vibha and Ngamjahao Kipgen. (2012). "We Can Live Without Power, But We Can't Live Without Our Land: Indigenous Hmar Oppose the Tipaimukh Dam in Manipur". *Sociological Bulletin, 61*(1), January–April, pp. 109–128.

Baro, Olympa and Kumar Abhishek. (2015). "Review Paper: A Review on the Tectonic Setting and Seismic Activity of the Shillong Plateau in the Light of Past Studies". *Disaster Advances, 8*(7), July, pp. 34–45. For images of such damages, see pp. 41–43.

D'Souza, Rohan. (2016). "Framing India's Hydraulic Crisis: The Politics of Modern Large Dams". *Monthly Review, 60*(3), July–August. Available at http://monthlyreview.org/2008/07/01/framing-indias-hydraulic-crisis-the-politics-of-the-modern-large-dam. Accessed 17 April 2016.

Kayal, J. K. (1998). "Seismicity of Northeast India and Surroundings-Development Over the Past 100 Years". *Journal Geophysics, xxIX*(1), pp. 9–34.

Khandaker, Mosharraf Hossain. (2016). *Sesimo-Tectonic Risk of Tipaimukh Dam*. Available at http://www.thedailystar.net/news-detail-215226. Accessed 11 August 2016.

Laskar, Anwarul Alam and Parag Phukon. (2012). "Erosional Vulnerability and Spatio-Temporal Variability of the Barak River, NE India". *Current Science, 103*(1), 10 July, pp. 80–86.

Linton, J. (2008). "Is the Hydrologic Cycle Sustainable? A Historical-Geographical Critique of a Modern Concept". *Annales of Association of American Geographers, 98*(3), pp. 630–649. See, p. 633.

Mazumder, Muhammed Khairujjaman, Freeman Boroa, Badruzzaman Barbhuiya, and Utsab Singha. (2014). "A Study of the Winter Congregation Sites of the Gangetic River Dolphin in Southern Assam, India, with Reference to Conservation". *Global Ecology and Conservation, 2*(2014), pp. 359–366.

Northwest Hydraulics Consultants (NHC), Edmonton, Alberta, Canada. (2006). *River Flooding and Erosion in Northeast India*, Background No. 4, October. Available at http://web.worldbank.org/archive/website01062/WEB/IMAGES/PAPER_4_.PDF. See, p. 12.

Pamei, Aram. (2001). "Havoc of Tipaimukh High Dam Project". *Economic and Political Weekly, 36*(13), pp. 1054–1148.

Yumnam, Jiten. (2016). *Oil Exploration: Boon or Bane for Manipur*. Available at http://epao.net/epSubPageExtractor.asp?src=education.Science_and_Technology.Oil_Exploration_Boon_or_Bane_for_Manipur. Accessed 12 August 2016.

Achieving Last Mile Delivery: Overcoming the Challenges in Manipur

Sukhreet Bajwa

INTRODUCTION

In the times of e-commerce where the goods and services are now being made available at our doorstep, there is a large section of population still deprived of basic services. While we book our tickets at one click in no time, people in Manipur have just one private bus plying to take them to the city. Consider the emergency of a pregnant woman in labour pregnancy or a disaster like flood or earthquake and the only road link is damaged cutting off people from the essential services. This pushes them to a world of bygones when we were striving for mere survival. Freedom means having a right for best possible development of oneself. In this mad race of achieving new heights, we have left a huge section of our population behind. Not moving ahead at the same pace has its cost of giving rise to difficult situations like conflicts which stall the common progress and happiness of humanity. This paper intends to explore the challenges arising to achieve the last mile delivery in Manipur.

Since earthquake can damage the only possible route to provide aid, it becomes imperative to study the last mile delivery challenges in a

---

S. Bajwa (✉)
Gujarat Disaster Management Authority, Gandhinagar, India

© The Author(s) 2018   229
A. Singh et al. (eds.), *Development and Disaster Management*,
https://doi.org/10.1007/978-981-10-8485-0_16

place like Manipur which is under Zone V of earthquake vulnerability. This puts focus on the logistics network of the state. Logistics serves as a link between disaster preparedness and response. The Indian State may respond effectively to reach aid in case of a flood or a tsunami because we have built an efficient and dedicated National Disaster Response Force. However, it finds it more challenging to ensure round-the-clock supply of water or electricity. Within logistics, we explore the last mile challenges to delivery of essential services. Transportation system forms an important part of the last mile network. Apart from earthquakes, there have been incidents of floods washing away the bridges which connect the villages to the main town. Situations like this were observed at Chakpikarong, Chandel district, where the village was left to fend for itself cutting off the daily essential supplies from the market due to destruction of the bridge. As if earthquakes and landslides were not enough to isolate the people from connectivity, Manipur is also prone to frequent landslides. All these hazards when they get combined with an already vulnerable population increase the risk of lives lost and other destructions to properties. The basic needs during disaster get accentuated. Basic needs include food, shelter, education and health. Due to a disaster, there is a disruption in the supply of these essential services. To minimize such risks, the state needs to focus on capacity building measures. Robust last mile delivery will ensure that the citizens are being rescued on time and are provided aid in times of crises. The various challenges to last mile delivery need to be identified and their alternatives put in place. It will ensure regularity of supplies of food, shelter, water, education and health. Thus, smooth and continuous delivery of essential services becomes critical.

DISASTER MANAGEMENT CYCLE AND LAST MILE DELIVERY ACTIVITIES

Disaster management process has been divided into five phases for the ease of planning, the various phases being—prevention, mitigation, preparedness, response and recovery. These phases, if integrated with various last mile delivery activities, can ensure the continuance of the essential services during a disaster.

Prevention Preventive measures include ensuring sustainable development. Sustainable development ensures that the existing hazards in

ACHIEVING LAST MILE DELIVERY: OVERCOMING THE CHALLENGES ... 231

nature do not turn into disasters. Living with nature has been the way of life. The existing livelihood options as observed in the villages of Manipur included cutting and selling timber. Such activities lead to magnification of hazards like soil erosion and subsequent landslides and floods in the area. The timber supply work must be accompanied by tree plantation along the river bank to build embankments and build safety against floods. Creation of alternate employment opportunities in allied and services sectors like fisheries, homestay tourism and marketing of the skills of the tribes like handicrafts can lead to prosperity without harming nature. Effective dissemination of information on link between disasters and development will build a culture of prevention.

Mitigation The risk can be minimized by limiting the impact of a disaster or reducing the probability of its occurrence. Various last mile mitigation efforts include construction of embankments, afforestation activities, reducing the dependence on timber trade, etc. Such activities if undertaken at the community level will be sustainable in the long term rather than if executed by top-level authorities. Communities if sensitized will engage in such mitigation measures and hence make their habitats safe.

Preparedness Preparedness means being ready to face the imperative disaster and hence minimizing the loss. Various preparedness measures include development of an early warning system, evacuation routes in place, creation of task forces to act quickly and construction of disaster safe shelters to take refuge in, in case of any mishap. Development and implementation of Village Disaster Management Plans is the most effective preparedness strategy. Such activities are best implemented by local-level governance structures. Community-based disaster management including last mile integration of policy, plans and execution has been the underlying theme of National Disaster Management Authority's plan on Disaster Management.

Response Disaster response requires quick rescue of the survivors so as to minimize the injuries and loss. After the rescue, the immediate needs of the survivors are to be met. This includes setting up of camps, provision of food, water aid and medical aid. Response planning would require identification of available health professionals in the vicinity and equipping them with necessary instruments. This also requires effective coordination with the Non-Government Organisations (NGOs) and Community-Based Organisations (CBOs).

Recovery Recovery entails reconstruction in a disaster-proof way. The disaster proofing should include the locally available resources so as to increase their acceptability. Post-Gujarat Earthquake in 2000, the owner-driven approach of reconstruction has been hailed as successful. The local government should build the capacity of the people by providing them adequate finance and knowledge about disaster safety measures and use the energy of the survivors to create a model disaster-proof reconstruction, i.e. 'Build Back Better' (Fig. 1).

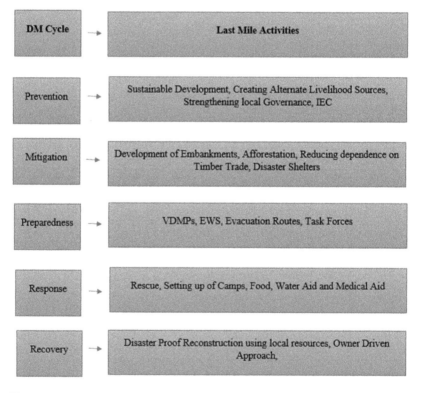

Fig. 1 Disaster management cycle and last mile delivery activities

CHALLENGES TO LAST MILE DELIVERY IN MANIPUR

Last mile delivery is to ensure that the last person standing gets his due rights. Manipur as a state is facing various issues in governance. It is also marred with constant internal conflict. The various identified challenges to achieve the last mile delivery are lack of infrastructure, inadequate administrative will, difficult terrain and financial constraints. Infrastructure includes access to schools, hospitals, police stations and safety shelters.

The survey in the villages of Manipur revealed that the villages had adequate access to education till primary level but faced problems in higher education as colleges are in the city and the infrequency in bus service makes it difficult for a person of meagre resources to access education. Similarly, in spite of the availability of police stations, the access is limited due to transport challenges. In case of health care, the primary health centres are functional but the access to secondary and tertiary health care is limited due to problem of connectivity. Disaster mitigation requires the construction of safety shelters which are absent in the community. Even after the major flood of 2015 and the earthquake of 2016, there is no provision of construction of safety shelters.

Since in case of infrastructure, the major problem is access being limited due to inadequate transportation network, it is imperative to study the obstacles in the existing transportation network. *First*, there is inadequacy of roads connecting the cities to the towns as well as villages. The villages mostly have kutcha roads which turn into muddy pathways during monsoon. *Second*, the state bus transport system in Manipur being dysfunctional, the vacuum has been occupied by private buses. Privately owned buses and wingers ply on roads. Being privately owned, they charge a disproportionately high amount of Rs. 100 from Imphal to Chandel district. Also, during monsoon, the transportation cost shoots up as there are few drivers willing to take the risk of driving through the muddy patches. Since Manipur gets constant rainfall beginning from March to October, it is often heavily flooded. The roads become inaccessible. Instead of buses, high powered vehicles like Tata wingers ply during that time. The Tata wingers charge an exorbitant mount of Rs. 5000 to travel during rainy days. It is a risk to the vehicle due to increased probability of landslides during that period. Many a times, the vehicles get washed away along with the road due to overflowing of the rivers. *Third*, the existence of bridges makes it extremely unsafe to drive through them. There have been instances wherein the whole bridge has been washed away due to rising river water level. One such instance is of Chakpi river bridge which was washed away during 2015 monsoon. It not only created problems of access but the wooden logs floating as debris accelerated the flood-like situation in the village. *Fourth*, the frequency of vehicles is insufficient with only one or maximum two buses in a day from the villages to the cities (Fig. 2).

Overcoming the Challenges

Goods and services which are normally needed after a disaster are not available to the people even in their normal lives. To increase the resilience, the delivery of basic and essential services must be made disaster-proof. Since disaster is a result of interaction of vulnerabilities and

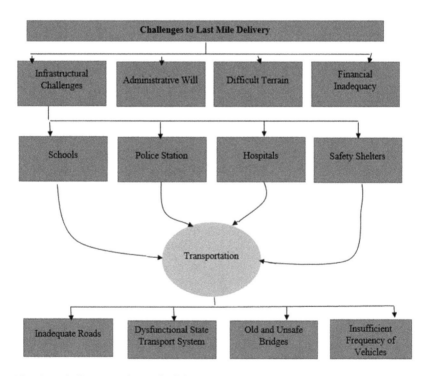

Fig. 2 Challenges to last mile delivery in Manipur

hazards, we can either reduce the vulnerabilities or build the capacity to decrease the disaster risk.

$$Disaster Risk = Hazard \times Vulnerability/Capacity$$

Keeping the hazard and vulnerabilities as constant, the paper focuses on the need to build the capacities to address the disaster risk. Building the capacity to deliver at the last mile through, the strengthening of transport network is an important intervention required in Manipur.

Right to Public Service Delivery

Transportation system is a basic public service delivery activity. A dent in this system would bring to halt various other service delivery activities of the government. However, in Manipur the dysfunctional state transport system is a major challenge to the state's development as well as capacity building efforts. The buses are plying in private sector but public sector needs to take responsibility of the same. Public service differs from private service as their intention is that of welfare rather than profit. Also, the sense of fiduciary responsibility of the state requires the state to ensure the delivery of services. Since the public services are funded by taxes, the government is duty-bound to ensure the delivery of services. The principles of *fairness* and *equity* are important in public services. Unlike private sector, public sector does not have the advantage of alternate supply in a competitive manner. This makes public sector less cost-effective. But public services are required to be continued in non-viable service sectors also and to geographically and financially disadvantaged groups. During a disaster, 80% relief effort is dependent on transportation service. Ready availability of aid and equipment would reduce the losses in a disaster.

Strengthening the Grassroots Governance

Grassroots governance and decentralization have been identified as the pillars towards achieving development in the most sustainable way. Developing local capacities remains the agenda of United Nations Organizations. Since governance is the overarching framework to build, create and maintain service delivery networks, there is need to identify and restructure the existing pillars of governance. The Sendai Framework for disaster risk reduction also gives priority to develop local capacities to build community resilience.

Manipur as a state has Hill Councils as the last tier of governance. There are six Hill Councils in Manipur according to the Manipur (Hill Areas) District Council Act, 1971:

1. Chandel Autonomous District Council
2. Churachandpur Autonomous District Council
3. Sadar Hills Autonomous District Council, Kangpokpi
4. Manipur North Autonomous District Council, Senapati
5. Tamenglong Autonomous District Council
6. Ukhrul Autonomous District Council

Essential services like primary education, primary health, sanitation, development of roads and bridges fall within the ambit of Hill Councils. Management of any forest, that is not a reserved forest, also falls under the Hill Councils. This gives enough scope to develop disaster preparedness and mitigation activities within the framework of local governance bodies. The research team visited Senapati, Chandel and Churachandpur districts as part of the field visit. It was observed that the villages were mostly governed by the village chief who was generally the eldest person in the community. The data collated reflect the absence of the essential services at local level and also non-integration of disaster management activities at the local level, the reason being the non-coordination of the concerned district's officials with the village level chiefs. The failure of governmental institutions and officials to provide adequate service in time during crises has led to erosion of faith and authority of these institutions. This has created a wide gap, and this gap has been filled by the emergence of various NGOs. These NGOs are either faith-based or ethnicity-based. Thereby, it is imperative now to coordinate and consult these community-based organizations in making disaster management plans. There is need for an institutionalized forum to bring the officials and the CBOs on a common platform. This will make the functioning of some of the CBOs and NGOs more transparent and will bring about professionalism within their functioning.

CONCLUSION

The major intervention required in Manipur is to build the capacity of state transport system. This would place in line other infrastructural services and also solve the problem of access. Strengthening the state transport system will ensure the reach of services like health, education and water in Manipur. Local level bodies need to engage in developmental activities which develop community resilience. Transportation service delivery can efficiently deliver the right product at the right time at the right place. Efficient service delivery will accelerate our progress towards the achievement of sustainable development goals. The disaster survivors need to be seen through the rights-based lens and not in the age-old altruistic manner. Being entrusted with electoral power, the state is duty-bound to provide these essential services. The delivery of these essential services is the solution to achieving legitimacy among the citizens which will also result in the fading away of social unrest. Humanity is very much capable of developing and exploring alternate routes and building connectivity. India definitely is endowed with the capability of engineering marvels. Thus, the policies must shift to distribution of resources to the historically neglected areas and bring all the citizens at par with development. We may not eliminate the inequalities but we can strive to diminish them by provision of basic and essential services needed for development.

BIBLIOGRAPHY

Disaster Risk Reduction Tools and Methods for Climate Change Adaptation, Inter-Agency Task Force on Climate Change and Disaster Risk Reduction, http://www.unisdr.org/files/5654_DRRtoolsCCAUNFCC.pdf.

Gyöngyi Kovács and Karen M. Spens. (2007). "Humanitarian Logistics in Disaster Relief Operations". *International Journal of Physical Distribution & Logistics Management, 37*(2), pp. 99–114, http://dx.doi.org/10.1108/09600030710734820.

Human Development in India: Analysis to Action, October 2010. http://www.gdrc.org/uem/disasters/1-dm_cycle.html.

http://www.prsindia.org/uploads/media/Citizen%20charter/Bill_summary_Citizens_Grievance_Redressal_Bill_2011.pdf.

Janet L. Bell. (1995). "Traumatic Event Debriefing: Service Delivery Designs and the Role of Social Work". *Social Work, 40*(1), pp. 36–43, http://www.jstor.org/stable/23718348.

Monica Escaleras and Charles A. Register. (2012). "Fiscal Decentralization and Natural Hazard Risks". *Public Choice, 151*(1/2), pp. 165–183, http://www.jstor.org/stable/41406921.

National Disaster Management Plan. (2016). A Publication of the National Disaster Management Authority, Government of India, May, New Delhi.

PART III

Accomplishing Community Resilience

A Study of Socio-economic Community Resilience in Manipur

Thiyam Bharat Singh

Vulnerability of Manipur to Natural Disasters

Over the last decade, the North-Eastern Region of India in general and the State of Manipur in particular have seen an increase in the occurrence of natural disaster in the forms of earthquakes, landslides, flood and drought causing people homeless and huge economic losses. The people of Manipur are now highly vulnerable exposing to risk of natural hazards. Manipur falls in Zone V, which is the zone most susceptible to earthquakes.[1] Though the region is endowed with a number of rivers, streams and freshwater lakes, the region is placed in a vulnerable situation due to frequent earthquakes, floods, drought, landslides and mudslides. Hilly areas are ecologically sensitive that needs special care while

[1] Manipur falls in Zone V, which is the zone most susceptible to earthquakes. The movement of the Indian tectonic plate towards the Eurasian Plate is taking place continuously. Earthquakes and heavy rains trigger floods and landslides which are a common feature during the monsoon season every year.

T. B. Singh (✉)
Centre for Study of Social Exclusion & Inclusive Policy (CSSEIP),
Manipur University, Imphal, India

© The Author(s) 2018
A. Singh et al. (eds.), *Development and Disaster Management*,
https://doi.org/10.1007/978-981-10-8485-0_17

241

taking up any development programmes. The recent earthquake that hit many parts of Manipur and surrounding states on the 4 January 2016 with its epicentre at Tamenglong district has impacted on the people's lives and their properties. Following this, another earthquake followed on the 13 April 2016 with its epicentre at Indo-Myanmar border worsens the already affected people, buildings and properties. Landslides are one of the common natural hazards causing loss of life and property.[2] Three national highways (NH-39, 53 and 150) running across the Manipur State are the only means of surface transport. Large-scale anthropogenic activities like construction and widening of road, quarrying for the construction material, fragile lithology, complex geological structures and heavy rainfall are the main causes of landslides in the area. All the major river systems in Manipur are vulnerable to flooding, as captured in the Vulnerability Atlas. Manipur receives heavy rain every year causing flash flood and breaching of embankments. It causes large-scale damage to people's livelihood and property. Most of the affected population took shelters in nearby relative and neighbouring village houses during flood.

It is reported that the rate of exposure of Asian and Middle Eastern countries to natural disasters and the damage inflicted by them has escalated in recent decades. According to a report by the United Nations ESCAP, half of the world's disasters in 2014 occurred in Asia and the Pacific, affecting 80 million people and incurring nearly $60 billion in economic losses.[3] The average number of natural disasters in the Middle East and North Africa (MENA) almost tripled since the 1980s, resulting in average annual losses of over $1 billion.

The objective of this present paper is to examine resilience of local community (socio-economic resilience) during disaster by selecting two districts of Manipur. The two districts are Imphal West (valley) and Senapati (hill district). Imphal West district has been selected from the valley districts to examine the vulnerability of its people exposing to risk of earthquake. Imphal Valley is particularly selected because the recent two earthquakes in Manipur have caused extensive damage to people's lives and major infrastructure including government offices, buildings, markets, schools, roads, bridges and houses. Senapati district has been

---

[2] M. Okendro et al. (2010).
[3] The World Bank (2014).

A STUDY OF SOCIO-ECONOMIC COMMUNITY RESILIENCE IN MANIPUR   243

Table 1   District-wise area, population, sex ratio and density of population (2011) (*Source* Economic Survey of Manipur 2013–2014)

S. No	District	Area (in sq. km)	Population (in '000)	Sex ratio	Density
1	Senapati	3271	479	937	146
2	Tamenglong	4391	141	943	32
3	Churachandpur	4570	274	975	60
4	Chandel	3313	144	933	44
5	Ukhrul	4544	184	943	40
6	Imphal East	709	456	1017	643
7	Imphal West	519	518	1031	998
8	Bishnupur	496	237	999	479
9	Thoubal	514	422	1002	821
10	Manipur	22,327	2856	985	128
11	All India	3,287,263	1,210,855	943	368

selected from the hill districts as people are exposing to risk of both landslides and earthquake. Over the past decades, the district has witnessed a rise in the occurrence of landslides. Data have been collected from the two districts by visiting the affected sites of both the districts. Questionnaires were used to collect data from the respondents. Qualitative data have also been generated by conducting group discussions and in-depth interviews with villagers, local teachers, Church leaders, SDMA officials, PDA officials, NGOs, legal experts, officials of Health Department, IFCDs officials and observations. Data were also collected from newspapers, research papers, books, journals, government reports and other available literature on disaster studies. The paper has been organized into seven sections. The first two sections present introduction, objectives and data. The following sections III and IV provide socio-economic demographic profile of Imphal West and Senapati districts. Sections V and VI discuss community resilience and flood. The last section provides the finding and conclusion of the study.

Table 1 presents data on district-wise area, population, sex ratio and density of population. It may be observed from the table that hill districts in Manipur recorded larger areas as compared to valley districts. In terms of population, Imphal West and Imphal East districts in the valley areas recorded higher figure compared to other districts. Valley districts recorded maximum figure in terms of sex ratio and population density as compared to hill districts. Table 2 provides data on the share of Central

244   T. B. SINGH

Table 2 Share in the disaster relief fund for Manipur, 2010–2011 to 2014–2015 (*Source* Economic Survey of Manipur 2013–2014)

	Share (Rs. Crore)		
Year	Central	State	Total
2010–2011	6.5	0.72	7.22
2011–2012	6.82	0.76	7.58
2012–2013	7.16	0.8	7.96
2013–2014	7.52	0.84	8.36
2014–2015	7.9	0.88	8.78

Table 3 District-wise people at risk during natural disaster in Manipur (*Source* Manipur State Disaster Management Plan Volume I & II (2013))

Districts	Population	Earthquake	Landslide	Flood/Flash Flood/Cloud Burst	Drought	Forest Fire	Industrial &CBR	Stampede	Road Accidents	Dam/Lake Burst
Imphal East	452661									
Imphal West	514683									
Bishnupur	240360									
Thoubal	420517									
Churachandpur	271274									
Chandel	144028									
Senapati	354972									
Tamenglong	140143									
Ukhrul	183115									

and State in the Disaster Relief Fund for Manipur State for the period 2010–2011 to 2014–2015. It is evident from the table that the total share marginally increased from Rs. 7.22 crore in 2010–2011 to Rs. 8.78 crore in 2014–2015, respectively.

Table 3 presents information on district-wise people exposing to risk of natural disasters such as earthquake, landslide, flood and drought in Manipur. It may be observed from the table that people of Manipur are exposing to risk of natural disaster such as earthquake, flood and drought. Landslide has affected only the hill districts of Manipur, and people in the hill are exposing to risk of landslide. It may be noted that valley people are exposing to risk of disaster in the form of flood. Sometimes, there are incidents of flood in the hill district due to disruption of river flows under bridge by timbers caused by mudslides.

IMPHAL WEST DISTRICT

Imphal West is one of the valley districts of Manipur. According to Census of India (2011), it has an area of 519 sq. km. Imphal district ranks seventh position in terms of its area covering only 2.32% of the total area of the state.[4] There are 124 villages of which 17 villages are uninhabited. The density of population per sq. km. accounted for 998 persons. Sex ratio of the district accounted for 1031, which is well above the state sex ratio of 985. The total literacy rate of the district recorded a figure of 86.1% which is higher than the state literacy rate with 76.9%. While the proportion of Scheduled Caste (SC) population in the district accounted for 3.2% which is marginally lower than the state average of 3.4%, the proportion of Scheduled Tribe (ST) constitutes only 4.6% of the total population against the state average of 40.8%. While the literacy rate of the ST accounted for 89.8%, literacy rate for SC population recorded 81.1%. The average size of the household in the district has been estimated at 4.6 persons, which is slightly lower than the state average of 5.1 persons. Work participation rate for the total population has been estimated at 41.2%, which is less than the state average of 45.6%. Main workers constitute 32.1% of the total workers.

Table 4 presents data on the amenities of Imphal West district. It is obvious from the table that safe drinking water accounted for 57.2% indicating more than 45% of the people is not available of safe drinking water. There are still many people who are not yet connecting electricity in their household.

The earthquake has hit many buildings in Imphal city. The building of the Manipur Press Club in Imphal has also been partially damaged. A massive crack was seen on the ground floor of the BSNL office adjacent to the Old Secretariat which houses ministers' offices. A road

---

[4]Government of India (2011), 'Census of India 2011', Manipur Series-15–B District Census Handbook-Part XII, Imphal West, Village and Town Wise Primary Census Abstract (PCA), Directorate of Census Operations, Manipur.

246 T. B. SINGH

Table 4 Distribution of amenities of Imphal West district (*Source* Census of India (2011), Transport & District Elementary Education Report Card, 2013–2014)

S. No.	Amenities	Numbers (percentage)
1	Total inhabited villages	124 (2011 census)
2	Total households	110, 672 (2011 census)
3	Safe drinking water facilities	57.2% (2011 census)
4	Electricity	78.2% (2011 census)
5	Primary and upper primary[a]	392
6	College	27
7	Accredited private health institutions	6
8	Primary health centre	50
9	Health facility (RIMS)	1
10	Post office	3
11	Bus services[b]	1646

[a]Based on District Elementary Education Report Card 2013–2014, Government of Manipur
[b]Government of Manipur (2013), 'State Transport Policy Manipur', Transport Department, June 2013

leading to the Leimakhong Army base developed cracks in many places. Two houses at Keishampat and Dewlahland in Imphal West collapsed completely. Thangal Bazar and Paona Bazar, Imphal's two busiest markets, were shut down after many buildings developed cracks. They were opened late in the afternoon after government engineers assessed the damage. The Minuthong Bridge that connects Imphal East and Imphal West developed huge cracks in two places. Ima Market, which is exclusively run by women, was also badly damaged and shut down after an inspection by state works minister. He said the State Government was assessing the damaged buildings.[5]

SENAPATI DISTRICT

Senapati district has a total area of 3271 sq. km, and it ranks 5th in terms of area size. The district is predominantly inhabited by ST that constitutes 87.5% of the total population. The proportion of SC population in the district is only 0.2% which is lower than the state average of 3.4%. There is no urban area in the district in 2011 Census except

[5] *The Sangai Express* (2016b), 'Over 1000 Buildings Damaged in Senapati', Imphal, 9 January 2016.

A STUDY OF SOCIO-ECONOMIC COMMUNITY RESILIENCE IN MANIPUR   247

one census town. The average size of the household in the district is 5.7 persons and is above the state average of 5.1 persons. The population the district constituted is 16.7% of the total population of the state and ranks 4th position among the 9 districts. The density of population per sq. km has been estimated at 146 persons against the state density of 128 persons. Sex ratio of the district is 937 which is slightly lower than the state sex ratio of 985. The total literacy rate of the district registered at 63.6% which is below the state literacy rate of 76.9%. The literacy rate of the Scheduled Castes has been estimated at 58.1% and the Scheduled Tribes at 62.7%. Work participation rate for the total population has been estimated at 48.7%. Main workers constitute 38.8% of the total workers. Cultivation is the main occupation of the population of the district and accounts for 74.9% of the total workers. The total number of villages in the district is 686 of which 17 villages are uninhabited. The 6.8 magnitude powerful earthquake that struck entire Manipur and its neighbouring states with its epicentre in Tamenglong district on January 4 has damaged as many as 3218 households besides claiming 10 lives and 120 injured in Manipur alone.[6] Senapati district has the maximum number of severely damaged Kutcha houses (601) and other infrastructures (55). Two people were killed in Senapati district during earthquake. Within Senapati district, Saitu and Kangchup-Geljang subdivisions were the worst affected.[7] Most of the damaged houses belong to Nepali families. Generally, their houses were built of mud and stone and they crumbled at the impact of the earthquake. Altogether 33 persons sustained injuries in Senapati district. The number of affected villages was 26 in Saitu subdivision, 31 in Kangchup-Geljang, 18 in Kangpokpi, 14 in Saikul, one in Island and two villages in Bungte Chiru. About 114 government buildings too suffered damages in Senapati district. In Senapati district, Molhoi and Chalbung villages were affected and affected people were staying in relief camps.[8] Most of the houses were completely damaged and would require reconstructing from the scratch. Most people were traumatized by the incident and also busy shifting and reconstructing their damaged houses to look for work.

[6] Eastern Mirror (2016), 'January 4 Earthquake Report', 10 February, Imphal.

[7] *The Sangai Express* (2016b), 'Over 1000 Buildings Damaged in Senapati', Imphal, 9 January 2016.

[8] Sphere India (2016), 'Joint Rapid Needs Assessment Report on Manipur Earthquake, 2016', Inter-Agency Groups-Manipur.

248   T. B. SINGH

Table 5 Distribution of amenities in inhabited villages of Senapati district (*Source* Census of India, 2001 & 2011[a])

S. No.	Amenities	Numbers (percentage)
1	Total inhabited villages	491 (2001) 686 (2011)
2	Total households	25,850 (2001) 36,000 (2011)
3	Safe drinking water facilities	182 (37.1%)
4	Electricity (power supply-domestic)	432 (87.9%)
5	Primary school	283 (57.6%)
6	Middle schools	73 (14.9%)
7	College	2 (0.4%)
8	Medical facility	33 (6.7%)
9	Primary health centre	4 (0.8%)
10	Primary health subcentre	27 (5.5%)
11	Post, telegraph and telephone facility	38 (7.7%)
12	Bus services	163 (33.2%)
13	Paved approach road	182 (37.1%)
14	Un-paved (mud) approach road	317 (64.5%)

[a]Cited in Vaiphei Thathang (2014), Impact Study of Mahatma Gandhi National Rural Employment Guarantee Scheme in Senapati District, Manipur, Working Paper, CSSEIP, MU, Vol. 004, November 2014

Government provided some ration like rice and dal, but the ration was not sufficient according the people. Data on amenities of Senapati district have been presented in Table 5. The table shows that more than 60% of the population is not getting safe drinking water in the district. It is evident from the table that there is acute lake of medical facility and roads are muddy. There is also acute lack of transportation and communication in the district.

COMMUNITY RESILIENCE

Khwairamband Bazar is popularly known as *Ima Keithel*.[9] It has been estimated that there are 3913 women vendors in the three markets. Purana Market has 1857 women vendors. Laxmi Market and New

[9]In Manipuri, Ima means mother and Keithel means market. So, it is known as the mother's market. The Ima Market (mother's market) is also known as Nupi Keithel (women's market). Women vendors obtain their licences from the Imphal Municipal Council (IMC) by paying monthly tax of Rs. 95, and they are currently selling vegetables, fruits, clothes, fish, ornaments, etc., at the three multistoried markets in the city.

Market have 778 and 1278 women vendors, respectively. The recent earthquake occurred on 4 January 2016 in Manipur has caused extensive damage to the buildings of two major women markets of the state. These markets are New Market and Laxmi Market. The tragic incident has displaced about 2056 women vendors. Another earthquake occurred on 13 April 2016 in Manipur has caused extensive damage on another multistoried building of women market called Purana Market or also called as Ima Market. About 400 women vendors have so far been displaced from the Purana Market, and still, 1457 women vendors remain seated inside the market which is highly vulnerable to disaster. It may be recalled that the multistoried buildings of the three women markets were inaugurated by Smt. Sonia Gandhi, UPA Chairperson in 2010. A temporary market shed has been constructed to accommodate the displaced women vendors. All the displaced women have been shifted here. When asked about the difference in the volume of sales between the present and the previous market, respondent said, *'Earlier they could sell as much as possible per day. They could sell about Rs. 10,000 per day in the previous market. Now the volume of sales has declined to about 75 per cent day. Majority of the women vendors have become indebted because each woman could not contribute to their daily collection (done by group of women). Their livelihood is now affected. All the women have now decided to go back to the previous market'*. Further asking one woman vendor who sells silk garments, earlier she could sell goods for Rs. 20,000–30,000 per day but now she cannot sell goods for Rs. 10,000. Asking about the preparedness of the disaster, the respondent replied that *'The design of the building does not provide escape route for people during the disaster. We want to remove the heavy objects hanging on the top of ceiling which has been constructed as decoration'*. Women vendors have blamed the State Government for the 'poor' quality of construction of the market complex.[10] Most of the buildings in Manipur have been constructed without consulting engineers and architects.[11] It is therefore highly needed to impart skills to construction workers so that they do not violate construction guidelines. School

---

[10]Works Minister of Manipur State Dr. Kh. Ratankumar said an expert team from the IIT Roorkee, the oldest technical institution in Asia, has provided the detailed assessment of the damaged all-women market to the National Disaster Management (NDM) which will determine whether to rebuild the market buildings or to repair them.

[11]Data collected from Chairman, Town Planner, MAHUD, Gitkumar Nepram.

250 T. B. SINGH

buildings are most vulnerable in Manipur because if earthquake occurs during school hour, a large number of students present in the school will be hit. Therefore, lifeline buildings like school, government offices, hospitals and community hall must be earthquake-proof buildings. As regard to Ima Market in Khwairamband Bazar, he pointed out '*there is lack of structural design in the construction. He has termed it as "soft storey". There are no walls on the ground floor of Ema market and only pillars stand but on the first floor, a lot of construction is taking place putting*

Table 6 Selected socio-economic indicators for two districts of Manipur (*Source* Data compiled from various sources including Economic Survey Manipur, Planning Department, Government and Census of India (2011))

S. No.	Socio-economic indicators	Imphal west	Senapati	Manipur State
1	Average size of the household (2011 census)	4.6	5.7	5.1
2	Per capita income in constant prices (1993–94) for the year 2000–01	8008	4275	6573
3	Literacy rate (2011 census)	86.1	63.6	76.9
4	Work participation rate (2011 census)	41.2	48.7	45.6
5	Cultivators (2011 census)	37,107	1,75,127	5,74,031
6	Agricultural labourers (2011 census)	12,870	11,210	1,14,918
7	Marginal workers (2011 census)	47,378	47,897	3,30,447
8	Families below poverty line	21,438	11,362	1,02,400
9	Net area sown (land use classification)[a]	32.49	17.44	Nil
10	Total crop area (land use classification)[b]	49.85	24.67	Nil
11	Expectation of life at birth 1991[c]	71.2 (male)	67.1 (male)	68.64 (male)
12	Expectation of life at birth 1991[c]	74.06 (female)	70.68 (female)	72.42 (female)
13	Forest cover (2011 census)	73 sq. km	2559 sq. km	17,219 sq. km

[a]It was based in terms of ('000 hectares) and the year based on 2014–2015
[b]Per Capita Income for the year 2000–2001 was estimated before Imphal was bifurcated into Imphal West and Imphal East
[c]Expectation of Life at Birth was taken from E. Bijoykumar Singh (2005), 'Development Discourse in Manipur, Hills vs. Valley', Eastern Quarterly, Volume 3 Issue, October–December 2005, Manipur Research Forum, Delhi

heavy on the ground floor'. He also said that there are two objectives of construction. One is quality and another is durability. As long as we keep in mind about quality and durability, earthquake will not make much damage to buildings.

Data on selected socio-economic indicators of two districts have been shown in Table 6. The table shows that Imphal West district has recorded an average size of 4.5 members per household and Senapati district has on average 5.7 persons per household. Work participation rate shows that there are more than 50% who are unemployed in both the districts. There are a large number of families living below poverty line in both the districts. A large number of cultivators are found in Senapati district. In terms of forest cover, Senapati accounted for 2559 sq. km. while Imphal West accounted for 73 sq. km.

As regard to Senapati district, local community and its neighbouring villagers are the first group of people, who will come and help the most affected villagers during disaster. Then, administrators and medical teams will follow their visit for inspecting the landslide-affected sites and for distributing relief materials. During disaster like landslide in the hill, volunteers, relatives, neighbouring villagers, Army, BRTF, Red Cross Society and Government officials have come to rescue the affected villagers. Landslides have caused extensive damage to their houses, roads, paddy field, school and drinking water. Since most of the roads in the hills are slippery during rainy season, it is difficult to take affected and injured people to the hospital. Sometimes, affected people are evacuated to neighbouring villages, relative's houses, community hall and government schools. Villagers lack awareness about the preparedness of disaster. A few households tried to cover some portion of vulnerable hill-slope with plastic to prevent water. Apart from that, there is lack of cooperation from SDMA for preparation of disaster. It takes at least five months to recover from the impact of disaster. Recently, training on disaster management was conducted by the Red Cross.

Disaster reduction work in Chakpikarong flood (Chandel district), flood in Thoubal district, cyclone in Mayang Imphal (Imphal West district) and earthquake-affected places in Senapati district by the Red Cross society shows that there is lack of coordination among elected members, groups and organizations in distributing relief materials to affected people. Most of the villagers are not aware of the logo/emblem of Red Cross

Society.[12] The most important thing in helping the disaster-affected person is giving psychological support. Highly needed materials like kitchen set, blanket, mosquito net, towel, sari and bucket were distributed to affected people during disaster.

In the hill districts of Manipur, Pastor, village chief or chairman, teachers, civil society organizations, clubs, youths and Church leaders are considered as educated group of the society.[13] Any kind of initiatives taken by the Church is successful particularly in the hills. During the period of disaster, the role of Church is multidimensional. They create awareness and training among the local villagers. They sympathize the sorrow and sufferings of the villagers. They provide places for gathering of people every Sunday in the Church. During disaster, people come to Church donating money in cash or kind. They give financial assistance or help in kind to affected villagers during disaster.

National Self Disaster Management Organisation (NSDMO) plays a vital role in creating awareness programmes and training on disaster risk reduction in various parts of Manipur.[14] Tamenglong, Senapati, Churachandpur, Imphal East and Imphal West are some of the districts in Manipur where awareness programmes have been conducted. They have been creating awareness among people about road safety, fire safety, earthquake, landslides, flood, etc. They conducted awareness programme on disaster management for 400 participants in Panchayat Bhavan at Porompat in Imphal East district. Further, they held awareness programmes for flood and cloudburst in Churachandpur district. Awareness programme for landslide in Tamenglong district has been conducted. Programmes for women on how they will prepare themselves during earthquake have been conducted. They held training for 50 villagers in Imphal West district. Lack of finance is major problem for the NSDMO.

When asked about the role of Municipal Council Members for disaster risk management, Ph. Baleshwor Sharma,[15] former Counsellor

---

[12] Data collected from T. Jinesh Singh, Team member of National Disaster Response Team (NDRT) under Red Cross Society, Manipur State Branch during field survey.

[13] Data collected from Father Dominic during fieldwork.

[14] Data collected from N. Sanajaoba Singh, Secretary of National Self Disaster Management Organisation (NSDMO), Manipur. The NSDMO office is at located at Canchipur, Manipur University Gate, Imphal West, Manipur.

[15] Ph. Baleshwor Sharma was elected three times (2001–2016) in Uripok Ward No. 5.

of the Uripok Ward 5, said, '*Imphal Municipality Council (IMC) is playing an important role for disaster mitigation programmes. We help in capacity building programmes among local community by conducting awareness programmes and training programmes among Self-Help Group (SHG) of Women. Usually, a SHG has about 10 members and we organised for 10 SHGs. There were around 100 participants. We invited resource persons from SDMA, Civil Defence, and Police Department came and delivered lectures*'.

It may be summed up that that landslide is mainly caused by deforestation, forest burning, Jhum cultivation, stone quarry and human greed. Planting vegetation with deep root systems is an effective method of preventing landslides, as the roots stabilize and anchor the soil. Native grasses and wildflowers are often useful for very steep slopes.

THE HAVOC OF FLOOD-PRONE RIVERS

Imphal Valley is drained by Imphal River and its tributaries (Laishram Sherjit 2015). The river rises in the highlands to the west of Kangpokpi and flows towards the north on the way. The river is joined by many tributaries like the Sekmai, the Thoubal, the Kongba and the Iril. Among them, the Thoubal and the Iril rivers are more important. The Iril river rises in the western part of the Mao and flows southwards to join Imphal river at Lilong. The Thoubal river rises in the Huimi hills of Ukhrul. It flows south-westwards and joins the Imphal river at Mayang Imphal. The Imphal river does not fall in the Loktak, rather it flows in the eastern part of the lake.[16] According to the study by Laishram Sherjit (2015), it is found that all the major river systems in Manipur are vulnerable to flooding, as captured in the Vulnerability Atlas. The confluence of the Imphal

---

[16] Khordak river drains the water of the Loktak to the Imphal river. In the south of the Loktak, the Imphal river is known as Manipur river, and it is joined by the tributaries, the Khuga and the Chakpi. The Khuga river rises in the Thinghat hills of Churachandpur district and flows eastwards to join the Manipur river at Ithai. The Chakpi river originates from the Laimatol hills of Chandel district. It flows south-westwards and joined the Manipur river at Sugnu. The Manipur river crosses the southern boundary through narrow gorges and enters the Chin hills of Myanmar. In the Chin hills, it is joined by the Myittha and finally it falls into the Chindwin river of Myanmar. The Nambul is another important river of Manipur that drains the Imphal valley. It rises in the Kangchup hills that lie in the western part of the valley. The river passes through the heart of Imphal city and follows a course in the western of Imphal river till it falls into the Loktak.

river with the Iril river at Lilong makes it voluminous and rapid, causing the breach of embankment in Chajing, Haoreibi, Samurou and Lilong. Khuga River flows through strong to steep slope. Hence, river velocity is moderate to high, creating flood problem in Kumbi and Ithai villages. The Chakpi River meets the Manipur River at a reverse direction causing flood in surrounding areas of Sugnu extending up to Wangoo during rainy season. Most of the embankments are poorly maintained. Many vulnerable points in Imphal River, Thoubal River and Iril River were identified. Further, the study shows that the rapid increase in the valley's built-up areas is also an important factor for the recent increase in flash floods in urban areas. Many natural groundwater recharging structures like ponds, tanks and swampy areas are greatly reduced due to urbanization and change in land use pattern. Breaching of river banks, heavy precipitation, inadequate drainage facility and siltation are the main causes of floods in Manipur. According to a Situation Report (2008) prepared by the Sphere India, Manipur received heavy rain during May in 2016, causing flash flood, breaching of embankments that has further caused damage to life and property. It has affected thousands of people by making them homeless in the valley districts of Manipur which are Thoubal, Imphal East, Imphal West and Bishnupur districts. Most of the affected population took shelters in nearby relative and neighbouring village houses. It is reported that 1671 households and 1302 households in Imphal West district and Bishnupur district were affected by the flood. Further, 70 households in Imphal West and 310 households Thoubal districts were affected. Shelter, non-food items (like hygiene kit, chlorine tablets, groundsheet and tarpaulin), safe drinking water and food are highly needed during time of flood. Flash floods due to incessant rainfall in the state have affected several houses in and around Pallel, Thoubal and Chandel districts headquarters. Similarly, large tracts of land at Sugnu, Nungu, Tangjeng and Serou have been submerged (*The Sangai Express* 2015).[17]

---

[17]The swirling flood water washed away the bridge built over Etok canal at Yairipok Chandrakhong under Heirok Assembly Constituency today. The bridge connects Chandrakhong Fangjakhong with other parts of Yairipok. With Sekmai river overflowing, several houses and agricultural lands located at Pallel in Thoubal district and some adjoining villages of Chandel district have been submerged. Houses and properties including livestock as well as agricultural crops were also destroyed. Local people have prohibited vehicular movement along Pallel's old bridge and have temporarily closed the old bridge. Most of the villagers of Island, Theimungkong, Sainem villages in Chandel district have shifted to safer places after their houses were half-submerged in water.

Flood in Manipur is mainly caused by several factors such as man-made factor, deforestation, three-time increase in siltation, encroachment on the river bank for construction and swallow of river bed.[18] The most important cause of flood in Imphal is the contraction in the areas of water bodies or water retention areas such as wetland. There have been large-scale constructions of office buildings, educational institutions, offices of club and organizations, government projects on the water retention areas in Lamphel in Imphal West district. Earlier in 1974, the time of run-off took 24 hours and there was not much flood in Imphal areas. By that time, the natural setting of water bodies was maintained. At present, the natural setting of water bodies has undergone significant changes. For example, the time of runoff from water bodies to rivers is zero causing instantaneous floods around Imphal area. Growing urbanization is another major responsible factor for causing flood in Manipur. Farming activities practised on the river banks are another problem. Apart from that, hindrance in opening Ithai barrage has been a major problem in controlling flood in Manipur.

When asked question to a Ng. Joykumar Singh, Uripok Khaidem Leikai, Imphal West, Manipur, about the preparedness and prevention of flood in Uripok, he said, My place is flooded very year. Basically, the water comes to my house due to over-flowing of Nambul River. We have laid our drainpipe up to the river bank and we used to shut the outlet of drain-pipe at the river bank. This helps in controlling the water flow. Now, the shutter has been removed and water runs quickly against the drain pipe. This has escalated the flood occurrence in my place. During flood, people are evacuated to their relative's houses, schools and community hall. Sometimes community hall in my locality is flooded and we cannot use it for evacuation. All the houses in my locality are flooded and we put important household goods in a higher place inside the room. Then, we moved out of the house to safety place like schools, community hall and relatives. Local clubs are taking vital role in helping the evacuation of affected people. They will report to the DC office and will collect relief materials. Club volunteers help in distribution of relief materials to the affected people. Sometimes, local club go to IFCD and collect materials bamboo, rope, mat, sand put in cement bag, for making embankment higher. Sometimes, State Government constructed

---

[18] Data collected from Chief Engineer, IFCD, G. Robindro Sharma.

retaining wall on the river bank but it is not effective in preventing overflowing of water. As regard to preparedness, household members will measure the level of water inside their house during flood. Depending on the level of water, they will try to increase the level of their house-floor with clay. Since there is inequality of income and wealth among the local communities, some well-to-do household will increase the level of their house-floor by buying truckloads of clay but poor household cannot afford to do it. Poor households remain flooded every year during monsoon. About the evacuation of livestock, local people know one day ahead before flood occurs and they will take their cattle on that night to some safe places.

It is also reported that IFCD is helpless and unable to take action as Flood Plain Zoning Act 1982 is not implemented in the state. The department is fully prepared to deal with any eventuality during rainy season. For this purpose, IFCD has set up a 24 hour control room. Moreover, a storeroom has also been provided from which the public can avail flood-fighting equipment round the clock. IFCD has taken up several schemes under flood management programme (FMP) of Union Ministry of Water Resources. Further, five new schemes for anti-erosion flood control works with a total cost of Rs. 626.63 crore have been proposed to the Union Ministry of Water Resources, River Development and Ganga for consideration during 2015–2016.[19] As regard to flood in Imphal West, a part of Uripok town is always flooded during monsoon every year due to overflowing of water from nearby Nambul River. Community resilience differs between valley and hill districts of Manipur.[20] There is strong community resilience in the valley districts because of availability of health care facility, better communication and transportation but still the high-rise buildings are highly vulnerable due to violation of building codes. On the other hand, there is weak resilience in the hills due to lack of health care facility, poor transport and communication. During this disaster, transportation is a major problem in the hill causing much difficulty in taking injured persons to hospital. There is lack of primary health centres. Another problem is related to the practice of Jhum cultivation in the hills causing soil erosion. As a part of capacity building process, SDMA

---

[19] *The Sangai Express*, Imphal Thursday 21 July 2016, pp. 1 and 8.

[20] Data collected from Jason Simray, Secretary, Relief & Disaster Management, Government of Manipur.

has been giving training to NCC, Scout, local bodies, Municipality Councils, representatives of local communities, stakeholders, teachers, etc., over the years.

FINDINGS AND CONCLUSION

The findings of the study show that local community lacks awareness about their responsiveness, preparedness, mitigation and recovery during disaster. They also lack training about their shelter, rehabilitation, relief entitlements, schemes and legal access during disaster. In this connection, local community should be able to identify their own vulnerabilities and capacities, public schemes, relief entitlements and special provisions for disabled, children, orphans or widows. During disaster in the state, rescue teams are not able to reach properly to the interior part of the affected villages due to poor infrastructure such as road and communication. This condition has in fact agonised and anguished the affected people. The study argues that we need to strengthen the local community so as to make a quick response when disaster occurs. As regards to earthquake, we should make our buildings safe by building based on earthquake resilient technology. We also need to ensure that public buildings specially are safe as these are the places where affected people will take shelter during disaster. In case of landslide, we need to protect our forests and other vegetation and to ensure that persons living in landslide-prone areas are shifted and rehabilitated to safer areas. The most vulnerable people in Manipur are agricultural and non-agricultural labourers and daily wage earners like small-scale vegetable vendors who earn their living on a daily basis.[21] Any hazard that disrupts normal day-to-day activity affects them most. Next vulnerable group is small and marginal farmers who get regularly affected by drought, floods, hailstorms and sometimes landslides. Most secure persons are government employees whose livelihood is assured. Poor people may also be affected by epidemics due to low immunity levels resulting from malnutrition. Women are generally more vulnerable to all the hazards because of the paternalistic society in Manipur. Even among women, pregnant and lactating mothers are more vulnerable. Old and infirm people, as well as small children, are more at risk than the general

---

[21]Government of Manipur (2013), '*Manipur State Disaster Management I & II*', Volume 2013.

population. Physically and mentally challenged people are more vulnerable to any disaster. As Manipur has a high ratio of HIV-positive persons, whose immunity has been compromised, they shall be much more vulnerable during any disaster. ST population is generally more vulnerable due to their poor socio-economic conditions.

REFERENCES

Eastern Mirror. (2016). *January 4 Earthquake Report*, 10 February, Imphal.
Elangbam Bijoykumar Singh. (2005). "Development Discourse in Manipur, Hills vs. Valley". *Eastern Quarterly, 3*, October–December, Manipur Research Forum, Delhi.
Government of India. (2011). *Census of India (2011)*. Manipur Series-15 Part XII—B District Census Handbook, Senapati, Village and Town Wise Primary Census Abstract (PCA), Directorate of Census Operations, Manipur.
Government of Manipur. (2013). *State Transport Policy Manipur*. Transport Department, June, 2013.
Government of Manipur. (2013–2014a). *Manipur State Disaster Management Plan Volume (I) & (II)*. Relief & Disaster Management Department, Government of Manipur.
Government of Manipur. (2013–2014b). *District Elementary Education Report Card 2013–14*. Government of Manipur.
Laishram, Sherjit. (2015). "Floods in Imphal Valley: Causes, Effects and Preventive Measures: Part-I". *The Sangai Express*, 22 August. Available at epao.net.
Okendro, M., R. A. S. Kushwaha, and O. P. Goel. (2010). "A Study of Landslides Along Part of a National Highway in Manipur, India". *International Journal of Economic & Environmental Geology, 1*(1), 48–50, January–June.
Situation Report. (2008). *Floods in Manipur 28th May 2016*. Prepared by Sphere India, Manipur, National Coalition of Humanitarian Agencies in India.
Sphere India. (2016). *Joint Rapid Needs Assessment Report on Manipur Earthquake, 2016*. Inter-Agency Groups-Manipur.
The Sangai Express. (2015). "Heavy Rain Washes Away Bridge". Imphal, 31 July.
The Sangai Express. (2016a). Imphal, Thursday 21 July, pp. 1 and 8.
The Sangai Express. (2016b). "Over 1000 Buildings Damaged in Senapati", Imphal, 9 January.
The World Bank. (2014). *Natural Disasters in the Middle East and North Africa: A Regional Overview*. Urban, Social Development and Disaster Risk Management Unit Sustainable Development Department Middle East, Global Facility for Disaster Reduction and Recovery (GFDRR), January, Washington.
Vaiphei, Thathang. (2014). *Impact Study of Mahatma Gandhi National Rural Employment Guarantee Scheme in Senapati District, Manipur*, Working Paper, CSSEIP, MU, Vol. 004, November.

'Living with Floods': An Analysis of Floods Adaptation of Mising Community—A Case Study of Jiadhal River

Tapan Pegu

INTRODUCTION

Floods are considered as one of the most significant natural disasters worldwide. The period between 1985 and 2008 evidently witnessed extreme rainfall which was responsible for more than USD 700 billion of damages in the world. Back in India, 59 years of statistical data from 1953 to 2011 revealed that the average annual cost of the flood damages in India is estimated to be more than Rs. 1800 Crore (Brahmaputra Board-CWD, India). Of late, the frequency and intensity of flood incidences have increased in the Himalayan region in the form of flash flood to be precise, the flash floods in Kosi Nepal on 5 May 2012, Uttarakhand, India on 4 August 2012, and Pakistan on 10 September 2012. The worst was the Uttarakhand flash flood that happened in June 2013 which due to its massive impact on the entire region of Kedarnath and other neighbouring districts was called 'The Himalayan Tsunami' (Das 2013).

T. Pegu (✉)
Jiadhal College, Jiadhal, Assam, India

© The Author(s) 2018   259
A. Singh et al. (eds.), *Development and Disaster Management*,
https://doi.org/10.1007/978-981-10-8485-0_18

FLOOD DISASTER IN ASSAM

Flood, river bank erosion and siltation are the most frequent water hazards in the Brahmaputra basin in Assam, especially during monsoon (Hazarika 2006). Siltation has become a yearly phenomenon and increasingly devastating since 1980s, especially on the northern banks of the eastern Brahmaputra Valley (Das et al. 2009). These kinds of hazards affect land and the community lives and livelihood. To be specific, it destroys homes and people get displaced, destroys crops, and damages public and private property. Above all, the cyclical journey of this hazard has brought on the question of survival for the community. For a state like Assam, these disastrous events have brought about a heavy toll on the state's economy as the loss was estimated in several thousand millions of rupees.

THE MISING COMMUNITY

The Mising[1] belong to the Tibeto-Burmese family of the Mongoloid race. Their original abode was in the upper courses of river Huang-Ho and Yangtse-Kiang in the north-west of China, and they entered India around 2000 BC (Doley 2014). In search of cultivable land, they migrated along the Siang River in groups in different periods of history. It is assumed that the first group of Mising migrated to Assam in between thirteenth and fourteenth century. The main cause of migration of the Mising to the plains of Assam was economic and also to explore better avenue for habitat (Doley 2014). The migratory habit of Mising people continued until independence and is associated with an itinerant life from agriculture that represents an adaptive strategy to hazards and an inherently unstable flood ecosystem (Crémin 2012). Here in the Brahmaputra Valley amidst the alien communities and faith, the Misings had adapted themselves to the new environment and changed social circumstances. After the migration, they settled in alluvial land of river Brahmaputra and its tributaries. The sociocultural assimilation affected their beliefs and practices, cultural arena and the society imbibed many cultural elements from the greater Assamese society (Doley 2014).

---

[1] The Mising officially recorded as Miri in the list as scheduled tribes of India under Constitution Order 1950 is originally a hill tribe of the Himalayan region of the Northeastern India. The name '*Miri*' was given to them, evidently by the plains people, but they always preferred to be known by their own name Mising.

The Mising people are known as 'river loving' people. After many years of settlement in the river bank of Brahmaputra and its tributaries, the Mising people have developed a deep relation with the river ecosystem. They adjusted with the frequent floods that occur and never failed to learn and develop ways on how to live in the floodplain areas with the help of traditional knowledge system. However, the Mising community is affected by flood every year, yet the losses due to floods fail to erode their love and deep connection with the river ecosystem and hence they have been living with the river despite the continuous adversities posed by the river.

FLOODS ADAPTATION AROUND THE WORLD: AN OVERVIEW

Flood risk can be reduced through mitigation and adaptation process. Mitigation refers to actions that seek to modify the source or pathway in order to reduce the probability of a flood occurrence. Adaptation refers to actions taken to reduce the impact of the potential hazard of flooding of the receptor area. Vulnerability to flood hazard closely relates to hazard exposure, susceptibility and resilience capacity of the community (UNESCO-IHE 2009). The flood resilience is related to community adaptation which acts as a positive aspect in response to hazard and reduces the vulnerability of the community using their traditional knowledge. Adaptation and coping measures identified as a short, medium and the long-term hazard, and success depends on the capacity of the community to adapt to the situation and capacity to adapt varies across region, societies and income groups (Katyaini et al. 2012).

There are numerous examples of community adaptation in the world. The people of Malaysia built stilt houses in the floodplain areas of *Rumah rakit*, they built 'Kampong', and 'Bajaus' houses in Sabha (Ayog et al. 2004) floodplains for adaptation. In floodplain areas of Bangladesh, communities use stilt houses for flood adaptation, which is economically viable too. They also use flood resilience agriculture and developed specific food habit in floodplain areas (Zaman 1993).

FLOOD ADAPTATION STUDIES
UNDER THE CULTURAL ECOLOGY

One can relate the study of community's adaptation in floodplain areas given the period of long-term habitation within the context of cultural ecology. When a community adapts to the environment using

262 T. PEGU

certain tools and techniques under certain geographical condition, it is known as a cultural ecological adaptation. Steward (2006) stated, 'Cultural ecology has been described as a methodological tool for ascertaining how the adaptation of a culture to its environment may entail certain changes'. The cultural adaption endorsed that the roots of decision-making in the face of potential hazard are deeply embedded in traditional knowledge. One's cultural background is, therefore, a structural influence which shapes one's perception as well as behaviour in response to hazard. There is different type of adaptation in response to flood hazard depending on the community and the environment. In this research, the community adaptation is articulated from three distinct aspects: cropping pattern, house architecture and food habit in the riverine areas. All these are discussed in the context of cultural ecology.

Reports on flood damage are often discussed in terms of loss of property and livelihood. The crop and agriculture get damaged due to flood. However, the impact of flooding on agriculture varies and depends on the tolerance of the crop or land use practices and the frequency, duration and seasonality of the event. For this purpose, the farmers developed resilience capacity with the help of various adaptive measures for their survival. Other two measures of adaptation help the community from the perennial flood with developing flood resilient housing architecture and adopting food habits to cope with flood. The community developed this adaptation measure from their age-long relationship with the local environment, an important feature found within the notion called 'traditional ecological knowledge' (Ramakrishnan 2000).

OBJECTIVE

The study aims to understand the long years of community resilience to flood. In this study, the natural phenomenon of floods was looked at through the parlance of community-based knowledge system and cultural ecological perspective of Mising people. It helps in understanding the flood mitigating measures adopted by the people with help of traditional knowledge system which ultimately reduces the impact of natural hazard on the community. The knowledge system developed over the years by the Mising people has also inspired other riverine people to cope with the natural hazard caused by floods. The strong resilience shown by the Mising community has been able to give hope and inspiration to other riverine people of the region.

METHODOLOGY

The study tries to understand as well as analyse how the Mising community people adapted in the floodplain areas of Jiadhal river basin. To achieve this goal, the qualitative research method was used as it describes the reality experienced by the community, groups and individuals (Ahuja 2012). Both primary and secondary data are collected. The tools for data collections were household survey, interview, questionnaire and various participatory rural appraisals (PRA) method.

The present study was carried out in Jiadhal river of Dhemaji district of Assam. The river Jiadhal is the northern sub-tributary of river Brahmaputra that originates in eastern Himalaya Mountains of Arunachal Pradesh at an altitude of 1247 m. above the mean sea level. The total catchment of the river is 1346 sq. km. from which 306 sq. km. is in Arunachal Pradesh and 1040 sq. km. in Assam (Fig. 1).

HISTORY OF STREAM CHANGE OF THE JIADHAL RIVER

Jiadhal, the word is derived from Assamese words 'Jia' and 'Dhal'. Locally, the word 'Jia' means *living* and 'Dhal' means *flood*. The Jiadhal was thus named after the river. The river has a long history of changing its stream over the years and the name of the river also changed along with time. From 1930 to 1935, the Jiadhal river flowed towards northeast side of Dhemaji town and it was known as 'Aradhal' (means left of the stream). In 1936, the river changed direction towards the eastern side of Dhemaji town, the place called 'Moridhal' (means Dead River) and it did not change till 1946. In between 1945 and 1950, the Jiadhal River again changed course and flowed through Jamuguri, Nahorbari and Nalamukh.

After the 1950 earthquake, the whole Subansiri sub-basin was devastated. After this disaster, the Jiadhal River divided into many streams and it continued until 1970. This tributary flows between Santipur, Ronganoi, Bishnupur, Raasepori, etc., and it covers 16 kilometre radius area. In 1970, there was a construction of embankment for flood control along the right bank of the Jiadhal River from the hill area of Arunachal Pradesh to Bishnupur Barman Gaon. Meanwhile, the soil conservation department planted trees between the embankment and the river and as a result it changed the direction of Jiadhal River towards western side and flowed towards Kathani, Jay Rampur, Borbhilaa, Salmari. The Jiadhal

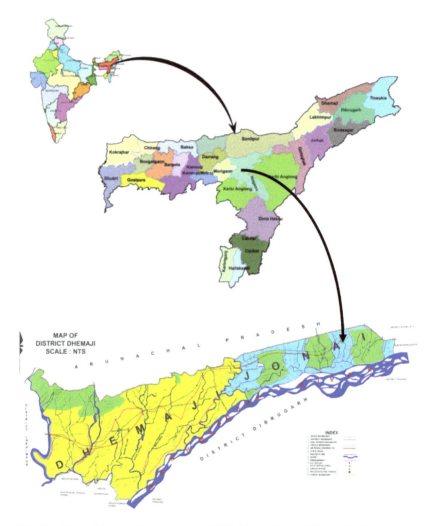

Fig. 1 Map of district Dhemaji scale: NTS (*Source* http://dhemaji.nic.in)

River then assimilated with Germer and fell into No-Nodi. The Jiadhal river flowed towards south of Gogamukh and finally met Subansiri. From 1977, it flowed through Kumtia after passing by Kaarchala, Abhaypur, Dihiri, and Khalihamari. In 1983, the river again bifurcated into two streams. One stream devastated the villages namely, Tekjuri Barman, Tekjuri Hajong, Baijayantipur, Barmathauri, etc. and the other stream took the route towards west of Barmathauri and assimilated in Kumotiya after crossing Dihiri village. In 1985, Tekjuri Hajong, Baijayantipur faced severe erosion. In the same year, construction of a new embankment was started along Kathanipur, Dharmapur, Dihiri, Khalihamari and Holodanga. But before the whole construction could be completed, another flood hit the area again and the whole embankment was washed away. The local people of Bordoloni witnessed the greatest havoc created by the river like never before.

The flood is continuing its devastation till today, and all government efforts seem to go in vain every time. Many people lose their homes due to this merciless flood. Many families are compelled to take shelter along railway tracks and other safer places during the calamity every year. The flood affected localities includes Tekjuri, Barman Gaon, Kekurichapari, Dihirichapari, Dihiri, Pachim Kekuri, Dihirikachari, Chomarajaan, Nepalikuthi, Kapahtoli, Rotwa, No. 1 Laumuri, No. 2 Laumuri, Kesukhanakachari, Kesukhanasapori, Kesukhana Koch, Misamari, Kahikuchi, Dhobabari, Gughuwa, Sakaladoloni, bordolata, Saru Dhekera, Sangmaibari, Powasaikia, Barmathani, Pehiyati, Loguwapara, Tinigharia, Bormathauri and many more villages.

FLOOD ADAPTION OF MISING COMMUNITY

Cultural ecological systems perform complex adaptation, where humans are basic components of the structure to reduce the vulnerability and increase resilience capacity with diverse management strategies (Chan et al. 2004). Berkes et al. (2000) described the socio-ecological system that makes it possible to explore environmental problems as a direct interaction or interplay between nature and society. This approach emphasizes that humans must be seen as a part of, not apart from, nature. Traditional knowledge on natural resource, ecosystem dynamics and associated management practices exist among the communities, and it benefits the ecosystem as well as livelihood of local people (Berkes et al. 2000).

TRADITIONAL SET UP OF MISING VILLAGE IN FLOODPLAINS

Generally, one would find a traditional Mising village settled along the alluvial embankment of rivers and wetland, with presence of forest patches around suitable land for the paddy cultivation. They have deep knowledge about the surrounding ecosystem and know proper uses of its resources. Their ingenuity and the urge to survive have created a perfect environment for enhancing knowledge system for living happily with all obstacles. The highlands are covered by forests and are used for grazing during high flood time. The forests are also the source of wild vegetables, fire wood, fodder, medicinal plant and domestic uses. The strategic location of the villages between the forest and paddy field makes it easier to access forest resources for daily use as well as continue their daily routine of agriculture. The low-lying area is usually chosen for the paddy field (Fig. 2).

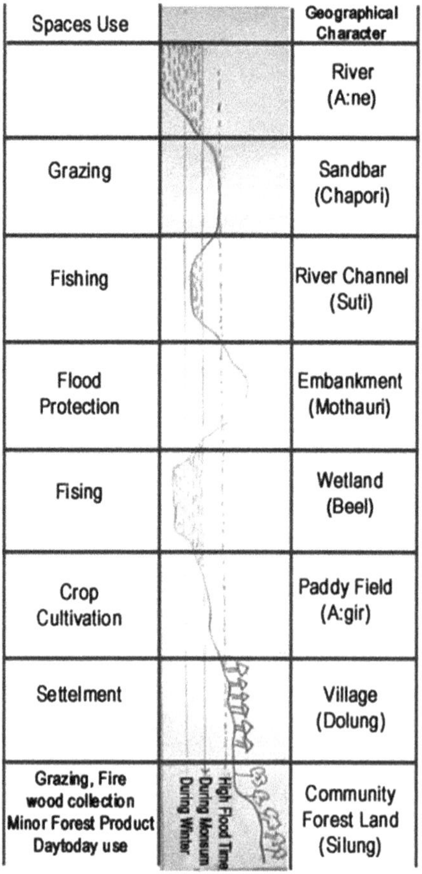

Spaces Use		Geographical Character
		River (A:ne)
Grazing		Sandbar (Chapori)
Fishing		River Channel (Suti)
Flood Protection		Embankment (Mothauri)
Fising		Wetland (Beel)
Crop Cultivation		Paddy Field (A:gir)
Settelment		Village (Dolung)
Grazing, Fire wood collection Minor Forest Product Daytoday use	High Flood Time During Monsum During Winter	Community Forest Land (Silung)

Fig. 2 Traditional Mising village set up in floodplains (*Source* Researcher's own findings [map not in scale])

AGRICULTURAL PRACTICE

Human societies on earth have undergone series of changes and transformation in the form of civilization and development. The Mising people are predominantly agricultural community. Originally Mising practised shifting cultivation in the hilly areas and continued it for a long time in the plains of Assam. During this time, they had to clear sprawling forest areas for this purpose and had plenty of arable land of the kind in their possessions all around their habitation (Taid 2008). Technology came from human needs, availability resource, innovation and influence from other culture. If there are any changes the technology will be changed and the cultural and environment will be affected (Sutton and Anderson 2013). Due to assimilation with plains, people in the valley, the Mising, gradually switched to settled cultivation. In the time of agricultural transition, traditional tools changed: for example, the hoe which was their chief tool in the hills for tilling the soil for cultivation was replaced by ox-driven plough of the plains (Pegu 2000). After settling along the floodplains of Brahmaputra and its tributaries, the Mising have adopted new natural resource exploitation techniques through interaction with people living on the plains of Assam (Fig. 3).

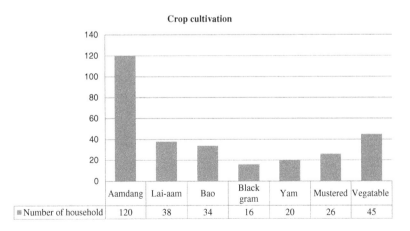

Fig. 3 Crop cultivation (*Source* Researcher's own findings)

268 T. PEGU

Traditionally, they practised two methods of rice cultivation Lai-aam[2] and Bao.[3] In addition to it, along with dry cultivation Mising began to practise wet cultivation more and more in the valley, growing different varieties of paddy. They began the wet paddy (Aamdang) cultivation for higher production that led to economic development. From the study, it is found that most of the people predominantly practise Aamdang method at present, as compared to traditional Lai-Aam and Bao.

The Mising community of river Jiadhal lost their flood resilience capacity in the absence of traditional knowledge system and changed into flood occurrence. The crop productivity of the area loses between 92 and 246 kg per hectare. The main reasons for sedimentation in the floodplain are shifting cultivation, rampant tree felling for commercial purposes and the extraction of boulders in the upstream mountain valleys. The practice of cultivating other cash crops like black gram, mustard, potato, pumpkin, peas, etc. has already decreased due to massive sand casting on agricultural land.

TECHNOLOGICAL ADAPTATION OF *CHANG OKUM*

The Misings have their own tradition of building houses which also reflects a beautiful picture of the lives of the community. Misings live in houses on stilts called *chang*[4] or *Kare Okum*. It has been found in the field that the houses are on piles about 4–5 ft. high from the ground and sometime more height based on previous flood experience and built in the east–west or north–south direction. The floor is raised using stilts made of wood or bamboo. On the stilts, they make the floor laying thick bamboo splits. This platform is the main house and walls are made using split bamboo. The main items used constructions of house are wood, bamboo and thatch. The walls are made using *Pi: muk* (Khagari), *Pi: ro* (Ikora) and *Dorbum* (nol), *iaa* (bamboo). These walls are not fixed by iron nails, rather tied together using *Jeing Ribv* (ropes made from cane). The roof is made from *Tase* or straw, *Ko: bung ekkam* (Toko leaves) and *Tara annc* (tal leaves). The complete house is usually 40–100 ft. in length and 10–14 ft. in breadth. Generally, the houses are made for

---

[2] Rain-fed cultivation.

[3] Deep water cultivation.

[4] The *Chang Okum* constructed using bamboo and wood.

two purposes: one is used as a granary for storage of various cultivated crops and other is used for residential purpose. The *Chang Okum* or stilt house is part and parcel of Tibet-Burma culture predominantly found in South and East Asian country. It is helpful where there is high rainfall and moisture both in the air and in the soil. For this purpose, the British officers adapted the mechanism of *Chang Okum* and they constructed their offices and houses called *Chang-bungalow*. In stilt house, the use of bamboo and cane was very popular among the Tibeto-Burmese origin. The Mising people continued this practice in central part of China, from where they migrated (Pegu **2000**).

IMPORTANCE OF '*CHANG OKUM*' IN FLOODPLAIN AREAS

Throughout the history, humanity has shaped nature and nature shaped the development of human society. The *Chang Okum* is an indigenously designed architecture, developed by the Mising for adapting to flood. It is a classic example of physical adaptation that has evolved through the experience of a riparian community (Das et al. 2009). The traditional house of Mising community is very suitable to adapt to flood because the raised platform keeps humans and as well as animals in a safe places. According to the villagers, the raised platform depends on the previous flood level and it keeps changing. The *Chang Okum*s are also made to provide protection against wild animals. According to villagers, the elephants generally do not attack *Chang Okum*; hence, granaries are protected. The grains in traditional granaries are saved from the moisture, insects and pests, and floods by its height and architecture.

CHANGED OF PATTERN OF *CHANG OKUM*

The traditional *Chang Okum* changed in terms of material use (Das et al. 2009) with of raw materials due to depletion of natural resources, the people shifted to the available resources. Most of the traditional Mising villages had community forests land near the village and supported the domestic needs. Due to the impact of flood and sand casting creating pressure on the agricultural land, the community forest land has been ultimately altered into agricultural field. The villagers used the materials for house making collected from the community forest but due to conversion of forest land into agriculture field there is resource scarcity now. At present, the people of Jiadhal depend either on Jiadhal Reserved Forest or hilly areas of Arunachal. Over-exploitation of natural resource in Jiadhal area by humans has led to the destruction of resources. So, the people of Jiadhal have replaced bamboo and wooden stilt with concrete pillars, tin sheets for roofs instead of thatch and concrete staircase for durability (Fig. 4).

FOOD HABIT OF MISING IN THE FLOODPLAIN AREAS

Food is a symbol of cultural identity of the community. Food habits also reflect the relation between community and nature. Indigenous agrarian societies developed their own food systems based on the ecosystem or landscape level by managing natural processes within the system. Food habit of the community is dependent on the availability of resources.

IMPORTANCE OF FISH IN THE COMMUNITY FOOD HABIT

The community have developed relations with river and fish products as they have lived their life living near the river bank for many years. The Mising people love to catch fish. Mising people use own traditional knowledge to catch fish with various traditional tools and techniques

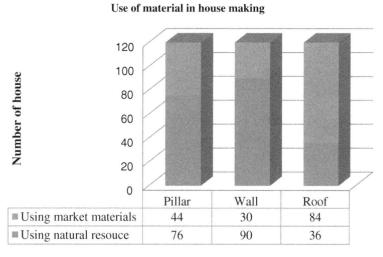

Fig. 4 Use of material in house making (*Source* Researcher's own findings)

such as Dingora, Porang, Uboti, Kholia, etc., and most of the tools are made from bamboo and cane (Dutta et al. 2012). They prepare delicious fish curry with various wild vegetables, bamboo shoot, cane shoot, etc. In their various functions, fish is used as a symbolic item of Lakhsmi.[5] The Mising people traditionally preserve the fish and use it during times of scarcity and flood. Mising traditionally use two distinct preservation techniques that help to retain the taste and nutritional content. The first method is drying fish over the oven in the kitchen. The dry fish is preserved under the 'Tuli'[6] and kept in the three-tier shelf on the bamboo platform over the oven. The other technique is *Namsing*. Soon after the drying, the fishes are mixed with some leaves and twigs and ground. The mixture is kept in Otung[7] and the mouth is sealed with clay soil. The mixture inside the *Otung* is called *Namsing*. For this purpose, the availability of fish is very important.

AVAILABILITY OF FISH IN JIADHAL

From the study, it is found that the fish stock is gradually decreasing. During survey, total 29 species fish were documented from the Jiadhal area. From the survey, it has been found that total 11 species of fish are not available now. Most of fish sizes have decreased; only few of them are of the same size as earlier. The destruction of habitat is the major cause of concern according to village people. Both anthropogenic and natural factors are responsible for the destruction. The other cause for reduction in fish population is over-exploitation due to rise in population. Reduction in the fish stock has ultimately affected food habit of Mising community of Jiadhal area (Table 1).

CHANGING PATTERN OF FOOD HABIT

Food sources from the aquatic environment are ample and unique for the macro- and micro-nutrients required in a healthy diet. In addition to the health benefits, fish is also an essential provider of a range of micro-nutrients, not extensively available from other sources in the diets

[5] Known as a God of Wealth in Hindu mythology.

[6] A bamboo pot.

[7] One side open hollow bamboo cylinder.

of the poor. It is also true that the small-size species of fish consumed in entirely including the head and bones can be an excellent source of many essential minerals such as zinc, iron, iodine, selenium, calcium, potassium, phosphorus, vitamins A, D and B complex.

IMPACT OF FLOOD ON FISH AVAILABILITY

Due to the depletion of fish species, daily diet of the community has been affected. It is affecting the whole community's food habit because most of them are below the poverty line and thus will directly lead to the decline in nutritional intake. It may cause loss of their traditional knowledge about fishing and its preservation.

Table 1 List of the fish availability (*Source* Researcher's own findings)

S. No.	Name of the fish	Local name of the fish	Availability of fish		Change in size
			Earlier	Present	
1	*Amblypharyngodon mola* (Ham-Buch)		Yes	Yes	Yes (small)
2	*Anabas testudineus* (Bloch)	MUWA	Yes	Yes	Yes (small)
3	*Bengala elanga* (Ham-Buch)	KIGANG	Yes	No	
4	*C. marulius* (Ham-Buch)	ELONG	Yes	Yes	Yes (small)
5	*C. orientalis* (Bloch & Schneider)	LIGAT	Yes	Yes	Yes (small)
6	*C. punctatus* (Bloch)	LIMAR	Yes	Yes	Yes (small)
7	*C. striatus* (Bloch)	LIGUM	Yes	No	
8	*Chitala chitala* (Ham-Buch)	TALI	Yes	No	
9	*Clarias batrachus* (Linnaeus)	CITAL	Yes	Yes	Yes (small)
10	*Colisa fasciatus* (Schneider)	PAVUR	Yes	Yes	Yes (small)
11	*Esomus danricus* (Ham-Buch)	KARMI	Yes	Yes	Yes (small)
12	*Gudusia chapra* (Ham-Buch)	JOBMUNG	Yes	Yes	Yes (small)
13	*Hara hara* (Blyth)	KOROTI	Yes	No	

(continued)

Table 1 (continued)

S. No.	Name of the fish	Local name of the fish	Availability of fish Earlier	Present	Change in size
14	Heteropneustes fossilis (Bloch)	GORUA	Yes	Yes	Yes (small)
15	L. calbasu (Ham-Buch)	NGMUK	Yes	No	
16	L. dyocheilus (McClelland)	MALI	Yes	No	
17	L. rohita (Ham-Buch)	TAGA	Yes	No	
18	M. cuchia (Ham-Buch)	ROW	Yes	Yes	Yes (small)
19	Macrognathus aral (Bloch & Schneider)	YUVI	Yes	Yes	Yes (small)
20	Mastacembelus armatus (Lacpede)	DANGKENG	Yes	Yes	Yes (small)
21	Mystus bleekeri (Day)	GERME	Yes	Yes	Yes (small)
22	Nandus nandus (Ham-Buch)	NARENG	Yes	Yes	Yes (small)
23	Notopterus notopterus (Pallas)	BEDELONG	Yes	Yes	Yes (small)
24	Ompok bimaculatus (Bloch)	PEMPELANG	Yes	No	
25	Orichthys aor (Ham-Buch)	NGOSER	Yes	No	
26	P. conchonius (Ham-Buch)	ARI	Yes	Yes	Yes (small)
27	P. sarana sarana (Ham-Buch)	NGERTAK	Yes	No	
28	T. progenius (McClelland)	GALI	Yes	No	
29	Wallago attu (Scheidner)	PITIYA	Yes	Yes	Yes (small)

Younger generation from the village know only of a few instruments used for fishing. This decrease in fish availability is making it difficult to preserve which is needed at the time of scarcity and emergency. Therefore, they eventually depend on the market products (Kumari and Dutta 2012). The decrease in fish production has affected the people and has reduced the self-sustaining village economy into a market-dependent economy and which is ultimately a challenge for food security (Table 2).

274 T. PEGU

Table 2 Time spent for fishing (daily consumption) (*Source* Researcher's own findings)

Month	Time for catching fish (before 30 years)	Time for catching fish (present)
January	30 minutes to 1 hour/per day	2–3 hour/per day
February	30 minutes to 1 hour/per day	3–6 hour/per day
March	30 minutes to 1 hour/per day	Not available
April	30 minutes to 1 hour/per day	Not available
May	30 minutes to 1 hour/per day	Not available
June	Nearly half an hour/per day	2–3 hour/per day
July	Nearly half an hour/per day	2–3 hour/per day
August	Nearly half an hour/per day	1–2 hour/per day
September	Nearly half an hour/per day	1–2 hour/per day
October	Nearly half an hour/per day	1–2 hour/per day
November	30 minutes to 1 hour/per day	1–2 hour/per day
December	30 minutes to 1 hour/per day	2–3 hour/per day

CONCLUSION

From the present study, it is found that the traditional Mising people are adapting to flood in the floodplains areas of Brahmaputra and its tributaries. They have adjusted with flood using their age-old traditional knowledge system. For this, they use stilt house in floodplains area, deep water and rain-fed cultivation for adjustment to flood and developed aqua culture for nutritional food. It is very helpful for the community settlers in riverine area, because the government policy of flood control has failed to safeguard the community from flood-related problems.

In this study, it has also been established that the community failed to adapt to the present changes in the nature of flood. The Mising people have somehow lost their resilience capacity against flood because of the present nature of siltation in the Jiadhal river basin. The changes in agricultural pattern have affected the livelihood of the community, who are now primarily dependent on paddy cultivation. The change in river hydrology and destruction of habitat has affected the fish granary and in the long run will affect the poor family's nutritional level. The catastrophic floods and heavy siltation have affected the community forest, which has finally affected their traditional house-making process.

Finally, change in the nature of river hydrology and flood hazard have affected the human settlement, caused displacement, loss of livelihood, destruction of public and private property and other socioeconomic effects. Due to destruction of educational institution and poor economic condition of parents, students are forced to leave school and work as domestic child labour. The younger generation suffer depression due to the flood which creates havoc every year affecting their living condition, and most of them are choosing to migrate to different industrial places for work.

From the present study, it can be concluded that through the age-old traditional knowledge system, the Mising community lived with flood. But with the recent changes in river hydrology and the nature of flood, the community is failing to adapt with the situation. At the end, the flood of Jiadhal has become a cause of concern for its adverse impact on the socioeconomic lives of the village people, abrupt human settlements, displacement and food insecurity.

ANNEXURE I:
SOCIOECONOMIC SURVEY OF HOUSEHOLD

Date:
Name of the respondents:

I. Name of Hamlet:
Block:
Tribe: Clan:
Religion: Economic status: APL/BPL

II. Details of Household Amenities (Use √-mark):
(a) Types of the house: Brick cement/Wood/Bamboo/Mud/Other.
(b) Electric supply: Own/Shared/None.
(c) Source of drinking water: Own pump/Pipe/Tape/Open well/ Tank/Pond/River.
(d) Toilet facilities: Sanitary/Pit/Open depiction.
(e) Cooking arrangement: Gas/Charcoal/Wood/Any other.
(f) Which one more problems faced during the flood: Man/Women/ Children.
(g) Rearing livestock: Yes/No.
(h) Using forest product: Yes/No.

276 T. PEGU

(i) Using firewood: Yes/No.

III. Land Use:

Use of land	Area (in Bigha)	Details of land use	Remarks
Total owned land			
Own crop land			
Flood-affected area			
Sand-deposited area			

IV. Agriculture Detail: (total area under cultivation = Hectare/Acre):

S. No.	Cultivated crops family	Season	Consumption (kg/year)	Quantity sold last year (kg)
1				
2				
3				

VI. Collection and Utilization of Forest Products.

ANNEXURE II:
SEMI-STRUCTURE INTERVIEW

1. What is the name of village and why the name is given? In which time people settled in this area?
2. What are the important institutional bodies of the village? Why it is important to have such body?
3. Is there is any unique traditional ways of using the resources, which is in a sustaining way? Give us a little explanation about it?
4. Who is the actual owner of the resources (lands/valuable property) in your household? Who took the final decision in your family?
5. How do you sustain your livelihoods, totally dependent on natural resources or have an alternative earning?
6. What is the name of your clans?
7. What are festivals celebrated?
8. What is the religious believe in the village before the present faith?

9. Does the conserved species or resources have anything to do with the culture, identity and the tradition of having in your community?
10. Why community people have a preference living near the river side?
11. Do you think it is important and necessary for your livelihood? What are the things support the livelihood of the people?
12. What is your perception about the flood, before and know?
13. Do you think flood is problem?
14. How do you prepare during the flood emergency?
15. What are the current threats coming from floods?
16. What are the changes you observed about floods, before and know?
17. Do you belief that present flood management system of government really helpful (embankments)?
18. What are traditional mechanism systems to adopt with flood?
19. What are the problems faced by women during the flood?
20. Is there any folk song or folk tales about the flood?
21. What types of materials are used traditionally for house-making process? Is there any specific trees not used for house making? Give reasons.
22. What are the fish found earlier? Still available or not? Is there fish size changed?
23. What types of instruments are used for fishing?

References

Ahuja, R. (2012). *Research methods.* Jaipur: Rawat Publications.

Ayog, J. L., Bolong, N., & Zakaria, I. (2004). *Human adaptation for survival against flood in Sabha floodplains areas: The past and present.* Retrieved 5 June 2016, from http://archive.riversymposium.com/2006/index.php?element=4.

Berkes, F., Colding, J., & Floke, C. (2000). Rediscovery of traditional ecological knowledge as adaptive management. *Ecological Applications, 10*(5), 1251–1262.

Chan, N. W., Zakaria, N. A., Ab Ghani, A., & Tan, Y. L. (2004). *Integrating official and traditional flood hazard management in Malaysia.*

Crémin, É. (2012). Wild life conservation and tribal livelihood in the Brahmaputra flood plain. The case of the Mising Tribe in the Fringe villages of the Kaziranga National Park (Assam, North-East India). *Nature,*

Environment and Society: Conservation, Governance and Transformation in India.

Das, A. (2014). *Living with (the politics of) flood: A study in Dhemaji district of Assam.* Retrieved 5 June 2016, from http://shodhganga.inflibnet.ac.in/bitstream/10603/19502/6/06.abstract.pdf.

Das, K. (2013). *Farm productivity loss due to flood-induced sand deposition: A study in Dhemaji, India* (SANDEE Working Paper No. id: 5337).

Das, P., Chutiya, D., & Hazarika, N. (2009). *Adjusting to floods on the Brahmaputra Plains, Assam, India.* ICIMOD Report.

Doley, B. (2014). *Glimpses on Mising folk cultures.* Jonai: Miro-Migang Publication.

Dutta, N. N., Borah, S., & Baruah, D. (2012). *Traditional gears used for capturing and preservation of fish by Mising community of northern bank of the Brahmaputra River, Assam, India, 12.* http://www.sciencevision.org/current_issue/dl/Baruah.pdf.

Hazarika, S. (2006). *Living intelligently with flood.* Retrieved 15 June 2016, from http://web.worldbank.org/archive/website01062/WEB/IMAGES/PAPER.

Katyaini, S., Barua, A., & Mili, B. (2012). Assessment of adaptations to floods through bottom up approach: A case of three agro climatic zones of Assam, India. *The Clarion, 1*(1).

Kumari, P., & Dutta, S. K. (2012). Changing eating pattern of Mising food culture. *International Journal of Humanities and Social Sciences, 2*(2), 211–219.

Pegu, N. C. (2000). *Mising hakalar itibritta aru sanskriti.* Guwahati: Bharati Publication.

Ramakrishnan, P. S. (2000). *Mountain biodiversity, land use dynamics, and traditional ecological knowledge.* Dordrecht: Oxford & IBH Publishing.

Steward, J. H. (2006). The concept and method of cultural ecology. *The Environment in Anthropology: A Reader in Ecology, Culture and Sustainable Living, 5*–9.

Sutton, M. Q., & Anderson, E. N. (2013). *Introduction to cultural ecology.* UK: Rowman & Littlefield.

Taid, T. (2008). *Traditional system of the Mising community: Change and continuity.* Seminars in Vivekananda Kendra Institute of Culture, Guwahati.

UNESCO-IHE. (2009). *Flood vulnerability indices.* Delft, The Netherlands: Institute for Water Education.

Zaman, M. Q. (1993). Rivers of life: Living with floods in Bangladesh. *Asian Survey, 33*(10), 985–996.

Website

http://dhemaji.nic.in/climate.htm. Accessed on 10 June 2016.
http://dhemaji.nic.in/demography.htm. Accessed on 10 June 2016.

http://dhemaji.nic.in/flood/flood_history.htm. Accessed on 15 June 2016.
http://dhemaji.nic.in/flood/rivers.htm. Accessed on 15 June 2016.
http://dhemaji.nic.in/geography.htm. Accessed on 18 June 2016.
http://dhemaji.nic.in/history.htm. Accessed on 20 June 2016.
http://dhemaji.nic.in/physiography.htm. Accessed on 25 June 2016.

Bibliography

Bhandari, J. S. (1992). *Kinship, affinity, and domestic group: A study among the Mishings of the Brahmaputra Valley*. New Delhi: Gyan Books (Print. [2]).

Gait, E. (1983). *A history of Assam*. Guwahati: LBS Publications (Reprint, 1997).

Kothari, C. R. (2004). *Research methodology: Tools and techniques* (2nd ed.). New Delhi: New Age International Publication.

Mipun, J. (1993). *The Mishings (Miris) of Assam, development of a new lifestyle, Mishing of the Brahmaputra Valley*. New Delhi: Gyan Books.

Nath, D. (1998). The Mising in the history of Assam. *The Mishings: Their history and culture*. Guwahati: Ayir publication.

Needham, J. F. (2003). *Outline grammar of the Shiyang Miri Language*. Dhemaji: Mising Agom Kebang.

Pegu, J. (2006). *Mising Samaj: Sanskritir aru parampara*. Guwahati: Ronganoi Publisher.

Taid, T. (2007). *Glimpses: A collection of articles and papers* (1st ed.). Dhemaji: Mising Agom Kebang.

Taid, T. (2010). *Mising Gompir Kumsung: A dictionary of the Mising language*. Guwahati: Anandaram Baruah Institute of Art and Culture.

Role of CBOs in Resilience Building: Good Practices and Challenges

Thongkholal Haokip

This paper examines the role of community based organisations in resilience building within the north-eastern states of India. There is a plethora of indigenous practices and value systems in the largely egalitarian societies of the North-East that help them in facing disasters without much assistance from the state. To demonstrate this, value systems as well as a practice among the Kukis known as *tomngaina* and *khankho* will be taken as a reference point to show how CBOs are obliged under these value systems to assist anyone within the community in the face of any kinds of disasters. The study assesses the historical past as well as from 2015 Manipur landslide, the 2016 Manipur Earthquake and hailstorms, particularly to prove or disprove the role played by community's cultural value systems and practices in building resilience. However, among the ethnically divided and antagonistic groups in the region they are so bound to restrict to themselves, within their own ethnic groups, despite their value systems. An attempt is also made to highlight the challenges that CBOs are confronted with in disaster resilience building.

T. Haokip (✉)
Centre for the Study of Law and Governance,
Jawaharlal Nehru University, New Delhi, India

© The Author(s) 2018   281
A. Singh et al. (eds.), *Development and Disaster Management*,
https://doi.org/10.1007/978-981-10-8485-0_19

BUILDING RESILIENCE TO DISASTERS

The Hyogo Framework of Action 2005–2015 defines resilience as 'the capacity of a system, community or society potentially exposed to hazards to adapt, by resisting or changing in order to reach and maintain an acceptable level of functioning and structure' (UNISDR 2005, p. 4). The Framework further adds that resilience 'is determined by the degree to which the social system is capable of organising itself to increase this capacity for learning from past disasters for better future protection and to improve risk reduction measures'. Disasters are primarily local phenomena and they first impact local communities, in which the initial emergency response is vital in saving lives. Therefore, there is a need to focus on improving the local communities' resilience to disasters. For this, the involvement of local resources in the preparation, response, and recovery is critical to reducing outside assistance and speeding post-disaster recovery. Local communities are the essential cornerstone in saving lives and livelihoods (Truesdale and Spearo 2014, p. 756; UNISDR 2007, p. ii). Resilience is, thus, an attribute of the community which would not only increase adaptive capacity and mitigate disasters but also reduce the risks.

Community based resilience building is a process of bringing people of the same community together and collectively managing a disaster by way of adaptation and utilising societal values. Community resilience building, thus, involves helping the community discover their culturally resilient values, stories, memories and connections in their life for the purpose of understanding their identity and becoming resilient. The activities also promote self-discovery and reflection, understand their own identity, manage change and transition, and build the skills necessary to become resilient.

India's North-East, particularly the hill areas, were a comparatively safer region and less prone to natural disasters. The relatively simplistic forms of life, eco-friendly livelihood and their traditional knowledge of the environment and its indigenous ways of protecting the environment had maintained the ecological balance. Until recently the region had not witnessed massive disaster since the 1950 Assam earthquake. The communities in the hills were known for their relative isolation and many of the cultural values they uphold are closely related to the environment and the understanding of it. With this distinguishing trait is the formulation of a distinctive policy by both the colonial and post-colonial Indian governments for the hill people. In their relative isolation the communities in the hill areas of the North-East continue to uphold the traditional social institutions and cultural values, which were also the survival strategies of the communities. With such cultural values and practices the different communities in the region were able to withstand and being resilient to disasters. Thus, the traditional cultural values become cultural resilience.

During the primordial past there were no modern forms of community based organisations. The various communities in Northeast India had different social institutional systems in the form of bachelor's dormitory in each village; such as the *morung* of the Nagas, *nokpante* of the Garos, *zawlbuk* of the Mizos, and *lom* and *som* of the Kukis. These bachelor's dormitories served their village and communities, and even beyond, in times of disasters. They are the first line of defence in conflicts and disasters and are responsible for the safety of the entire village. With the coming of the British and the advent of formal education along with Christianity, these social institutions gradually lost their relevance and finally disappeared by the middle of the twentieth century. The British created identities and elites among such groups to established compact with such elites so that the legitimacy of colonial governance was not challenged. Administrative units are loosely created around such identities (Haokip 2016, p. 2). With the emergence of new elites in each ethnic group, organisations came to be organised either along ethnic lines or based on religious denominations. Ethnic boundaries became rigid.

OF CULTURAL VALUES AND DISASTERS

Cultural values are intrinsic to communities in their existence as a distinct social group. These cultural values instruct the youth and their social institutions to act in times of disaster, conflict, war and other calamities. This value system is known among the Mizos in the Indian State of Mizoram as *tlawmngihna* and by their ethnic cousin, the Kukis, as *tomngaina* and *khankho*. To the Kukis in Northeast India and North West Myanmar, *tomngaina* and *khankho* are the 'code of ethics' and universe of all philanthropic activities. Both the terms are related to the 'norms' in a cultural society. *Tomngaina* is basically 'altruism (and being) hospitable, kind, unselfish and helpful to others' (Sanga 1990, p. 6) and the doctrine 'revolves round the need to renounce the idea of individuality and self and thereafter be brave and firm in the practice of this doctrine'

(Pillai 1999, p. 129). On the other hand, the breakup etymology of *khankho* is: *khan* = grow, deal and behave; and *kho* = village, land, or lifetime. The conjoined word *khankho* means 'the way a person should grow up in the village or land'; or 'the manner in which a person should conduct himself/herself in life'; or 'the norms one adheres to while living in the village or land' (Chongloi 2013, p. 224). Thus, this cultural value informs all able-bodied to act for the well-being of those in need of help when disaster strikes an individual, a family or any group of people. They are trained to become skillful people, with every woman having the knowledge of weaving and spinning and meeting all domestic needs of clothes in the family. While men also have skills in handicraft and produce beautiful products of bamboo, cane, woods, iron and brass (Gosh 1992, p. 204).

During the pre-modern days the traditional socio-cultural life for the youth was centered on two social institutions: *lom* and *som*. *Lom* is basically an informal labour group in which all able-bodied youths are a part of it. Besides others, *lom* also functions as a mechanism to bring about welfare and development of the village by way of helping the poor and needy, and also safeguarding the village from external and internal dangers. On the other hand, the primary purpose of *som* is to safeguard the village from external dangers such as attacks and invasions from enemies as well as internal dangers such as conspiracies and disputes, fire, theft, gambling and quarrelling and other untoward incidents which may bring disintegration in the village (KIM 2003, p. 39). *Som* members offer their free and compulsory services to the village. The main purpose of this traditional institution was to defend the village from external danger. Besides its main purpose *som*, in times of peace, rendered every possible service to the village community whenever required (Gangte 2008, p. 70). Chongloi (2012, p. 15) aptly remarked what prevailed in the past: 'Healthy and friendly competition prevailed among the *Som* members in various fields, be it sports, helping Meithai/Chaga (widow/orphan), defending the village, etc.... *Som* basically is an institution to mobilize the young for village services. It manifests the spirit of free service i.e. *tomngaina*. The members of *Som* do not expect to receive any reward for the help they rendered. It is believed that *tomngaina* were born out of *Som*'.

In the traditional institution of *som*, the elders of the dormitory (Som Upas) teach *som* boys discipline, responsibility and leadership. They also imparted them 'various traditional practices and systems, folk songs, folk lores, legends and myths of different kinds and also various social norms and manners along with the spirit of *Khankho*' (ibid., p. 16). As a consequence of the inculcation of cultural values and their introduction into a distinct life world through *som* 'the youth acquired the most important qualities of one's life i.e. *khankho* thereby rendering their free services to Meithai/Chaga (widow/orphan) in particular and society in general' and 'as a result it was a centre for competition in rendering services to the society among the members' (ibid., pp. 20, 17). Regard this, a sociologist C. Nunthara (1996, pp. 72–73) states that everyone in 'zawlbuk was always ready for any emergency', 'and tried to surpass his friend in chivalrous act and bravery depending on the situation'. He continues his statement that: 'In the event of death, the tribal community would year for one's readiness to be the first one available to help their people in need. It was a healthy competition which even kept many members of the community awake at night so that they could help those in need. T. T. Haokip (1991) adds the altruistic activities of *som*: which continued as an institution of providing help to the economically strained or marginalized. They provided house material for repairing community houses and also looked into many welfare schemes to see their fellow-beings more at peace and comfort.

However, as much as it exists today, conflicts were prevalent, though not ethnicised in the past and it was regarded as one of the most disastrous situation. In such situation of inter-tribal warfare the *som* boys were disaster mitigator or preventer. T. S. Gangte (1993, p. 132) encapsulates the role of *som* in mitigating disasters as the able bodied amongst them could be available during crisis and calamities. They have to offer them free and compulsory services to the village'. Thus 'Lom and Som were the institutions that stood for the total well being of the community, serving the community as centre for learning discipline and manner, social etiquette, fellowship of oneness, spirit of *Tomngaina* and *Khankho*, unity, reconciliation of differences in opinion and habit that binds the community together' (Haokip 2013a, p. 191).

What can be drawn from the above discussion on *tomngaina* and *khankho* is that these cultural values cement communities and bring cultural resilience. Imparting such altruistic values and inculcating the spirit of healthy competition among the youth for community service, particularly in stressful times and disasters, through the traditional youth institutions of *lom* and *som*, thus, foster community resilience and reduce the risk of disaster in the past.

DISASTERS AND CBOs

Though India's North-Eastern Region had not witnessed massive natural disasters since the 1950 Assam earthquake, the region is frequented with floods and lesser intensity earthquakes and landslides throughout the region. This naturally peaceful zone in recent years has increasingly witnessed natural disasters. Unfortunately unlike in the past where sustainable livelihood practices and traditional yet earthquake resistant houses made them disaster prepared coupled with the cultural values of altruism that made them disaster resilient, rapid modernisation and development have changed this disaster resilient community into a largely unprepared society. Traditional earthquake resistant houses are rapidly replaced by concrete structures and buildings devoid of any disaster-resilient policy and the communities slowly becoming almost oblivious of the altruistic past. Furthermore, with the coming of Western education in the early decades of the twentieth century traditional societies and their social institutions in the North-East were gradually transformed or subsequently disappeared. They were replaced by community based organisations. In this section, the role of Kuki community organisations, Kuki Khanglai Lawmpi (KKL), Kuki Inpi, Manipur (KIM) and the Kuki Students Organisation (KSO), will be examined in the light of the 2015 landslides in Joumol, and 2016 Manipur earthquake and hailstorms.

The Kuki Khanglai Lawmpi (KKL) and Kuki Inpi Manipur (KIM)

The KKL is a philanthropic organisation based in Churachandpur district of Manipur with a motto 'Panpi, Ngaite, Panpi', meaning 'help those who needed help'. It has six blocks and a zone in Jiribam. One of the main aims and objectives of KKL is to help widows, orphans, poor and destitute, and those who are in need of help (KKL 2000, p. 4). T. Jamkhothang Haokip (2002, p. 167) views that the Thadou–Kukis have changed the old ways with new ideas, and that *Lompi*, which was once very important, has become very important again. He contends that the way *lompi* has been used and its importance may not be the same as before; now that *lompi* is known as Kuki Khanglai Lawmpi (KKL) and it is actively involved in philanthropic works among the Kuki people.

Incessant rains during the monsoon season in 2015 caused floods and landslides in several parts of Manipur and other states of the North-Eastern Region. Located in Dingpi ridge in Chandel district of Manipur,

Joumol, a small hamlet with 21 houses and a population of 96 people as per the decennial census of India in 2011 (Census 2011), on 1st August was totally buried under mud by a massive landslide. When the heavy rainfall induced disastrous landslide took place only nineteen people were in the hamlet, in which nine survived. Among the ten dead, five dead bodies are untraceable 'as they lie buried along with the debris of the village. The only thing that remains today is the site of the mudslide, approximately 2 km in breadth and 3 km in width, as testimony to a forgotten village'. It was only after three days since the tragedy occurred that relief and evacuations reached the village (Lunminthang 2015). The KKL distributed rice, clothes and blankets to the landslide survivors of Joumol.

Prior to this, since its establishment in 1998, the KKL has been doing enormous philanthropic works. In times of conflicts, which are common in the region, it played a major role in looking after the affected people by providing relief and rehabilitation. For instance, some Kuki families that fled Assam's Karbi Anglong district after arson and killings by Karbi militants found shelter in a makeshift relief camp set up by KKL at Tuibuong in Churachandpur district. About 15 quintals of rice and other essential items have been donated to the camp by various organisations (*Telegraph* 2003).

Massive hailstorms occurred in 2015 affecting thousands of people. It not only destroyed standing crops, but also houses. The KKL not only distributed rice, crockery items, clothes and blankets, and also distributed aluminum sheets and built about thirty houses of the affected families in Churachandpur district. It mobilised youth clubs in the rebuilding of houses destroyed by hailstorms; assessed the extent of damage caused by the hailstorms and submitted a report to the government (Figs. 1 and 2).

Fig. 1 Hail damaged galvanised iron sheet roofings

Fig. 2 Distribution of galvanised iron sheet roofings

The workers of KKL are ready 24×7 for any kinds of help needed from accidents, deaths, facilitating blood donations, and transportation of serious patients to better hospitals through free ambulance service. The service rendered by KKL goes beyond ethnic boundaries. They provided the needful help to anyone seeking from other communities also. Such is the case of patients from Chandel district of Manipur visiting Churachandpur town. They also bravely provided many social services in times of conflicts and bandhs/blockade which other kindred ethnic groups are afraid to tread, especially at the dead of night. Though limited by the territory of India KKL also provided assistance by way of dropping poor patients from Myanmar till the Indo-Myanmar border town Moreh who visited Churachandpur for treatment.

The Kuki Inpi, Manipur (KIM), an apex body of the Kuki tribes, appealed and collected relief material such as rice, money, tin roof, blankets and clothes not only for Joumol villagers, but also for the whole affected people during the 2015 flood and landslides (Figs. 3 and 4).

The 4th January earthquake measuring 6.7 on the Richter scale that occurred at 4.37 a.m. with epicentre near Noney in Tamenglong district of Manipur was the most powerful earthquake ever recorded in several generations. It had shaken denizens for months. There were several reports of loss of lives, over 100 injured, buildings developed cracks and collapsed in Saikul, Noney and Imphal, and geological changes in

Fig. 3 Joumol village after the landslide (Photo: Michael Lunminthang)

Fig. 4 Collapsed women market in Saikul

Tamenglong district (*Indian Express* 2016). The KKL provided service by transporting several victims of the January 2016 earthquake to hospitals and drop the dead to their villages. They also visited and provided assistance to houses that had collapsed in the earthquake in Tuibuong and Twitha (Figs. 5, 6, 7, and 8).

Kuki Students Organisation (KSO)

The KSO is a community based student organisation which has its presence in the nook and corner of India with its general headquarters in Manipur. It has state level (Manipur, Assam and Nagaland), district level,

Fig. 5 Pillars of the office of the Saikul Hill Town developing cracks after 4th January Manipur earthquake

Fig. 6 An earthquake affected house in Noney

block level units and several city branches. Besides its main objective of promoting education among the students, the KSO also strives for scientific advancement of kuki students and their integration to achieve greater participation in community activities. (see Khongsai 2013, pp. 130–131).

Fig. 7 A new spring developed after the 4th January Earthquake at Noney

Fig. 8 Rescue efforts by a joint community based organisations at Joumol

In the 2015 heavy rain induced landslide, particularly Joumol, the KSO (General Headquarters) in collaboration with its units—KSO Chandel district and KSO Khengjoi block, along with Kuki Chief Association (KCA), Chandel open a relief donation centre at KSO Imphal office in New Lambulane, Imphal and also sent a rescue team to Joumol. They also formed Natural Disaster Relief Committee to tackle such natural disaster in future. Heavy floods also damaged roads and bridges. In order to restore connectivity the KSO, KCA, Kuki People Forum (Kana Area) and its kindred Anal Chiefs Association mobilised hundreds of volunteers and constructed a road from Serou till Chakpikarong with the help of four excavators and 30 Shaktiman trucks (*Sangai Express* 2015). Students from Manipur in University of Hyderabad collected donations and submitted Rs. 32,000 to relief committees in Manipur (Figs. 9, 10, and 11).

The Imphal and Sadar Hills district units of KSO collaborate with a Non Governmental Organisation (NGO) and distributed relief materials such as first aid kits, solar lamps, rice and money to the affected families of the 4th January earthquake. During the 2016 hailstorms, KSO Tengnoupal block pursue the Manipur State Government through the local MLA (Member of Legislative Assembly) for providing relief materials and distributed galvanised iron sheets to the affected households.

Fig. 9 Relief donation centre and collection of relief materials

Fig. 10 Transportation and distribution of relief materials

Fig. 11 Distribution of relief materials under the banner of Natural Disaster Relief Committee

Apart from the formal community organisations, several community based social network groups in whatsapp and facebook also, from time to time, helped people-in-need through crowdfunding. A whatsapp group named 'Khulkon Post' collected a sum of Rs. '23,450 and 20 bags of rice for survivors of Joumol and surrounding areas.

'TOMNGAINA', 'KHANKHO' AND CIVIL SOCIETY ORGANISATIONS

Every community's culture and their location are unique. The unique culture and location of a community have to be understood for a common disaster preparedness and resilience building. To make India's North-east disaster resilient there needs to be a deeper understanding, particularly, of local communities' cultures and also their relationship. Among the Kuki-Chin-Mizo communities in India, Myanmar and Bangladesh, who settle in a contiguous area though ethnified into different countries, they uphold common unique cultural values of *tomngaina* and *khankho*. These cultural values were abandoned in the process of Christianisation and modernisation during the first half of the twentieth century. Yet there were attempts to revive and propagate *tomngaina* though the social institutions of *Lom* and *Som*, and Mizo's *zawlbuk* cannot be revived in its pristine form. In the Lushai hills (Mizoram) it was 'around the '50s and again around the '70s, that the Church took a fresh look at Tlawmngihna and now not only accepts it as a vital part of the moral life of the Mizos but actively propagates its principles' (Pillai 1999,

p. 132). Chhetri (2013) gave a vivid description of contemporary *tomngaina* in times of disaster, suggesting that the existence of Tlawmngihna has no parallel in any other society. Tlawmngihna means a Mizo code for Dharma that puts the welfare and interest of others above self-interest and personal welfare. In times of catastrophes, Mizos go out braving all odds to help rather than shoutingfor help from the government. The spirit of Tlawmngihna has survived through the changesin the mizo society as demonstrated in many natural disasters (Chhetri 2013).

During disasters in the last two years the cultural values of *tomngaina* and *khankho* were imbibed by the Kuki civil society organisations, making them ever more to clamour for self-sacrifice and contribute for those in need of help. These cultural values have become cultural resilience and the community returns back to normalcy without much outside assistance, as shown above in the activities of various Kuki civil society organisations during disasters.

LIMITATIONS

Almost every community in Northeast India is as compassionate as others and altruism is the core of their cultural values. However, as communities that had existed together as a hill people now emerged as hostile groups who had inherited a history of antagonism. British colonial governance transformed inter-village feuds into inter-ethnic conflict between hill people who were grouped as the Kukis and the Nagas (Ningmuanching 2010, p. 107; Haokip 2013b). In post-independent India, the North-Eastern Region is one of the most enduringly conflict-ridden regions. It has witnessed several ethnic conflicts in the past and many of the conflicts perpetuate till today. The communities are interlocked on several irreconcilable issues. This psychologically separates each ethnic community from the other antagonistic neighbouring community/communities. Even in times of need each community is reluctant to seek help from the other communities, though they will be very much willing to help, leading to ethnic confinement. Ethnic conflicts have made communities inhibited and more vulnerable to the risks of disasters. On the other hand, communities at peace with one another can share each of their indigenous knowledge of disaster prevention, mitigation and resilience building and become more prepared of the eventuality of any disaster. As the world rapidly changes due to advancement in science and technology, and industrialisation with the resultant environmental change, natural disasters are on the rise. This along with the rise in conflicts among communities makes society more vulnerable to disasters. There is an urgent need to reduce conflicts and bring back communities to normalcy so as to build resilience.

CHALLENGES

The prolonged inter-community conflicts and struggle against the Indian State has left the North-Eastern Region underdeveloped, and a huge chunk of resources are diverted to maintain law and order situation. As a result almost all policies framed are seen through the lens of security. In such a situation of perpetual latent conflict, as Twigg (2009, p. 15) underscores: 'conflict often undermines community resilience, for instance, by breaking down social cohesion'. One of the other challenges of disaster management is fostering resilience, and 'enhancing community resilience is considered the key to preparing for, responding to, and recovering from disasters and other crises' (Thornley et al. 2015, p. 23). In order to increase community resilience there must be an effort to understand the local community more. Apart from understanding the cultural values of communities there is a need to integrate local and indigenous knowledge, observations, and practices related to disaster risk reduction and scientific knowledge (Hiwasaki et al. 2014). Furthermore, in an ethnically diverse and divided society, there is also a need to understand ethnic relations, and the conflict in such relations in multi-ethnic and multi-cultural society. As modern societies are increasingly confronted with minority groups demanding recognition of their identity, and accommodation of their cultural difference (Kymlicka 1995, p. 10); to reduce conflicts among social groups there is a need to recognise ethnic and cultural diversity and ensure minority rights. This 'challenge of multiculturalism' is to accommodate national and ethnic differences in a stable and morally defensible way (Gutmann quoted in Kymlica 1995, p. 26).

Multiculturalism is a policy response for coping with social and cultural diversity in society. Thus, the concept of multiculturalism as a democratic policy in responding to cultural diversity has to be promoted to ensure recognition of minority rights. The importance of the rights of ethnic minorities has been recognised and promoted by international organisations (Inglis 1995). Article 27 of the International Covenant on Civil and Political Rights of 1966 protects the cultural, ethnic, linguistic rights of ethnic, religious or linguistic minorities. The state has to adopt multiculturalism as a social policy and not only recognise the rights of minorities but also effectively address their problems. This will reduce socio-political conflicts and is an effective way for minority development. The concept particularly needs to be concerned with the advancement of the rights of the oppressed minorities. When conflicts are reduced through state intervention and other non-state initiatives by way of recognising multicultural principles, it can bring about a cohesive society among multi-cultural societies leading to better resilience building among diverse communities.

CONCLUSIONS

The traditional cultural values and indigenous knowledge and practices are unexplored in the government efforts to bring community resilience. Thus, there is a need to integrate indigenous knowledge of cultural values and other good practices 'into scientific research, policy making and planning' (Hiwasaki et al. 2014, p. 16), as 'societal cultures' influence 'educational technologies' and vice versa, with the two entities engaging in a politically reciprocal relationship (Al Lily et al. 2016, p. 205). Such integration would help the indigenous communities in understanding scientific knowledge better and develop an integrated approach to community based disaster risk reduction and resilience building resulting in the local communities acquiring the basic skills of preparedness and survival.

In the wake of the 2015 landslide, 2016 earthquake and hailstorm that had resulted in disasters unseen before particularly in the Indian State of Manipur, it has also triggered people's concern about the safety of the houses and the surroundings in which they live. It is a wake-up call for reckless urbanisation and construction of buildings without strict adherence to National Building Code of India 2005 and local building by-laws in India, earthquake-resistant building codes and other relevant safety codes, particularly the North-Eastern Region. What is more alarming is that most of the newly constructed buildings could not even withstand 6.7 magnitude of earthquake on the 4th January 2016. An increased public awareness is required on disaster risk and vulnerability as an aware community is better prepared to face any disaster when they occur and to minimise the loss of lives, injury and loss of assets, property and infrastructure through their conscious actions (Menon 2009). This can be achieved with public education which aims to achieve an integration of indigenous with scientific knowledge. Disaster resilience is achievable relatively better in the north-eastern states than in the rest of India.

ROLE OF CBOS IN RESILIENCE BUILDING ... 297

Acknowledgements The photos in this paper are collected from Kuki Students Organisation, Michael Lunminthang and my own photograph collections.

Interviews and Telephonic Conversations
Some of the data of cultural values during pre-modern times are drawn from the following interviews:

1. Interview with Mr. Mangkhosei Haokip, aged 95 (Indian freedom fighter) at T. Champhai Village, Churachandpur district, Manipur on 27 June 2016.
2. Interview with Mr. Songkhothang Kipgen, aged 87, on 11 July 2016 at Gangpijang Village, Sadar Hills East, Manipur.
3. Interview with Mr. Thangkhosei Haokip, aged 87, at Khokon Village, Sadar Hills East, Manipur.

Data during the 2015 landslide, 4th January 2016 Manipur Earthquake and the hailstorms were drawn from the following telephonic conversations:

1. Telephonic conversation with Mr. Kamthang Haokip, Kuki Khanglai Lawmpi (KKL), Churachandpur, on 2nd September 2016 on the activities of KKL.
2. Telephonic conversation with Dr. Satkhokai Chongloi, General Secretary of Kuki Organisation for Human Rights (KOHR), on 3rd September 2016.
3. Telephonic conversation with Mr. Seiboi Haokip, General Secretary of Kuki Students Organisation-General Headquarters (KSO-Ghq), on 3rd September 2016.

REFERENCES

Al Lily, A. E., Borovoi, L., Foland, J. R., & Vlaev, I. (2016). Who Colonises Whom? Educational Technologies or Societal Cultures. *Science, Technology & Society, 21*(2), 205–226. https://doi.org/10.1177/0971721816640624.

Census. (2011). *Joumol Population—Chandel, Manipur.* Accessed September 3, 2016. http://www.census2011.co.in/data/village/270992-joumol-manipur.html.

Chhetri, P. (2013). 'The Landslide of Aizawl: Hope Amidst Despair'. *Eastern Panorama.* Accessed September 9, 2016. http://easternpanorama.in/index.php/component/content/article/105-2013/june16/2462-the-landslide-of-aizawl.

Chongloi, H. (2013). Integrating Christian Faith and Kuki Khankho Towards Cultural Renewal. In T. Haokip (Ed.), *The Kukis of Northeast India: Politics and Culture* (pp. 219–228). New Delhi: Bookwell.

Chongloi, L. (2012). Som: A Decaying Traditional Institution of the Thadou. *Journal of North East India Studies, 2*(1), 13–21.

Gangte, M. P. (2008). *Customary Laws of the Meitei and Mizo Societies.* New Delhi: Akansha Publishing House.

Gangte, T. S. (1993). *The Kukis of Manipur: A Historical Analysis.* New Delhi: Gyan Publishing House.

Gosh, G. K. (1992).*Tribal and Their Culture: Manipur and Nagaland, Volume III.* New Delhi: Ashish Publishing House.

Haokip, T. T. (1991). *Kuki Polity with Special Reference to Village Administration.* M.Phil. dissertation, NEHU, Shillong.

Haokip, T. (2013a). 'Reinculcating Traditional Values of the Kukis with Special Reference to Lom and Som'. In T. Haokip (Ed.), *The Kukis of Northeast India: Politics and Culture* (pp. 177–193). New Delhi: Bookwell.

Haokip, T. (2013b). Essays on the Kuki–Naga Conflict: A Review. *Strategic Analysis, 37*(2), 251–259. https://doi.org/10.1080/09700161.2012.7557 85.

Haokip, T. (2016). 'Spurn Thy Neighbour: The Politics of Indigenity in Manipur'. *Studies in Indian Politics, 4*(2), 1–13. https://doi. org/10.1177/2321023016665526.

Haokip, T. J. (2002). *The Concept of God in Traditional Thadou Kuki Religion.* Dissertation, Manipur University.

Hiwasaki, L., Luna, E., & Shaw, R. (2014). Process for Integrating Local and Indigenous Knowledge with Science for Hydro-Meteorological Disaster Risk Reduction and Climate Change Adaptation in Coastal and Small Island Communities. *International Journal of Disaster Risk Reduction, 10,* 15–27. https://doi.org/10.1016/j.ijdrr.2014.07.007.

Inglis, C. (1995). 'Multiculturalism: A Policy Response to Diversity'. Management of Social Transformations (MOST)—UNESCO Policy Paper No. 4. Accessed September 5, 2016. http://www.unesco.org/most/pp4.htm.

Indian Express. (2016, January 4). '9 Killed, Over 100 Injured as Strong Quake Hits North-East'. Accessed September 3, 2016. http://indianexpress.com/article/india/india-news-india/ earthquake-measuring-6-8-magnitude-strikes-india/.

Khongsai, L. (2013). A Study of the Birth of Kuki Students' Organisation in Manipur. In T. Haokip (Ed.), *The Kukis of Northeast India: Politics and Culture* (pp. 115–140). New Delhi: Bookwell.

KIM. (2003). *Kuki Customary Law.* Imphal: Kuki Inpi Manipur.

KKL. (2000). *The Constitution of Kuki Khanglai Lawmpi.* Churachandpur: Kuki Khanglai Lawmpi General Headquarters.

ROLE OF CBOS IN RESILIENCE BUILDING ... 299

Kymlicka, W. (1995). *Multicultural Citizenship: A Liberal Theory of Minority Rights* (Reprint 2013). Oxford: Clarendon Press.

Lunminthang, M. (2015, August 23). Have You Heard of Joumol? *The Hindu.* Accessed September 2, 2016. http://www.thehindu.com/features/magazine/michael-lunminthang-on-the-manipur-army-ambush/article7565450.ece.

Menon, N. V. C. (2009). Earthquake Risk Management in the North East. *Dialogue, 10*(3). Accessed September 11, 2016. http://www.asthabharati.org/Dia_Jan%2009/N.%20vin.htm.

Ningmuanching. (2010). *Reading Colonial Representations: Kukis and Nagas of Manipur.* Unpublished M.Phil. dissertation, Centre for Historical Studies, Jawaharlal Nehru University, New Delhi.

Nunthara, C. (1996). *Mizoram: Society and Polity.* New Delhi: Indus Publishing Company.

Pillai, S. K. (1999). Winds of Change in the Bamboo Hills: Learning From a Mizo Way of Life. *India International Centre Quarterly, 26*(2), 125–137.

Sanga, R. R. T. (1990). *Administrative Development in Lushai (Mizo) Hills Up to 1972.* Dissertation submitted to North-Eastern Hills University, Shillong.

Sangai Express. (2015). 'Relief Committee Formed, Road Construction Begins'. Accessed September 4, 2016. http://www.thesangaiexpress.com/relief-committee-formed-road-construction-begins/.

The Telegraph. (2003, November 26). Kukis Find Shelter in Manipur. *The Telegraph* (Calcutta). Accessed September 1, 2016. http://www.telegraphindia.com/1031126/asp/northeast/story_2611938.asp.

Thornley, L., Ball, J., Signal, L., Aho, K. L., & Rawson, E. (2015). Building Community Resilience: Learning from the Canterbury Earthquakes. *Kotuitui: New Zealand Journal of Social Sciences Online, 10*(1), 23–35. https://doi.org/10.1080/1177083x.2014.934846.

Truesdale, S. B., & Spearo, J. P. (2014). Contemporary Community Resilience: Success, Challenges, and the Future of Disaster Recovery. In A. Farazmand (Ed.), *Crisis and Emergency Management: Theory and Practice* (2nd ed., pp. 733–760). Boca Raton: CRC Press.

Twigg, J. (2009). Characteristics of a Disaster Resilient Community. Accessed August 30, 2016. http://discovery.ucl.ac.uk/1346086/1/1346086.pdf.

UNISDR. (2005). Hyogo Framework of Action 2005–2015: Building the Resilience of Nations and Communities to Disasters. In *World Conference on Disaster Reduction*, 18–22 January 2005, Kobe, Hyogo, Japan A/CONF.206/6. UN Office for Disaster Risk Reduction.

UNISDR. (2007). *Building Disaster Resilient Communities: Good Practices and Lessons Learned.* UN Office for Disaster Risk Reduction.

Floods, Ecology and Cultural Adaptation in Lakhimpur District, Assam

Ngamjahao Kipgen and Dhiraj Pegu

INTRODUCTION

Floods are the most common and widespread of all natural disasters, and India is one of the highly flood-prone countries in the world. As per the National Flood Commission report, around 40 million hectares of land in India is prone to floods.[1] The State of Assam located in the North-Eastern Region of India is prone to natural disasters like flood and erosion which have a negative impact on overall development. Floods, flash floods, riverbank erosion and deposition of sand are the most frequent water-induced hazards in the eastern Brahmaputra basin in Assam (Das et al. 2009), in the monsoon period each year. These have not only affected land, lives and livelihoods of communities living in the region but also due to the changing landscapes, people living near such river banks have to adapt with the changing dynamics of the environment by

---

[1] TNAU Agriculture Portal, Disaster Management. Retrieved 30 August 2014, from http://agritech.tnau.ac.in/agriculture/agri_majorareas_disastermgt_flood.html.

N. Kipgen (✉) · D. Pegu
Department of Humanities and Social Sciences,
Indian Institute of Technology Guwahati, Kamrup, Assam, India

© The Author(s) 2018   301
A. Singh et al. (eds.), *Development and Disaster Management*,
https://doi.org/10.1007/978-981-10-8485-0_20

302 N. KIPGEN AND D. PEGU

Table 1 Damage caused by riverbank erosion due to flood (Government of Assam)[a]

Year	Area eroded (in hectare)	No. of villages affected	No. of families affected	Value of property including land loss (rupees in lakh)
2001	5348	227	7395	377.72
2002	6803	625	17,985	2748.34
2003	12,589.6	424	18,202	9885.83
2004	20,724	1245	62,258	8337.97
2005	1984.27	274	10,531	1534
2006	821.83	44	2832	106.93

[a]Retrieved 15 September 2014, from http://assam.gov.in/web/department-of-water-resource/flood-and-erosion-problem

means of altering their livelihoods and ways of living. Such resettlement and shifting of villages because of the changing landscape are common in this flood-prone areas.

During the post-independence period, Assam faced major floods in 1954, 1962, 1972, 1977, 1984, 1988, 1998, 2002, 2004 and 2012. The average annual loss due to flood in Assam is to the tune of Rs. 200.00 crores and particularly in 1998, the loss suffered was about Rs. 500.00 crores and during the year 2004 it was about Rs. 771.00 crores. As assessed by the Rastriya Barh Ayog (RBA), the flood-prone area of Assam stands at 31.05 lakh hectares against the total area of state 78.523 lakh hectares, i.e. about 39.58% of the total land area of Assam. This is about 9.40% of total flood-prone area of the country. As assessed, the annual average loss of land is nearly 8000 hectares. The riverine fertile agricultural lands of the state are reducing due to erosion, which has a very negative impact on the rural economy of the state (see Table 1).

This study is premised upon the theories put forward in ecological studies such as 'environmental determinism' (Andrew 2003) and 'possibilism' (Sahlins 1976; Steward 1955). 'Environmental determinism' studies how environment is the dominant force over humans and how it mechanically dictates a culture to adapt. Whereas 'possibilism' rejects the environment to have a controlling influence, rather claims that culture is determined by social conditions or human action. And in dealing with these challenges is an intimidating but essential task of human, and culture makes the choice of what possibilities to employ (Stutton and Anderson 2010). Culture, a uniquely human attribute, is something which man interposes between himself and his environment in order to ensure his means of security and livelihood. Herskovits (1956) defined culture as the man-made part of environment, and White (1949) defines culture as a specific and concrete mechanism employed by a particular animal organism in adjusting to its environment. In the same manner, Piddington (1950) states that culture is essentially an adaptive mechanism, making possible the satisfaction of human needs, both biological and social.

Human and cultural ecologist also emphasizes on the aspects of culture and environment—how cultures adapt to their physical environment; and how they understand the environment (see Stutton and Anderson 2010; Steward 1955). The ecologist Woodbury (1954) recognized that social life is an adaptation for efficient use of time and space upon the earth. Adaptation is a set of actions resorted to over the long term to reduce the adverse impacts of floods on people's lives. Some of these practices have become part of the lifestyle or culture when the benefits accrued has become time-tested. Adaptation practices evolve or are acquired over a long-time span, and these are effective as short-term and long-term measures of response to impacts of floods. To sum up, any human adaptation partially involves historical influence inherited down the generations and technological evolution or innovations, practices and knowledge which help human to survive even in the harshest of the environment. All people belonging to certain specific groups or culture may share same basic but have a unique pattern of learned behaviour. This is because each cultural group has their own distinctive ecological adaptation.

In this context, this paper tries to understand the traditional coping and adaptation practices of the riparian[2] communities against natural disasters and the extent to which cultural norms and traditions make people vulnerable or resilient. This study looks into the impact of floods on the cultural behaviour and ecological landscape of the Mising community inhabiting the Brahmaputra plains in Lakhimpur District of Assam. It also examine—whether the traditional coping capacities and adaptation strategies are still useful in dealing with the changing nature of water-induced disasters. The study also focuses on the impact of floods on the changing livelihood patterns, social lives and cultural adaptive mechanism of the local people.

---

[2] A riparian zone or riparian area is the interface between land and a river or stream.

The Setting

The areas undertaken for the study are Matmora and Laiphulia village under Dhakuakhana subdivision in Lakhimpur district of Assam, which are affected by recurrent floods affecting lives and livelihood. The Mising (also called Miri), one of the major ethnic tribe of Assam predominantly inhabits these two villages. The population of Mising in Assam is counted to 587,310 (Census 2001). Recent survey done by the Mising Organization shows the Mising population as more than 1 million approximately, i.e. 1,250,000 persons. The number of households in the village of Matmora is 111, and the number of households in Laiphulia is 80. The rationale for selecting this area are based on the following criteria: (1) Vulnerability of the villages to regular flood predominantly due to embankment breaching causing destruction to land, properties and affecting livelihood, (2) Floodwaters have engulfed the fertile farm fields leaving villagers homeless and poor, and (3) Large chunks of lands are gradually eroding away affecting the agrarian and riparian culture of living.

In this study, we used both qualitative and quantitative methods of data collection. The samples in both villages were randomly selected in which 30 households from each village were surveyed using questionnaires and 20 in-depth interviews were conducted with elderly person within the age group 50–65 and above. Questionnaire for the study was designed with a focus on the impact of flood on the livelihood and culture of the Mising people. The questionnaires for household survey was intended to gathered the socio-economic profile, impact of flood—ecology and livelihoods, adaptive mechanism, social support system during floods and post-flood experiences (Fig. 1).

Ethnographic Profile of the Mising Tribe

Linguistically, the Misings come under the Tibeto-Burman speakers of the greater Sino-Tibetan groups. The Mising people observe both Hinduism as well as their traditional religious practices.[3] They live in village settlement and have traditional institutions such as *kebang*, the

---

[3] See Census of India (1901) and Assam District Gazetteer (1905), by B. C. Allen reports—about the religious beliefs and practices of the hill tribes including the *Misings* as 'Animists'.

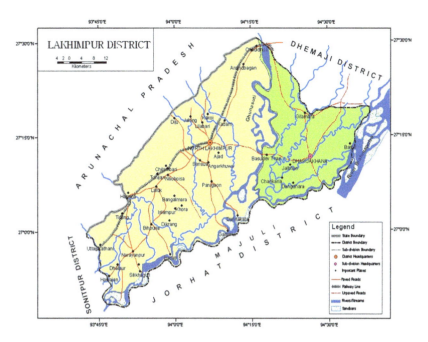

Fig. 1 Map of Lakhimpur district, site of the field study (*Source* http://lakhimpur.nic.in/)

village committee which looks in matters of social life, and *Mimbir-Ya:me* another committee comprising the village youths. The *murong* is another institution that imparts social training and other duties to the younger generation.

The sociocultural life of the Mising circles around agricultural practices which is their basis of livelihood. There are two main festivals namely, *Ali-Aye-Ligang*[4] and *Po:rag*. *Ali-Aye-Ligang*, the main festival of Mising community marked the onset of sowing of crops and beginning of agricultural calendar.[5] The season of 'Ahu' cultivation is marked by the

---

[4]The term *Ali-Aye-Ligang* means *Ali*—roots, *Aye*—fruits and *Ligang*—sowing.

[5]'Ahu' paddy and 'Bau' paddy are the principal product cultivated both through *Jhumming* and wet-rice cultivation. Rice is cultivated twice or thrice a year—*Āhu (Lāi)* in the month of December–January, *Bāo* in the month of May–June and *Sāli (Āmdāng)* in the month of June–July.

306   N. KIPGEN AND D. PEGU

Table 2 Educational profile of Matmora and Laiphulia

Educational qualification	Villages	
	Matmora	Laiphulia
Below 10th	14	12
10th	6	3
10+2	4	8
Graduate	6	5
Postgraduate	–	2
Total	30	30

celebration of *Ali-Aye-Ligang*.[6] In this festival, the head of every household offers their prayer to the mother earth for good agricultural season and harvest. The seeds are sown chanting the following incantation:

> On this auspicious day, oh forefathers – *Sedimelo, Karsing-Kartak, Donyi Po:lo*, ants and alike, all of you bear witness; today we are sowing the seeds into the womb of mother earth; let mother earth be fertile, capable of bearing abundant crops. Let there be good harvest! (*Field notes*, 18 February 2015)

The *Po:rag*, another important festival of the Mising community, is primarily organized by the youths (*Mimbir-Ya:me*) of the village. Owing to its exclusive and extravagant engagements, this festival is celebrated at an interval of 5 years.

Literacy rates in both the villages are very low, although Laiphulia has higher literacy rate (see Table 2) than Matmora as the young people in Matmora are engaged in daily wage and labour jobs to sustain them. The literacy level has gone down over the years with the youths opting for jobs to earn extra income for the family over education. The village which had produced eminent personalities in the past is now one of the most backward and poverty-stricken village among the Mising communities. The only primary school in *Janji-Arkep* area is in a dilapidated condition with only few students.

Economically, the Mising people are one of the poorest sections in the region (Pegu 2012). The Misings are generally known to

---

[6]The present-day celebration of *Ali-Aye-Ligang* reflects the sociocultural identity of the Mising people along with a definite role in the cultural convergence within the greater Assamese society.

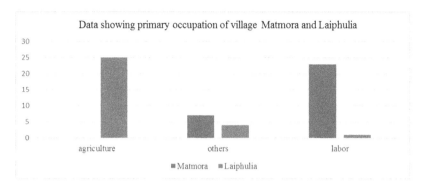

Table 3 Primary occupation of Matmora and Laiphulia villages

be hard-working people as they are engaged in agricultural practices and livestock rearing such as cattle, pigs and poultry etc. (Kuli 2012). Traditional methods of farming techniques are used for agricultural productions. They generally cultivate rice, mustard seeds, black pulse, jute potatoes and other vegetables. Usually rice is produced for domestic consumption and mustard and black pulse for commercial purposes. Presently, along with traditional tools and implements, some commercial farmers have also adopted advanced machines and production procedures. In some areas, the economy of the Mising have gradually transformed from subsistence economy to surplus economy (see Table 3).

The Mising women are generally known to be hard working with extensive participation in the agricultural works. Women also work as labourers in road and embankment construction. Besides, the Mising women are known for their dexterity in weaving (see Fig. 2). Weaving of traditional cloths was generally meant for domestic use within the family but with limited income from daily-wage labours, they have started to weave cloths for commercial purposes.[7] They weave traditional cloths like *ri:bi* (a sheet with narrow stripes, wrapped to cover the lower garment and the blouse), *gāsèng* (used for the same purpose as that of a *ri:bi*, but having broad stripes of contrastive colours), *Gonro Ugon* (dhoti), *Gero* (a sheet, wrapped round the waist to cover the lower part of the body, or

---

[7] Other than the income earned privately, the Mising women have no right to claim or inherit any family properties (for details, see Doley 2012).

Fig. 2 A Mising woman seen here is engaged in weaving cloths at Matmora

round the chest to cover the body down to the knees) and among their woven products the *Gadu* or *Miri Jim* (blanket) is well known.

Comparatively, the Mising women of Matmora are less impacted than the non-Mising women as they are more comfortable in the water. They can swim, manage boats and rafts, and they live in houses on stilts that are more convenient than the temporary raised platforms of non-Mising households.

Of late, livestock rearing for commercial purposes has increased since the damage caused by the recent floods has resulted to lack of income from other sources. In some instances, it is also observed that womenfolk are engaged in production and selling of local beverage known as *apong* (rice beer). Setting up such small stalls for selling *apong* along with other drinks is common and a thriving source of income in Matmora.

USING TRADITIONAL KNOWLEDGE SYSTEMS TO ADAPT WITH FLOODS

The Mising community has a rich repertoire of traditional knowledge as they have a prolonged history of survival and coping strategies with floods for generations. The Misings have developed their own ways of adapting physically to floods that have withstood the test of time by

FLOODS, ECOLOGY AND CULTURAL ADAPTATION ... 309

Fig. 3 Stilt houses (*chaang ghar*) in Laiphulia

evolving in response to changing environmental and social conditions. Different traditional knowledge systems that are used as early warning of flood by the local communities play an important role in traditional disaster management resulting to lessen the lost of lives and livelihoods. Some of the helpful indicators such as traditional knowledge systems (TKS), cultural traits, housing type, social relationship, food and seed storage that played a significant role in the local communities preparedness to flood are discussed in this section. Though consider primitive, the Mising people still rely on traditional knowledge of flood control and other precautionary measures.

Physical Adaptation Strategies

Houses—Housing plays a key role in protecting communities from floods in the study sites. Misings have a unique style of housing. They live in thatched houses raised on stilts called '*chaang ghar*' generally built facing the river, a type of house ideally suited to adapt to flood waters (see Fig. 3).[8] Usually, the heights of the stilts vary from six to eight feet

---

[8] *Chaang ghar* is an 'eco-friendly' house, built from locally available resources such as bamboo, cane, reed, wood and thatch due to the fact that they are cheap and easily available and would minimize the damage during floods.

above the ground, conforming to the highest flood level of the area based on long-term observation and experience of past floods. In a sense, raising a homestead is a traditional method of coping with floods. The floor of the house made of bamboo and wood is adjustable and can be raised to cope with rising flood waters. Widths of the house generally vary from 10 ft to 15 ft while lengths of the houses vary depending on the size of family since they prefer to live in a joint family. The *chaang ghar* is an ingeniously designed, multipurpose house that makes it possible for the inmates to stay protected amidst flood waters and which allows enough light and air into the house. It has facilities for a kitchen, living room and food store storage of essential household goods in the basement. Below the raised structure, they keep their domestic animals.

Chaang ghar is a classic example of physical adaptation that has evolved through the experience of a riparian community. It is interesting to examine how cultural norms and traditions determine the way people build their houses and influence their physical adaptation to floods. The Mising communities in Matmora live in their *chaang ghar* following their age-old tradition.

Under the influence of modern housing styles, the traditional *chaang ghar* has undergone changes in terms of building materials and style. The use of bricks and concrete is an exception as those who can afford used concrete pillars in place of bamboo or wooden stilts and corrugated iron sheets as roofing material to enhance durability and even concrete staircases have also been added to the floor.

Food storage systems—The Mising community has a well-defined traditional practice of storing food and seeds on a three-tiered bamboo platform with the shelves placed at various heights over the fireplace (*meram*) in the kitchen. The fireplace is used for drying raw fish and meat as well as paddy in rainy season as objects on this shelf get direct heat. The shelf above *meram* is called *rabbang*, where pitchers filled with *apong* are kept during winter to keep the wine warm and prevent it from becoming sour. In summer, the pitchers of *apong* are placed on the floor to keep it cool. The uppermost shelf is called the *kumbang*, where vegetables are stored. Another storage place called *sansali*, a mat made of bamboo placed in the living room where rice and seeds are kept so that floods do not damage them.

Social relationship—The Mising's have a strong clan affiliation. The members of the village are known to be cooperative in nature and are widely seen in cooperative activities such as *rikbo-ge'nam and dagle'k-alek*. For example, whenever a member of the village is unable to work or cultivate due to some reasons, he may call for help and cooperation (*rikbo-ge'nam*) of the villagers to which the members of the village may respond according to their time and capacity.

In both villages, the social relations are strong and people help each other in times of need. They are generous about sharing their boats with other families, and those having boats help others selflessly while rescuing and evacuating people. Those families having well-built stilt houses give shelter and food to a number of other families, which are either homeless or have dilapidated houses that cannot withstand the current and flood waters.

The local youth organizations namely, the Takam Mising Porin Kebang (TMPK) and the All Assam Mising Students Union (AAMSU) engage in providing help in relief operations in these areas. They help the local people during the floods by helping them to move to safer place temporarily. They also engage in collecting donations from different sources which they use in providing relief to the flood affected people by distributing basic necessities such as food and shelter.

Prediction of Rains and Floods Through Traditional Beliefs

The use of traditional wisdom and beliefs facilitate the Misings to forecast about the impending occurrence of rains and floods. These predictions are interpreted through the observance of animal behaviour and signs of celestial bodies.

1. On the eve of *Magh Bihu* (harvesting festival), the last day of Assamese lunar calendar (usually mid-January)—if the cattle are reluctant to move out from their shed in the morning, it indicates an impending flood.
2. In the morning of *Goru Bihu* (New Year festival)—if the cattle behave abnormally and frantically when they are brought for bath, it is believed that the flood is ahead.
3. The howling of fox can be of two types; when the fox howls irritably at higher place it indicates a forthcoming prolong drier season and when it howls from a low-lying location it indicates flood.
4. Birds also indicate an impending flood. For instance, a dove cry monotonously before a forthcoming flood and a mysterious bird

locally called *Melong* cries before a devastating flood. Also, a local bird called *Chatok* chirps sadly before flood.

5. Sometimes, insects like grasshopper and locust come out of their shelter and fly randomly and enter houses. This behaviour is indicator of a sudden change in weather condition, more occasionally a heavy rain and flood. Also when ants shift their shelter to higher places with their eggs and foods; it indicates certainty of forthcoming flood.
6. Uninterrupted frog calls imply heavy showers and possible flooding.
7. If there are no fish or few fish during the early monsoon season, there will be floods.
8. Flowering of the mango tree indicates impending floods, whereas flowering of jack fruit implies good productivity of paddy and soil.

Prediction Based from the Observation of Cosmic Bodies

1. If the moon inclines towards south it implies a forthcoming devastating flood.
2. If the cloud gathers in the south-west direction it indicates a probable storm, in south-east direction it indicates a rain and flood and in north-west indicates normal rain.
3. Massive bamboo flowering with the onset of summer season indicates a forthcoming devastating flood.

Apart from these indicators, information about the nature of the river is passed on through verbal communication by watching the level of the rising water. In the distant past, the Mising community used traditional instruments namely *Le-long* and *Mabong*, which have sounds like cymbals, are used to warn people within a distance of three to four kilometres about advancing and rising waters. The practice however has disappeared with time, so have the instruments.

Upholding the traditional knowledge and beliefs is diminishing nowadays, especially among the younger generation. The elderly people are of the opinion that some of these folk beliefs are no longer operational since the nature of the rainfall, weather and floods is altering. Currently, the indicators mentioned in folk beliefs, especially those dealing with animal behaviour or observation of celestial bodies are no more reliable though found to be useful in the past.

FLOODS, ECOLOGY AND CULTURAL ADAPTATION ... 313

Erosion, Embankment and Shifting of Village

The primary flood-prone areas of Assam are remote and quite underdeveloped and lack basic amenities such as electricity, proper transport and communication. Both the two villages Matmora and Laiphulia villages are prone to flood (*barikha or aasi aanam*), and they have experienced major floods in the year 1998 and 2012. These have caused damage to their agricultural lands including livestock, stored grains and other properties. With recurrent floods, it has been observed that the villages under study have shifted 10 kilometres inwards with the constant widening of the river and westward movement of the river towards Dhakuakhana. It is observed that Matmora village has been resettled seven times due to river bank erosion and sand deposition.[9] Relocated and resettled several times, the village once a very fertile land, now have nothing but a mere sandy field besides the embankment. As a result of this, the agriculturalist were rendered landless and forced to look for other works such as weaving of traditional clothes for commercial purposes, selling of local liquor, daily-wage labour and fishing.

People had to shift their villages periodically, depending on the movement of the river and consequent cycle of erosion and deposition. As a result, the villagers kept moving backwards (westwards along the line of advancement of the river) and resettled on the river bank, retaining the name of the village each time. The history of floods for the villagers is also the history of the rise and fall of the embankment intertwined with progressive shifting of old villages and migration to new locations. The Matmora villages (along with neighbouring villages like Janji, Arkep and Modarguri) self-initiated their own resettlement and rehabilitation, and these have become an integral part of their lives and an important adaptation mechanism with respect to the changing course of the river.

In recent years, the construction of embankments emerged as the most sought after method of controlling the recurrent floods.[10]

---

[9] The original village site of Matmora was located approximately 6 km outwards to the river from the present embankment, and the river has advanced towards Dhakuakhana town by at least 10 km in last decade.

[10] The constructions of these embankments are funded by the Central Government through the State Government departments such as Public Works Department (PWD) and Water Resources Department (WRD). Labor works in the construction of embankments, and geobanks provide opportunities to the people of Matmora. However, it is observed that young men from the villages are migrating to nearby cities and towns in search of job opportunities, as the pay grade in the construction of embankments is significantly low.

Fig. 4 The ongoing construction of the new embankment in Matmora where geotube technology is being used for the first time in India

Embankments are the only mode of any major flood disaster mitigation in the area of Matmora. Embankments to some extent would help prevent water spilling over to the paddy fields. The construction of embankment was constructed according to the eroding lines of the Brahmaputra using geotextile tubes by filling up the tubes with sand and water from bank of the river. Presently, along the banks of the river, using the same technology as geotextile construction of geobanks is going on (see Fig. 4). Dumping of sand bags along the bank of the river is used for this construction. Although the present embankments are built using state of the art technology by using geotextile fabric, it has been breached in 2012 causing catastrophic damages primarily in Janji area (SANDRP 2014).[11] The locals mentioned this breaching as the slow pace of construction of the embankments.

---

[11] Matmora (Assam) Geotube Embankment on Brahmaputra: State Glorifies, but No End to Peoples' Sufferings after Three Years of Construction. *South Asia Network on Dams, Rivers and People (SANDRP)*, 6 May 2014. Retrieved 10 December 2014, from https://sandrp.wordpress.com/2014/05/06/matmora-assam-geo-tube-embankment-on-brahmaputra/.

The overall impact of floods on the landscape, lives and livelihoods has also brought about significant changes to the socio-economic conditions and cultural milieu of the villages in both the study sites. From a state of self-sustenance, the people have been reduced to impoverishment. The once fertile paddy fields are now totally unfit for any type of conventional agriculture. In the absence of agriculture, people have resorted to other work, virtually anything they can find to earn enough to survive. People engage in daily-wage labour, fishing and sale of dried fish, sale of driftwood and country liquor, and migrating to other places for menial labour such as pulling rickshaws and hand carts and working in factories.

The floods immensely disrupt their social, economic life disrupting daily lives and livelihoods. Both the villages are completely inundated by water including their agricultural land. Their livelihoods are extremely hampered as they are confined to live in camps on the embankment, and there is no possibility of finding any work as movements are hampered. During this time, ferry services are also closed down due to the dangerous nature of the river. Womenfolk are the most affected as they are completely dependent on their male counterparts for food and livelihood as they can't either weave cloths or rear livestock on the embankments. Social life is also affected, as they have no capacity to perform any social functions or ceremonies mainly due to the lack of movement facilities hampered by the flood.

Post-flood, the people from both villages are concerned about assessing the damages caused by the flood and repairing or either reconstructing there houses. Besides, materials for constructing house are hard to find naturally and the cost of procuring bamboo is not possible due to lack of income after the flood. Thus, migrations to nearby places are often seen in the post-flood from this area mainly where their relatives dwell who helps them in setting up in the nearby places.

SUMMARY AND CONCLUSION

From the study, it is seen that both the villages have been extremely affected by recurrent floods. Matmora have been more affected in terms of livelihood and cultural change. Due to loss of agricultural land, the people of Matmora once a self-dependent village now has to look for other sources of livelihood to which they were not much acquainted. Women earlier used to weave cloth only for domestic uses now started weaving for commercial purposes. It is seen that rearing livestock for domestic consumption has gone down in favour of earning extra income for the household. With recurrent floods, the people of Matmora are finding it hard to find a sustainable means of livelihood.

In regard to cultural changes, one significant finding is that they no longer possess any land holdings even to perform the rites and rituals.[12] The *Po:rag* festival known for extravagant requires huge amount of preparation and money—is now almost redundant in the Matmora area primarily due to huge cost incurred and inability to arrange feast for the guest. Ceremonies such as *dodgang* (funeral) which also requires huge amount of expenditures are performed but within a very moderate budget. Some of the festival is now revived at the district level as a platform to express and assert their cultural identity. But with no land holdings it is foreseeable that the importance of rites and rituals could be soon forgotten which is a very sensitive issue for the Mising people. Such important events will remain a distant memory for the people of Matmora soon.

It also observed that with dwindling youth population in the villages due to migration in search of sustainable jobs, an important Mising social structure *Mimbir-Ya:me* is hardly visible today. The womenfolk of the villages are so vigorously engaged in earning extra little income by weaving of cloths night and day that they barely have time to spare and involved in activities of the youth organisation. Thus, the tradition of *Mimbir-Ya:me* is diminishing. Earning their income as daily-wage labour has a colossal impact on their culture which has seen various changes such as not being able to perform social functions such as *Ali-Aye-Ligang*, *Po:rag* or following the traditions such as *Mimbir-Ya:me*.

To sum up, going back to the theoretical perspectives deliberated, both 'environmental determinism' and 'possibilism' can be situated in the context of the Mising people. As discussed, recurrent floods and wide change in landscape as a result of erosion and sand deposition have rendered the people of Matmora landless and forced them to look for other forms of livelihood. The area of Matmora, once a flourished village with fertile agricultural lands, is now a barren land due to sand deposition. Thus, it can be deduced that environment has mechanically dictated the people of Matmora to adapt to the changing landscapes to other forms

---

[12] This can be attributed to the fact that people in Matmora have no land holdings as floods had washed them away. During the good old times, mainly in the Arkep area the villagers land holdings ranged from 12 *bighas* to 200 *bighas* and above.

of livelihood. On the other hand, traces of 'possibilism' can also be seen in the changing cultures of the Mising communities in order to adapt to the changing environment. The possibilities of different forms of livelihood although not sustainable exist in the form of fishing or even selling of local liquor, wage labour available in construction of embankments or other nearby places. The impact of floods has forced the people to consider every opportunity available to make a subsistence living.

REFERENCES

Andrew, S. (2003). Neo-Environmental Determinism, Intellectual Damage Control, and Nature/Society Science. *Antipode, 35*(5), 813–817.

Allen, B. C. (1901). *Census of India*, Vol. IV: Assam.

Allen, B. C. (1905). *Assam district Gazetteer*, Vol. III. Calcutta.

Census of India. (2001). Directorate of Census Operation. Assam.

Das, P., Chutiya, D., & Hazarika, N. (2009). *Adjusting to floods on the Brahmaputra plains, Assam, India*. Guwahati: ICIMOD.

Doley, D. (2012). Role and status of women in Mising society. In J. J. Kuli (Ed.), *The Misings: Their history and culture* (pp. 91–97). Dibrugarh: Kaustubh Prakashan.

Herskovits, M. J. (1956). *Man and his work: The science of cultural anthropology*. New York: Alfred A. Knopf.

Kuli, J. J. (2012). *The Misings, their history and culture*. Dibrugarh: Kaustubh Prakashan.

Pegu, L. N. (2012). Agrarian economic system of the Misings: Past and present. In J. J. Kuli (Ed.), *The Misings, their history and culture* (pp. 198–203). Dibrugarh: Kaustubh Prakashan.

Piddington, R. (1950). *An introduction to social anthropology*. Edinburgh: Oliver and Boyd.

Sahlins, M. (1976). *Culture and practical reason*. Chicago and London: University of Chicago Press.

SANDRP. (2014). 'Matmora (Assam) Geo-tube embankment on Brahmaputra: State glorifies, but no end to peoples' Sufferings after three years of construction', May 6, 2014. Retrieved 15 April 2015, from https://sandrp.in/2014/05/06/matmora-assam-geo-tube-embankment-on-brahmaputra/.

Steward, J. (1955). *Theory of cultural change: The methodology of multilinear evolution*. Urbana: University of Illinois Press.

Stutton, M. Q., & Anderson, E. N. (2010). *Introduction to cultural ecology*. Maryland, USA: Altamira Press, A Division of Rowman and Littlefield Publishers.

White, L. A. (1949). *The Science of Culture, a Study of Man and Civilization.* New York: Grove Press.

Woodbury, A. M. (1954). *Principles of general ecology.* New York: McGraw-Hill Book Co.

Disaster and Resilience Building among Women in Manipur

Mondira Dutta

INTRODUCTION

Time and again one observes how the disaster risk reduction strategies fail to address the issue, endangering lives at a greater risk with the most basic needs missing from the system. Women and girls have been found to be at a much greater risk in terms of reproductive health problems, sexual abuse, forced marriages and other forms of gender-based violence including death. It is estimated by the UN that worldwide, women and children are 14 times more likely to die than men in a disaster. In fact, women across the world are charged with responsibilities to handle all crisis management at home simply because men are busy due to their obligations at work outside the home premises. Ushering in safety for women and girls in terms of maternal mortality, safety from unintended pregnancy, safety during humanitarian crisis and protection from violence will automatically ensure safety during natural disasters or even during the man-made conflicts.

M. Dutta (✉)
Jawaharlal Nehru University, New Delhi, India

© The Author(s) 2018
A. Singh et al. (eds.), *Development and Disaster Management*,
https://doi.org/10.1007/978-981-10-8485-0_21

319

The region of Northeast India and the neighbouring parts of Myanmar, China, Bhutan and Bangladesh encompass one of the most active seismic regions of the world. This region frequently witnesses earthquake, floods, landslides and droughts. Some of the earthquakes in this region are among the largest in the world impacting large chunks of the regions nearby leading to infertility of land, lack of livelihood options, health crisis and epidemic, and most severe is emphasizing violence against women.

DATA AND OBJECTIVE

The present paper is based on both primary and secondary sources of information. Primary data[1] have been collected from field visit drawn to the regions of Chakpikarong, Thoubal and Imphal in Manipur. Two sites were identified for an in-depth investigation. These were the village of Wangjing Tentha, of Thoubal district, and the village of Rungchang, of Chandel district. In order to understand the role of women, a special meeting was held with some of the women who were available (about 15 women) to understand their problems and the methods applied by them to mitigate disasters and contribute towards risk reduction. Another ten to twelve women entrepreneurs from the Ima market were also interviewed for an understanding of their role and contribution to the society. Data were collected with the help of a structured questionnaire. The methodology involves a secondary desk review through information based on media reporting, NGOs, government reports and local contacts. In addition, a field visit was undertaken by the Disaster Research Programme (DRP) team of Jawaharlal Nehru University, New Delhi, and other experts from academia, and research institutes in collaboration with Manipur University and National Institute of Disaster Management (NIDM), Ministry of Home Affairs and Government of India conducted field investigation for the sudden flooding of Thoubal and Chandel districts of Manipur State in July–August 2015. Other members of the research team included research scholars from Manipur University, NSS volunteers of Manipur University, Civil defence volunteers of Manipur State.

---

[1] The field visit was undertaken on 10th April 2016 to assess the causes and consequences of the sudden flooding of Thoubal and Chandel districts of Manipur during July–August 2015.

It attempts to use some of the field-based case studies from Manipur and examine a series of instructional discourses and practices that have been implemented by the women to mitigate disaster. The case studies from Manipur will disseminate the literature on disaster preparedness and disaster mitigation and establish the fact as to how women emerge as the main agents of crisis management. Directing women's attention towards 'disaster readiness' within the interior space of her home will create safety and security for the community at large and enhance the preparedness for all present and future emergencies. It needs to be understood that merely being a woman does not lead to higher vulnerability, rather it is the socially constructed gender norms, such as gender stereo type roles assigned to the women, responsibilities and 'appropriate' behaviours stipulated for woman and man, along with unequal distribution of resources that become the main reasons of vulnerability rather than biological differences. Such indicators of vulnerability affect the women manifold through the perceptions of risk and disaster, creating gaps in the differential capacities on the various processes of mitigation and responsive channels. The objective of this paper is to explore the complex dynamics surrounding a women and their role in bringing back normalcy at home particularly mitigating when the hazard strikes

DISASTER IN MANIPUR

In Manipur, the north-eastern state of India, river flooding is a regular feature witnessed by the state. Over the last few years, however, this hazard has manifested into disasters as a result of man-made situations leading to drainage failures, increased run-off loads in hard surfaces, large-scale deforestation leading to increase in the erosion capacity of flowing water, encroachment of river channels and illegal and engineering defective construction activities, creating barriers in the free flow of excessive water owing to heavy rainfall in catchment areas and so on. About two-third population of Manipur State (1.79 million out of 2.72 million population according to Census of India, 2011) lives in the valley area surrounded by mountain hills (Manipur valley constitutes only about 8.2% of the state area). Rivers from these mountain hills flow into valley and very often lead to flash floods and landslides during rainy seasons in the mountain areas and floods in the plain areas. All the major river systems of Barak River in the state are vulnerable to flash floods/flooding during rainy season, as depicted in the Vulnerability Atlas of Manipur State.

Manipur was hit by the worst flood in 200 years, due to unprecedented continuous rainfall with overflow of all major rivers that wreaked havoc washing away bridges, cutting up embankments while pushing villages away from the mainland in August 2015. Though the magnitude of the destruction in terms of number of exact villages and population affected cannot be estimated, the grim situation seen even today display large habited areas in Thoubal and Chandel district that continue to live in the absence of livelihood options. Their fields are silted with muddy water and sand, their cattle are lost, and vast scales of land are simply inundated leaving no scope for agriculture. In addition, a major landslide in Chandel district on 1 August 2015 claimed 21 human lives while washing off a complete village. Most of the affected population had to take recourse in schools, market places and vacant spaces as shelter and safety. Erosion and landslide pose as the greatest threat to their lives and safety. Moreh, commercial towns on the India–Myanmar border and Jiribam have been cut off from the rest of the state due to landslides caused by flash floods.

As per estimate, six out of the nine districts of Manipur such as Bishnupur, Imphal East, Imphal West, Churachandpur, Thoubal and Chandel districts were severely affected (Sphere India 2015). Three of these, namely Thoubal, Churachandpur and Thoubal districts, were the worst affected by floods, and the Joupi village and Chandel district were hit by severe landslides. In terms of population, Chandel district has 144,028 lakhs of people while Thoubal district has 420,517 lakhs of people as per 2011 Census. Some of the worst affected damages were recorded in Kairembikhok, Wangjing, Salungpham, Heirok, Wangbal and the surrounding villages under Wangjing Tentha and Heirok Assembly constituencies, in KakchingKhounou block, Serou, Wairi, Chumnang villages of Thoubal district. In Chandel district, areas worst affected include Pallel, Island, Theimongkung, Molmon, Rungchang, Chapikarong Village, Peace land, Novokom, Haralal Churachandpur: Samulamlan and Sangaikot Blocks, and another 70 villages.

GENDER AND DISASTER MANAGEMENT

Statistically, women are the single largest group who are the worst affected by natural and human-caused disasters. Therefore, all efforts need to concentrate to mainstream gender within disaster management.

They are primarily the ones also who require maximum support in the post-disaster scenario. Women have been mostly accepted as the main caregiver for the young, elderly, sick and the differently abled persons. Women are therefore the prime target group for relief creativities. 'In addition to the overall impact of the disaster on the general community, women are uniquely burdened by the breakdown of infrastructure, displacement and isolation, collapse of familial and social support networks. It has been observed that in the post-disaster communities, women are at a greater risk of sexual and domestic violence. The loss of the male heads of household, also the chief bread winners, complemented with livelihood loss also contributes to increasing women's burdens and responsibilities' (Chingtham 2014).

The roles and responsibilities of a woman seem to over flow naturally, in the advent of a disaster or any other eventuality that may touch her family or her loved ones. It is a marvel how those very women, who are considered un-skilled, illiterate, inexperienced, suddenly get transformed in carrying forward Herculean tasks during the disaster. Women who are mostly concentrated in the informal sector are the worst affected due to disasters. Therefore, post-disaster a woman tends to remain more invisible and uneconomic and automatically get deprived from the employed status of the country's Census documentation.

WOMEN IN MITIGATING DISASTERS

The two villages of Wangjing Tentha of Thoubal district and the village of Rungchang of Chandel district clearly displayed during the field visit how women approached the problem. When floods hit the village, women were the first to rush to protect their cattle, food products, grains and some domestic necessities that would help them to carry forward their family and children. They had to leave their homes and get on the streets which were at a higher elevation. During nights, they had to sleep under the sky on the streets along with their cattle. All the men slept on one side of the road, while the women slept together on the other side. Women disclosed how they were subjected to discomfort when it rained heavily and all men and women had to huddle together in the limited space to sleep throughout the nights. Women and girls thus become particularly vulnerable to violence—sexual violence being the most immediate risk. Young and old female confirmed their increased risk of violence during disaster and stress that lack of access to safe and private wash and sanitation facilities is an additional worry. During disaster and when displaced, women often waited until dark, to use wash and sanitation facilities. Moving at night in the search of privacy can make her even more vulnerable and at risk. This emphasizes the stark reality to the fact that how women and men experiences and responses to disaster varies.

Women from these villages stated that they would return back to their huts during the day wading through the water to try and cook for the community a meal. They are the ones who manage the food distribution where four to five families pool together and manage the show, taking care of the children and the old and other senior citizens of the area.

'When food was less, we [women and girls] ate after everybody had finished eating', revealed a young school girl. This implies that in the event of food shortages, women and girls are less likely to eat when compared with men and boys. Some of the NGOs such as the Red Cross and Integrated Rural Development Service Organization (IRDSO) came forward with help like providing rice and pulses. Governmental organizations were not visible for days on end. No minister visited them either. The village community came forward and was the most helpful in organizing community participation. They could not save much of their cattle, and their fields were full of mud water which converted their agriculture fields into silted barren areas devoid of cultivation for the next 5–6 years.

WOMEN AND ENTREPRENEURSHIP IN MANIPUR

Women have always remained actively involved in economic activities in addition to their role in the society. The *Ima* (meaning mother) market is one such live example where it is a complete women set-up, all managed and produced and sold by the local women. The market has long been a part of Manipuri tradition with evidence suggesting that it dates back to the sixteenth century, known as mother's market or 'Ima Keithal'. In Imphal which is the capital of Manipur State, this market is one of its kind in Asia or even the world at large. Women sit in rows on high platforms, in the market building with their products displayed in front of them. There are extremely cordial relations among them, and there has never been any aggressive behaviour from either the sellers or buyers. This is because it is an all women affair and they feel free and secure in the privacy of their space (Fig. 1).

This was a market that was badly damaged and destroyed severely by an earthquake of 6.7 magnitudes in January 2016. The earthquake

Fig. 1 Ima Market, Imphal, Manipur, 9 April 2016

is believed to have killed at least 5 people and injured over 40 others. The reconstruction of the building as seen in the field was found to be defective and not safe for the women who are more than 2000 sprawled in and all around the building. So women were not allowed into one of these huge structures. As a result, women occupy the neighbouring building and also sit along the roads and around the building. A section of the textiles is at a little distance away in another part of the same city. This is believed to be one of the few entirely women-run markets in the world. The Ima Market in the capital city of Imphal in Manipur exemplarily shows the power of women working together. Manipur unfortunately faces a high level of violence, bandhs and disruption of daily life on a regular basis and has few employment opportunities. This market depicts not the story of any single woman but 'this is about the many ordinary women of Manipur, who live with courage and dignity through life in a conflict zone' (Singh 2012) (Fig. 2).

'Ima Market consists of many small sections put together, each dealing with its own specialty; one set of vendors does only phanek (the wraparound skirts worn by women in this part of the world), another does phi, the shawl that goes over the blouse, and yet another sells only kitchen implements. Bamboo products, western clothes, fresh produce, trinkets, packaged foodstuffs, and even snacks to eat when you are tired of the shopping – Ima Market has something for

Fig. 2   Ima Market, Imphal, Manipur, 9 April 2016

everyone'.[2] Most of the products cater to domestic needs, and also some artisans display marvellous items of artistic grandeur. These women are always smiling and satisfied with whatever they earn with such hard work and following a strict religious faith.

Most of the women vendors stated that they take loan from the private sector with high rate of interest. If the government takes an initiative in financing the women vendors their economical status would improve to a large extent. Majority of the women vendors are devoid of a savings account or are simply not aware of it. As a result, their money tends to be stolen or lost often. The self-help programme, insurances and other national government schemes need to be organized by the government as well as the private sector. Majority of the women vendors do not have any idea about the marketing system and trading. They are needed to give proper awareness programme in order to improve trading and marketing. Women vendors who sit along the roadside also give tax at the same rate as those who are sitting under the shed. If attention was given to these women by the government through proper policy and programme for the roadside women

[2] Ibid.

vendors, the problems that are faced on a regular basis both by the women vendors and the public will be largely reduced. Government agencies need to take greater initiatives and cooperation in their role for the development of women vendors.

POLICY RECOMMENDATIONS

It would be appropriate to suggest certain policy recommendations in the backdrop of gender-sensitive approaches to disaster management.

While research and analysis is certainly a part of this process, the local inputs are seldom sought. An inclusive disaster management plan would make a significant difference at times of crisis. It is important to align disaster management with best practices in gender inclusive governance, domestic as well as global. There are several programmes of the government which provides a platform to community-based women's groups for interacting with local government officials. The *Indira Awas Yojana, MGNREGA, the Ujjawala* scheme and several others need to be interwoven at the micro-level with the local governance. Sometimes many such schemes remain unknown to women who face serious crisis in disaster mitigation. Proper training in this regard will reduce women's vulnerabilities manifold during a crisis in the post-disaster scenario. Many examples exist where women have been instrumental in contributing towards restoring and transforming infrastructure.

This fact was repeatedly stated in the national conference held at Manipur in April 2016. One of the recommendations stated 'A dedicated multidimensional livelihood generation programme should be launched to enhance adaptive and coping capacity of people through skill development and capacity building for local resources management. These programme should integrate all community resources such as water bodies, forests, indigenous knowledge, local crafts, community medicines and also their animals who sustain their subsistence economy besides providing companionship during crisis'.

- It is therefore important to usher in a greater interaction with communities at the micro-level during the entire planning process. The existing government schemes need to be imbedded into a comprehensive programme and shared among the Panchayat and community folks and made aware on how to go about utilizing the same during times of crisis.

Mainstreaming gender-specific concerns is also an essential part of disaster management. Special attention ought to be paid particularly to women's needs such as at times of menstruation, easy and safe access to toilets, availability of wheelchairs, private areas for women to change clothes/bathe and special provisions for pregnant and nursing women, are all an integral part of effective disaster management.

- All gender disaggregated assessments need to be incorporated before policies and programme are framed and disseminated in the field.

In the field of disaster management and relief efforts, inclusion of more women into the process would undoubtedly enhance the efficiency of the process. It is well known that many women have inhibitions in sharing bed with strangers at shelter centres, and hence, pre-disaster evacuation efforts tend to be met with resistance when women refuse to vacate the premises of her safety zone.

- All such disaster mitigation programme must necessarily employ more of female relief workers for micro-level dissemination process such as rescue, evacuation and rehabilitation.

Gender-sensitive disaster management cannot be complete without making it gender inclusive. Women's contribution towards relief and rehabilitation ranges from their traditional knowledge and skills that can be used to manage natural resources, aid in addressing to the needs of the injured and sick, prepare community meals, and nurse displaced infants and children during reconstruction and recovery processes. Moreover, with adequate training, one could capitalize on the stereotypical role of a woman as the emotional nurturer of survivors. Following the initial phase of relief operations, the involvement of affected local women becomes an added advantage for management of refugee camps, distribution of food, etc. Since these women belong to the affected community, they enjoy a degree of trust that relief workers normally have to work hard to achieve.

- It is recommended that greater involvement of particularly the affected women in relief efforts/planning will go a long way in gaining the trust of the women folk for further implementation.

Although the relief camps cannot be stated as 100% safe from sexual abuse, violence and sexual harassment, safety has always been a question mark in the lives of all displaced women. A secure area implies safe sleeping arrangements, adequate lighting and accessible toilets, and choice of safe location for camps. Designating spaces for 'women-only' is mostly ignored. Effective management will be harmed if such spaces are not provided particularly when a woman needs to change, during times of menstruation, and while nursing her baby. During times of disaster mitigation, relief should not be restricted under strict compartmentalization of widowhood, single women, female-headed households, elderly, differently abled and so on. At such a time, the entire affected community needs aid, rescue and rehabilitation with the same swiftness.

- It is important to imbibe safety and security measures for women as the prime focus points of all post-disaster mitigation/rehabilitation schemes including the centres for temporary shelter.

References

Chingtham, T. (2014, June). Women and Human Resource Development at Ema Market: A Case Study. *Women and Human Resource Development, 3*(1) (Voice of Research). ISSN 2277-7733.

Singh, A. (2012). *Ema Market Manipur: Women Working Together*. http://www.thealternative.in/business/ema-market-women-working-together/, 14 December 2012.

Sphere India. (2015, August). *Floods in Manipur-2015* (Data Analysis Report). National Coalition of Humanitarian Agencies in India.

Community Resilience Building and the Role of Paitei Tribe of Churachandpur in Manipur

M. Pauminsang Guite and Langthianmung Vualzong

INTRODUCTION

The Framework of Third UN World Conference on Disaster Risk Reduction (WCDRR) in Sendai, Japan 2015, has emphasized to determine the specific roles of stakeholders that included the community-based organization to participate in collaboration with public institutions, to, inter alia, provide specific knowledge and pragmatic guidance in the context of the development and implementation of normative frameworks, standards and plans for disaster risk reduction,[1] and noted that indigenous peoples, through their experience and traditional knowledge, provide an important contribution to the development

---

[1] UNISDR. (2015). *Sendai Framework for Disaster Risk Reduction 2015–2030.* UNISDR, United Nation, p. 23.

M. P. Guite (✉)
Jawaharlal Nehru University, New Delhi, India

L. Vualzong
Centre for the Study of Law & Governance, Jawaharlal Nehru University, New Delhi, India

© The Author(s) 2018   331
A. Singh et al. (eds.), *Development and Disaster Management,*
https://doi.org/10.1007/978-981-10-8485-0_22

and implementation of plans and mechanisms, including for early warning.[2] The indigenous tribal communities of Northeast India, who were observed to be bind by their strong cultural entities and egalitarian component, could serve a great role in community resilience in disaster risk management. Their participation as facilitators of resources in terms of traditional knowledge and cultural significances in response to natural disaster could contribute as a great significance in disaster risk management.

THE STUDY AREA AND THE PAITEI COMMUNITY

Manipur State, located in the north-east of India, is one of the most vulnerable and prone to natural disasters such as recurrent flood, landslides, windstorm and hailstorm which are commonly occurring disasters in this area. Besides, it is also located within the Seismic Zone V: Very High Damage Risk Zone of Seismic Zones of India Map.[3] Flood and 'earthquake', which were the highest-ranking natural events,[4] are also frequently occurred in the study area that needs to be addressed urgently (Fig. 1).

The study area of natural disaster incidents is concentrated in the main town of Churachandpur districts of Manipur, which is widely accepted by the local residents as *Lamka*.[5] It lies on the south-western part of Manipur between 24.0°N and 24.3°N latitude and 93.15°E and 94.0°E longitude. Churachandpur is a hilly geographical area of 4570 sq. km, which is the largest district of the state with a total population of 274,143 and 60 persons per sq. km. As per Census of India 2011, the district has a high literacy rate of 82.8% and about 92.9% belongs to different Scheduled Tribes. The sex ratio of the district is 975 females per 1000 males and about 52% are cultivators (Fig. 2).[6]

---

[2] Ibid., p. 23.

[3] Survey of India, GOI; *Seismic Zones of India Map IS: 1893–2002.* Retrieved from http://www.hpsdma.nic.in%2FResourceList%2FMaps%2FEqIndia.pdf.

[4] Weiner, J. M., and Jr. J. J. Walsh. (2015). Community Resilience Assessment and Literature Analysis. *Journal of Business Continuity & Emergency Planning, 9*(1), p. 86.

[5] Literally, *Lamka* derives from the Paitei tribe word where '*lam*' means '*road*' and '*ka*' means '*divergence*' as the main road from Imphal capital diverge into two road at *Lamka* town, where one road, i.e. *Tedim Road* leads to Myanmar and another one '*Tipaimukh Road*' leads to Mizoram State.

[6] Census of India. (2011). *District Census Handbook Churachandpur*, Village and Town Wise, Primary Census Abstract (PCA), Directorate of Census Operations, Manipur.

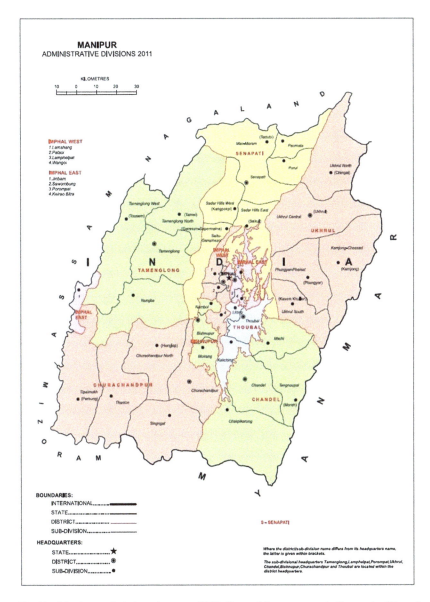

Fig. 1 Manipur administrative map 2011 (http://www.censusindia.gov.in/Data/Census_2011/Map/Manipur/00_Manipur.pdf)

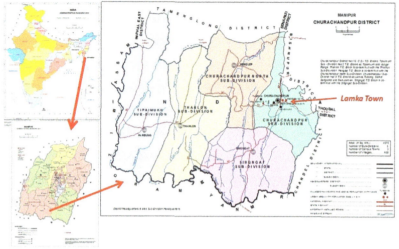

Fig. 2 Location map of Lamka town in Churachandpur district, Manipur (merge of various maps from Census of India 2011)

The Paiteis as a community in Manipur are 55,542 in numbers, which is just 1.94% of the total population[7] but they (are) found heavily concentrated in Churachandpur district[8] of Manipur. Paitei language can be considered as the '*lingua-franca*' of the main town since they are the dominant tribe group. Besides Paitei tribe, *Lamka* is inhabited by other small tribal groups of the same Tibeto-Burman family, where each one asserts for their own distinct cultural identities mainly based on dialect and languages leading to diverse ethnic composition.

Officially, the Paiteis are recognized Scheduled Tribe of Manipur (1956)[9] and Mizoram States (2002)[10] and they have their own

[7] Census of India, 2011.
[8] Siamkhum, Th. (2013). *The Paiteis: A Study of the Changing Faces of the Community*. Chennai: Notion Press, p. 7.
[9] Ibid., p. 195.
[10] Ibid., p. 204.

organized bodies like Paite Tribe Council, custodian of the customary laws[11]; the Young Paite Association (*hereafter* YPA), founded on the principles on 'altruism'; the Paite Literature Society, guardian of language and literature of Paite; and SSPP (Siamsinpawlpi) the students' welfare body and different church denominations. The Evangelical Baptist Convention (EBC) Church, which comprised almost only the Paitei community, alone recorded that the EBC is composed of 148 local churches with a membership of 51,201 according to 2008 (EBC church) census[12] across the globe.

STAKEHOLDERS OF DISASTER MANAGEMENT

Under Disaster Management Act 2005 (*hereafter* DMA 2005), each district is supposed to have their own District Disaster Management Authority and the Deputy Commissioner/Magistrate, who is already engaged with various administrative responsibility, is the chairman of the committee to tackle the disaster-related problems. But, in reality, the locals have observed in many incidents that there used to be, if at all any action were taken, inadequate and timely response by the officials. This also led to further lessening the confidence of the people on government and its governance, which, in turns, augment the involvement at community level in many responsibilities that are supposed to be taken by the government. The involvement of many programmes or activities took up by many NGOs also further strengthened the need for local communities in the case of disaster management too.

The YPA with its underlying principle on 'altruism' with selfless service to the society and responsible for the social upliftment of the Paitei society as a whole, and those who are affected by natural calamities, etc.,[13] has been engaging in the many occurrences of natural disasters in the pasts too. The cultural factor of language and religion is also a great factor in dissemination of information within the society. Paitei being the local '*lingua-franca*' in *Lamka* town, the role of Paitei community is not just important but indispensable to communicate the local people.

---

[11] Paite Tribe Council. (2013). *Paite Customary Law & Practices, 2nd Amendment, 2013.* Lamka, Manipur: Lamka Super Printer.

[12] Retrieved from http://ebc-india.org/about/history.html.

[13] Ibid., p. 93.

In the face of adversity, people of faith shows connection with the sacred and their religious way of life[14] and the Paiteis with 98% of them are Christians[15] tend to connect to their respective church during these disasters. Disasters were also deeply embedded in the causal perceptions…consequently, some people responded to the disaster by praying.[16] Church as a community platform also has been the driven force for many social outcomes by providing their spiritual, emotional and financial needs through different church programmes, which is very often among all communities in *Lamka* town. So, it also has a great role to play in the events of natural disasters which affect the members of their church.

During the past years, the vagaries of the seasonal Indian monsoon have caused various natural disasters in *Lamka* town and its surrounding areas causing destruction of houses, properties and lives due to heavy hailstorm, flooding of Tuitha River and Lanva River. In May 2013, the district official recorded that about 100 houses were totally damaged and 300 houses were partially damaged in the Churachandpur subdivision alone, i.e. *Lamka* town, due to windstorm and shower of hailstorm.[17] In July 2015, the heavy rainfall has caused in inundation of residential area by Tuitha River and Lanva River leaving to disaster-induced displacement of many families settling along the low-lying areas of the river bed.[18] Later within a month, twenty people of Joumol village in Chandel Districts with twenty households lost their lives due to a landslide caused by heavy rain in the Joumol village.[19] Recently in June 2016, many places in Manipur including *Lamka* town are again inundated with the arrivals of monsoon accompanied by torrential rainfall (Figs. 3 and 4).

[14] Pargament, I. K., and J. Cummings. (2010). Anchored by Faith: Religion as a Resilience Factor. In J. W. Reich, A. Zautra, and J. S. Hall (eds.), *Handbook of Adult Resilience*. New York, NY: Guilford Press, p. 207.

[15] Census of India, 2011.

[16] Misanya, D., and A. O. Oyhus. (2014). How Communities' Perceptions of Disasters Influence Disaster Response: Managing Landslides on Mount Elgon, Uganda. *Disasters, 39*(2), p. 403.

[17] Retrieved from http://kanglaonline.com/2013/05/ccpur-locals-reeling-from-effects-of-natural-furies-of-last-two-days/.

[18] Retrieved from http://pvtlimlak.blogspot.in/2015/07/lanva-leh-tuitha-khan-ziak-in-mipi.html.

[19] Retrieved from http://www.ndtv.com/india-news/20-killed-in-landslide-caused-by-heavy-rain-in-manipur-police-1202854.

Fig. 3 One of the many houses in Khominthang area of Lamka town damaged by hailstorm in May 2013

Unlike the seasonal monsoon effects, the unpredictable geological tectonic effects of earthquake are also a common occurrence. During the past one year, there are 31 occurrences of earthquakes of more than 4 Richter scale[20] with not much severity of risk and damage to the people except that hit Manipur in the early hours of 4 January 2016 with a magnitude of 6.8 Richter scale.[21] Despite the frequent occurrences of earthquake in the past and present, there were no noticeable damages and loss of lives and properties in and around *Lamka* town.

It is also noticed that most of the affected families from the recurrent flood of Tuitha River and Lanva River within *Lamka* town are also from economically poor background. The low socio-economic level compels them to settle in a place where there are affordable house rents or land at cheaper rates, which are near the river bed. They have migrated from

[20] Retrieved from http://earthquaketrack.com/p/india/manipur/recent.
[21] Retrieved from http://indianexpress.com/article/india/india-news-india/earthquake-measuring-6-8-magnitude-strikes-india/.

Fig. 4 Rescue works at Joumol village in Chandel district by Assam Rifles in the first week of August 2015 which was hit by a major landslide on 1 August 2015 (*Picture Credit* PRO, HQ IGAR(S), Imphal. Retrieved from http://www.e-pao.net/epGallery.asp?id=5&src=News_Related/Calamities_News_Gallery/Joumoul20150814)

rural villages and settled in those vulnerable peripheral areas during the past two to three years, for better education and better source of livelihood. As abundance of monetary resources reduces the potential impact of a given hazard,[22] it is not just the ignorance about natural disasters, but it is also the socio-economic status of the household that could be linked with the study area. Meanwhile, many opine in social network sites by blaming the people for encroachment of the river bed during the dry season and thus, the issue for river zoning arose during the different debates.

The natural phenomenon of Indian monsoon becomes the major common disaster risk that could result in loss of lives and properties which affect the livelihood and survival of the common people of the

[22]Cutter, S. L., et al. (2008). A Place-Based Model for Understanding Community Resilience to Natural Disaster. *Global Environmental Change, 18*(4), p. 601.

Fig. 5 A market shed of New Lamka bazar was flooded in July 2015

Fig. 6 A footbridge damaged by Lanva River in July 2015

Lamka town. The risk imposed by these natural calamities has become a challenge and threat to the survival, growth and development of the inflicted Paitei community. Thus, the needs to combat, prepare and manage the prevailing uncertainty of the natural disaster were inevitable for the community, since long.

This paper was aimed to find out how the Paitei community in Manipur deals with natural disaster that affects the people of *Lamka* town in general. Qualitative analysis was done using primary and secondary data. Field study with random interaction with disaster victims and different community leaders of the Paitei tribes was done through prepared questionnaire, in June 2016. Interaction with government official and people representatives was also done during the field study. Opinions of youths were also drawn from the social network sites (Figs. 5 and 6).

ROLE OF PAITEI COMMUNITY

Vulnerability and Awareness of the Community

As discussed, the people inhabiting *Lamka* are highly vulnerable to different natural calamities. Neither a survey nor research has been done on the study areas for managing and responding natural disaster. It was found from all the affected respondents and church leaders of the community that no one was aware about the existing of DMA 2005. However, one of the YPA leader responses that he vaguely knew about the DMA 2005 and under his leadership, the YPA contributed an amount of Rs. 30,200/- to the Deputy Commission of Churachandpur for District Disaster Response Team[23] (*hereafter* DDRT) for Victims of Natural Calamities within Churachandpur District, which seems to be the other way round. This could either signify the incompetency of district administrations in response to natural calamities or the ignorance of the local community about the DMA 2005. In a surprising observation, one of the junior officers of the district administration was also not aware about the existence of DMA 2005. The benefits of these programmes to a large extent depend on the level of awareness of the people about the programme.[24] The overall ignorance about the disaster risk management, thus, enhances the vulnerability to the people and amplifies the chance of

---

[23] Retrieved from https://thestmt.blogspot.in/2016/06/ypa-relief-fund-to-dc-ccpur-manipur.html.

[24] Das, O. K. (2009). *Baseline Survey of Minority Concentrated Districts District Report CHURACHANDPUR*, Institute of Social Change and Development, Study Commissioned by Ministry of Minority Affairs, Government of India, p. 33.

COMMUNITY RESILIENCE BUILDING AND THE ROLE OF PAITEI TRIBE ...   341

greater damages from natural calamities. In overall, not much initiative has been taken up by the government or local community leader in making aware of the vulnerability and risk from natural calamities in the area.

Disaster Mitigation and Preparedness

The recent frequent occurrence caused by the flood must have awakened the official for the need for preparedness and a sensitization seminar on 3-Day Disaster Management Course in Comprehensive Landslide, Cyclone Spin and Effect of Deforestation was held at Rayburn College Auditorium, New Lamka, under the flagship of Relief and Disaster Management, and Civil Defence, Government of Manipur, in mid-June 2016.[25]

The YPA has responded that they are imparting basic training on disaster management in collaboration with different agencies like Indian Red Cross Society, Civil Defence, various government departments of Manipur from time to time.

However, the responses from the affected people show their unawareness on how to prepare for natural calamities as they reportedly said that they do not expect any natural hazards. This also shows that there is a wide gap between the community leaders who were trained or participated in disaster management and the people, who are always at high disaster risk.

At all strata of the society, the people of *Lamka* claimed that they are ready to work with the government officials on disaster management as and when the need arises. However, the community as a whole felt that the response from the government was slow. Thus, the long procedural process of the government does not only meet the need of the people, but also indicate the unpreparedness of the government officials in the event of any disaster. The slow help from the government was also acknowledged by the chairman of the autonomous district council of Churachandpur.[26]

Largely, we can counsel the people of *Lamka* and its surrounding areas that they are always in a graved danger of any natural calamities due to their unpreparedness and poor mitigation, which need to embark upon by the community, immediately. This unpreparedness could also

---

[25] Retrieved from http://www.zogamonline.com/index.php/news/lamka-post/item/2650-the-lamka-post-%7C-June-15-16-2016.html.

[26] Retrieved from http://kanglaonline.com/2013/05/ccpur-locals-reeling-from-effects-of-natural-furies-of-last-two-days/.

Fig. 7 YPA relief team for Joumol landslide flag-off by the Deputy Commissioner of Churachandpur on 12 August 2015 (Retrieved from http://www.virthli.in/2015/08/hmasawnna-thar-13-august-2015.html)

pose a large-scale impact on loss of lives and damages of properties on the occurrence of any natural disasters, even on a small-scale calamity. The unmanaged large-scale damages caused by the recent flood along Tuitha river and Lanva river could also be linked with the unpreparedness of the community at large.

Relief

When a disaster occurs, most organization centre their efforts on the delivery of emergency supplies such as food, blankets and tents with the intention to relieve the effects of the impact.[27] Collecting of relief fund

---

[27] Serna, L. (1983). *Disasters and Development: Same Mistakes All Over Again?* [Review of the book *Disasters and Development* by Cuny, Frederick C.]. New York: Oxford University Press, p. 154.

COMMUNITY RESILIENCE BUILDING AND THE ROLE OF PAITEI TRIBE ...   343

Table 1   Relief materials collected by YPA Lamka Block for Joumol landslide and flash flood, Dated: 8 August 2015 (*Source* Retrieved from YPA general headquarters' record file handwritten in Paitei)

	In cash	In kind
1. Central Lamka unit	Nil	2 nos. bags of rice
2. Lanva unit	Nil	2 nos. bags of rice
3. Tuithapi unit	Rs. 1720/-	4 nos. of plates and 1 no. of carry bag
4. Headquarters unit	Rs. 1000/-	nil
5. Bungmual unit	Rs. 200/-	9 nos. bags of rice
6. Mission compound unit	Nil	1 no. 1/2 bags of rice
7. Mata unit	Rs. 600/-	2 nos. bags of rice
8. G Mualkoi unit	Nil	4 nos. bags of rice
9. Pearsonmun unit	Rs. 3328/-	4 nos. bags of rice
10. Lamka unit	Nil	2 nos. 1/2 bags of Rice
11. Zenhang Lamka unit	Rs. 5000/-	3 nos. bags of rice, 4 nos. of clothing, 1 kg of salt
12. D. Phailian/Chiengkon	Nil	3 nos. 1/2 bags of rice, 5 kgs of dal and 2 kgs of sugar

Table 2   Details of contribution to Joumol landslide (*Source* Record of YPA general headquarters)

1. Rs. 2000/- each to affected families
2. Rs. 1000/- each as condolence to families of 9 landslide victims
3. 45 nos. of 1/2 bags of rice
4. 15 nos. of bags full of clothing and utensils

and materials such as food, clothing, utensils, etc., and distributing it to the affected families were the prime activities of the district officials and at the community level, so far. It is also notably observed that the Paitei community also adopted collecting and distributing of relief materials to tackle and cope with any natural calamities in *Lamka* town and its close surrounding areas (Fig. 7).

In 2015, the YPA Lamka Block collected relief funds and materials through their subunit and eleven of the YPA units have responded positively (see Table 1). The collected relief materials were then transported by a truck hired by the YPA and finally, flag off by the Deputy Commissioner of the Churachandpur district for delivery at relief campsite for Joumol landslide (see Table 2), which is adjacent to Chandel district. As already mentioned, some amount of cash was also contributed to District Commissioner towards District Disaster Response Team for victims of natural calamities through the apex General Headquarters of YPA[28] in May 2016. The help from the YPA is very much applaudable, but it is far from adequate to meet the need of the affected family as community resilience. This unit-level contribution by the Paitei community through their community organization also shows that there are great capabilities of the community in disaster management if proper measures on systematic awareness and training are done.

However, the common mistakes organizations make include concentrating on the rapid delivery of material aid, treating the victims as aid recipients, supplementing local coping mechanisms and being accountable only to donors.[29] As a community, YPA is no exception to the common trend in disaster responses and it should also be accountable to the victims, not only to the donors.[30] By this way, disaster relief will be much more effectively and efficiently delivered to the affected people.

An official head from EBC church that largely comprises of the Paitei community also told that through their respective local church, the affected families and victims of the windstorm and Lanva river were given meagre sum of financial help. And, the concerned local pastor consoles and comforts the affected families through spiritual practice like prayers and counselling. In the meantime, other members of the church lend a helping hand during their difficult ordeal like relieving, temporary relocation and rebuilding of their houses.

During the windstorm and hailstorm in May 2013, about 30 affected families of Khominthang village were rehabilitated at two temporary relief camps, one at PT Sport Complex and the other at SSPP Residential School in New Lamka by the district officials. Also, those families of Zoumunnuam village affected by the flooding of Tuitha River were sheltered for two weeks at Zoumunnuam Community Hall. All these temporary relocation processes were done with the assistance and support of the local community. The community role of relief and rehabilitation is notable but shortlived and it does not endure long run impact.

[28] Retrieved from https://thestmt.blogspot.in/2016/06/ypa-relief-fund-to-dc-ccpur-manipur.html.

[29] Ibid., p. 154.

[30] Ibid., p. 154.

Fig. 8 Members of the Paitei community on their altruistic activities at Khominthang area, Lamka that was affected by hailstorm in May 2013

Community Resilience

In spite of the risk, majoring of the respondents has decided to return to their earlier respective resident when the level of waterfall back or will return after monsoon season, when it is liveable. One reason being that there is no alternate safe location for rehabilitation and no choice left, but to return to the same place, due to poverty. Other, the community are always ready to give support in rebuilding their damaged houses. During the rebuilding and rehabilitation processes, YPA at their different unit-level plays a great role in rebuilding houses by providing building materials, involving youth and skilled labourers of within their same community without charging a single penny, on the principle of 'altruism'. In contrast to other activities of the YPA, where there are equal participations of men and women, the response to natural disaster mainly involved men (Fig. 8).

And during various hardships due to natural disasters, religion can be a powerful force for resilience among people grappling with the most traumatic experiences in life.[31] The local churches in *Lamka* town played

---

[31] Ibid., p. 207.

a vital role by providing the spiritual, emotional and psychological assurance through prayers and counselling for the victim families. House visiting to the affected families by the local pastor of their concerned church and counselling with prayers is a common practice. The members of the church were also comforting social factors during their adversity. The faith-based beliefs are an important source of resilience for people in disasters.[32] Also, the social component of kinship is still very strong, the continuous support of their relatives as community resilience is notable in the present context. The assist and support of the YPA and other organization are shortlived and lasted less than a month in general; so at present, the community resilience on disaster management needs to be further amplified to restore and speed up all the damages back to normalcy and stability.

Recently, three days International Workshop on 'Earthquakes and Landslides: Addressing Administrative and Technological Challenges in Building Community Resilience' was held at Manipur University Campus from 9 April 2016 to 12 April 2016 by Disaster Research Programme (DRP) of Jawaharlal Nehru University, New Delhi; National Institute of Disaster Management (NIDM), New Delhi; and Manipur University, Imphal, Manipur. Various community leaders from all the nine districts of Manipur had participated this workshop and an '*Imphal Declaration*' on building community resilience was affirmed, which is expected to bear positive outcome in the near future.

CONCLUSION AND WAY FORWARD

Though there is still large vague in the proper and systematic procedure in responding the natural calamities in the study areas, the willingness of the Paitei community to tackle disaster with their own ways and means could also be undeniable facts. The people are aware about the natural disaster but not the need for proper management. This is so because of the ill initiative on part of the government and the ignorance of the local community leaders who are responsible to train the common people and disseminate about disaster management through awareness campaign, which is the need of the hour and also frequently. Poor socio-economic status of the community leads to low level of mitigations and high damages to properties which could, otherwise, be minimized or controlled.

[32] Ibid., p. 390.

The Paiteis as a community through the YPA and the church play a significant role in response to the disasters, which could be prominently observed through the collection and distribution of relief at their unit grassroot level of the organization. A sub-department such as Disaster Response Team could be set up by Paitei community under the banner of YPA, and they could organize awareness campaign on disaster risk management time to time and disseminate information about disaster risk management thoroughly within the whole community. The spiritual assurance given by the church leaders is also one prominent approach of the Paitei community for disaster-affected people that contributed to community resilience, presently.

Though the great initiative taken by the Paitei community is highly commendable, the powerful natural forces are unmatchable for a single community to manage it. In a town like *Lamka* where multiethnic are inhabited, the cooperation of all ethnic community is extremely necessary to deal with the ethnic-less risk imposed by natural disasters. This gigantic task of disaster management could be much more efficiently handled, if there is strong inter-communities coordination. It could also enlarge the required source of resources in term of manpower, finance and coordination at local level.

Many church buildings and community halls could be also utilized as short-term relief camps, and if it equipped with the requisite disaster-related instruments, tools, relief materials, etc., these relief or safe camps could well serve as a safety evacuation centres.

Also, there is no proper coordination of the community with the district government officials, which weakens the efficiency of the DDRT at the local community level. Local government officials and community leaders should be more sensitized to disaster management. Preparedness and mitigation of the present common availing disaster due to flooding could be much more easily dealt through the government initiative of the flood zoning in consultation and compensation with the concerned affected community. The DDRT should also include representatives from disaster-prone communities as its core member.

Learning from the past experiences, identifying the disaster-prone areas, vulnerable inhabitants and working out on disaster risk reduction are some immediate feasible steps that could be taken at the community level and instantly. So that, this will also ensure early warning system before the arrival of monsoon rain especially to those families residing along the river bed. With all the experiences and observations, this study could relate that the Sendai Framework 2015 has strong relevance and needed in implementation for disaster risk reduction in the context of this study area.

BIBLIOGRAPHY

Census of India. (2011). *District Census Handbook Churachandpur*, Village and Town Wise, Primary Census Abstract (PCA), Directorate of Census Operations, Manipur.

Cutter, S. L., et al. (2008). "A Place-Based Model for Understanding Community Resilience to Natural Disasters". Global *Environmental Change, 18*(4), pp. 598–606. https://doi.org/10.1016/j.gloenvcha.2008.07.013.

Das, O. K. (2009). *Baseline Survey of Minority Concentrated Districts District Report CHURACHANDPUR*, Institute of Social Change and Development, Study Commissioned by Ministry of Minority Affairs, Government of India.

Misanya, D., and A. O. Oyhus. (2014). "How Communities' Perceptions of Disasters Influence Disaster Response: Managing Landslides on Mount Elgon, Uganda". *Disasters, 39*(2), pp. 389–405. https://doi.org/10.1111/disa.12099.

Paite Tribe Council. (2013). *Paite Customary Law & Practices, 2nd Amendment, 2013*. Lamka, Manipur: Lamka Super Printer.

Pargament, I. K., and J. Cummings. (2010). "Anchored by Faith: Religion as a Resilience Factor". In J. W. Reich, A. Zautra, and J. S. Hall (eds.), *Handbook of Adult Resilience*. New York, NY: Guilford Press, pp. 193–211.

Serna, L. (1983). *Disasters and Development: Same Mistakes All Over Again?* [Review of the Book *Disasters and Development* by Cuny, Frederick C.]. New York: Oxford University Press, pp. 153–156.

Siamkhum, Th. (2013). *The Paites: A Study of the Changing Faces of the Community*. Chennai: Notion Press.

UNISDR. (2015). *Sendai Framework for Disaster Risk Reduction 2015–2030*. UNISDR, United Nation.

Weiner, J. M., and Jr. J. J. Walsh. (2015). "Community Resilience Assessment and Literature Analysis". *Journal of Business Continuity & Emergency Planning, 9*(1), pp. 84–93.

Urban Risks to Hazards: A Study of the Imphal Urban Area

Sylvia Yambem

Urban risks to hazards pose serious challenges to disaster governance in Manipur. Hazards are not new to the state; however, increasing population growth in the Imphal urban valley, particularly in the Imphal West district, and the lack of planned urban development, along with the inability of the state to institutionalize effective disaster risk reduction, have aggravated the effects of hazards such as floods, earthquakes in the urban areas. Urban risk not only presents a wider social, economic and environmental challenge but also affects inclusive development. This is because though hazards affect all areas equally, whether urban or rural, however it becomes particularly threatning in built-up densely populated urban areas (Twigg 2015, 245). The effects of natural disasters are thus worse in cities than in other environments (Wamsler 2006). This imperative to integrate disaster risk reduction strategies in urban planning procedures such as land use policies, building codes, standards, rehabilitation and reconstruction practices, etc., has also been emphasized in the Hyogo Framework for Action 2005–2015.

In the Imphal Valley, the impact of hazards in the urban areas can be seen in the interaction between the valley's naturally existing

---

S. Yambem (✉)
Department of Political Science, Manipur University, Imphal, India

© The Author(s) 2018 349
A. Singh et al. (eds.), *Development and Disaster Management*,
https://doi.org/10.1007/978-981-10-8485-0_23

geophysiologically predisposition to hazards, with the man-made phenomena of increasing population growth, unplanned development and lack of disaster risk reduction strategies. These factors combined together magnify the impact of disasters on human lives, livelihood and property. Importantly, this has created an urban system that is not disaster resilient.

THE IMPHAL URBAN AREA

Imphal, the capital of Manipur, comprises the municipal areas of Imphal West and Imphal East districts, urban outgrowth towns and contiguous census towns.[1] Collectively, this area is also known as the Imphal urban agglomeration. The Imphal urban area has experienced continuous expansion, both in terms of its geographical spread and population size. In the 1991 census, while the Imphal urban area covered a total geographical area of 33.30 sq. km and a population size of 198,535 persons, by the 2001 census the Imphal urban agglomeration increased to an area size of 41.62 sq. km and a population of 250,234. A brief area profile of the Imphal urban area is given in Table 1.

In 2001, the Imphal municipal area experienced a decadal growth of 12.61%. While in the Imphal urban agglomeration for the same period, a decadal growth of 23.37% was observed (2001 Census). Population density in the area of the urban agglomeration is 6012.3 persons per sq. km, while in the municipal area it is 6651 per sq. km. As compared to the overall population density of 103 per sq. km (2001 Census) and 122 per sq. km (2011 Census) for the state, population growth and density in the Imphal urban areas is therefore much higher. However, by 2011 population growth in the Imphal urban area increased to 418,739 people (2011 Census). Thus the urban areas comprising only 0.13% of the total area of the state accounts for more than 50% of Manipur's urban populace.

---

[1] Imphal urban agglomeration comprises of the 27 ward areas of the Imphal West and East district, the outgrowth towns of Bijoigovinda, Kongkham Leikai (Part), Langthabal Kunja, Langthabal Mantrikhong (Part), Naorem Leikai, Oinam Thingel, Porompat Plan Area, census towns of Chingangbam Leikai, Khongman, Khurai Sajor Leikai, Kiyamgei, Laipham Siphai, Lairikyengbam Leikai, Lamjaotongba, Langjing, Naoria Pakhanglakpa, Porompat, Sagolband (Part), Takyel Mapal, Thongju, Torban (Kshetri Leikai) and the nagar panchayats of Lamshang (Core Town), Lilong (Imphal West-Core Town) and Lilong (Thoubal).

Table 1 Imphal urban area (*Source* Directorate of Census Operations, Manipur)

Area	Size		Population		Population density		Decadal growth
	1991	*2001*	*1991*	*2001*	*1991*	*2001*	
Imphal Municipal Area	33.30	33.30	198,535	221,492	5962	6651	12.61
Outgrowth		4.12		6927	1015	1681.3	
Bijoy Govinda (Imphal West)		2.13		3710		1741.8	
Takyel Mapal (Imphal West)		1.12	1137	1370	1015	1223.2	20.49
Porompat Plan Area (Imphal East)		0.71		1057		1488.7	
Kongkham Leikai (Imphal East)		0.16		790		4937.5	
Census Town		4.20		21,815		5194	
Naoria Pakhanlakpa (Imphal West)		0.74		6631		8960.8	
Porompat (Imphal East)		1.54		5160		3350.6	
Khongman (Imphal East)		1.06		5464		5155.7	
Torban Kshetri Leikai (Imphal East)		0.86		4559		5301.2	
Imphal Urban Agglomeration		41.62	198,535	250,234		6012.3	23.37

This is because Imphal the capital is the only Class I city in the State of Manipur, and the largest town in the North-Eastern Region of India. This increasing urbanization of the Imphal West district, however, has wider implications to urban risks. The oval-shaped valley, home to the four valley districts of Imphal West, Imphal East, Thoubal and Bishnupur, is surrounded by rugged hill and mountain ranges that comprises of the five hill districts of Senapati, Ukhrul, Tamenglong, Chandel and Churachandpur, lies at an elevation of 790 metres above sea level. The four rivers, namely Imphal, Nambul, Kongba and Iril along with its numerous tributaries that flows through Imphal, have filled the valley with thick sediments (Gahalaut and Kundu 2016, NIDM). This geophysical characteristic makes the Imphal valley vulnerable to both earthquakes and floods; while the soft marshy soil amplifies the wave motions from tremors and quakes even from far away, heavy rainfall in the hill districts which constitute about 70% of the upper catchment areas drain in the Imphal valley making the valley districts most susceptible to flash floods (NIDM).

The Imphal urban area is thus situated directly atop of river bodies or marshy pits or swamps locally known as pats. The centre of the urban area is Kangla or Kanglapat and Keishampat while in the north lies Lamphelpat and in the east Porompat, and in the south the institutional areas of Takyelpat (Map). These constitute some of the most densely populated areas of Manipur. For instance the areas around Kanglapat and Keishampat is home to the state's main political and commercial centers—the Raj Bhavan, Chief Ministers Office and bungalow, Kangla Fort, Manipur Secretariat, Police Headquarters and the main commercial centers of Ima market, Khwairamband bazar, Thangal and Paona bazar. The areas of Kanglapat and Keishampat are also surrounded by the Nambul and Imphal River, in a sort of oval-shaped demarcation, making the areas vulnerable to flash floods. Lamphelpat and Porompat are home to Manipur's (and even the NER's) primary medical establishments Regional Institute of Medical Sciences (RIMS) and Jawaharlal Nehru Institute of Medical Sciences (JNIMS), respectively. Lamphelpat also has the commercial spots such as Sanakeithel, Supermarket complex and other important offices and departments such as the CRPF camp, DC Imphal West office, Election office, etc., while Porompat is the centre of the DC Imphal East office and others. The Imphal urban areas thus being the political, administrative and economic capital account for over 90% of employment opportunities, making the area attractive to migration. Except for the border trade town of Moreh, that is located in the hill district of Chandel, Imphal has more than 3389 commercial establishments, of which almost 80% are located in the Imphal urban area making the area attractive to migration (City Development Plan). Consequently, even though the urban population in the state has increased from 25.11% in 2001 to 30.21% urban growth is higher in the Imphal Valley (2011 Census). In the Imphal West district, the urban population has increased from 55.51% in 2001 to 61% in 2011, while for the same period in Imphal East urban population increased from 27.43 to 40.28% (Fig. 1).[2]

IMPHAL'S UNPLANNED URBAN DEVELOPMENT

The Imphal urban area poses a larger threat to human lives, livelihood and property. This is because urbanization of the Imphal urban area is in the nature of unplanned development, characterized by increasing number of urban poor, and incompetent urban infrastructure and low-quality public services. Imphal (along with the neighbouring city of Kohima in Nagaland) in the north-east region has the highest number of people living below the poverty line (City Development Plan: Imphal). It is estimated that at least 26% of the urban populace lives below the poverty line, of which 23% do not have access to safe drinking water facility and are largely dependent upon stand-posts water services, 3% rely on tank supply water, 59% do not have access to proper toilets, and 66% live in kuchha houses (Ibid). The lack of basic services, however, affects all citizens of the state. For instance, field survey revealed that in the Imphal urban area at least 78% complained of irregular and highly unreliable water services (Yambem 2013).

The Imphal urban area thus lacks access to quality infrastructure such as power, transport, water, sewerage, disaster resilient building

---

[2] In the hill districts, urbanization has been relatively slow. In 2001, except for the hill district of Chandel with an urban population of 12.64%, all other hill districts were predominantly rural. In fact, in the other four hill districts, there was no urban populace, and there was no urban demarcation. It is only in the 2011 Census that some form of urban growth has been registered in all the five hill districts; Senapati—2.10%, Churachandpur—6.40%, Chandel—11.74%, Tamenglong—11.22% and Ukhrul with 14.32%. In contrast, the urban growth in the other two valley districts of Thoubal—35.48% and Bishnupur—36.73% is rapidly increasing.

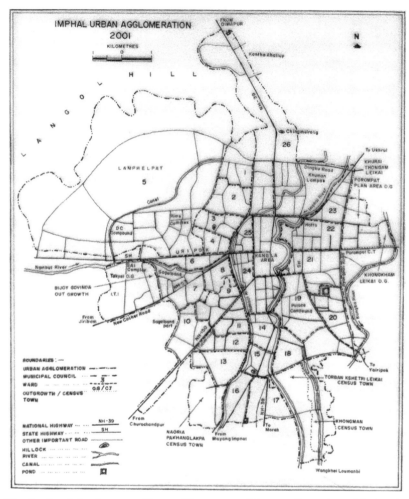

Fig. 1 Imphal urban area (*Source* Directorate of Census Operations, Manipur)

and basic public goods and service such as waste and garbage disposal system, well-planned drainage, etc. Take the case of the garbage disposal system. Since 2008, garbage collection in the Imphal urban area has been outsourced to four Non-Governmental Organization (NGO) the Thangmeiband Assembly Constituency Development

Forum (TACDEF), Seven Security Force (SSF), Centre for Research on Environmental Development (CRED) and Workers Union Manipur (WUM). The 27 wards in the Imphal Municipal area have been accordingly divided into 4 zones. TACDEF oversees Zone I comprising of the ward numbers 1, 2, 3, 5 (partial areas of Sanakeithel, Lamphel super market), ward number 25 (partial areas in the western side of Thangal bazar, the northern side of MG Avenue) and ward number 26 partially covering the western areas of Chingmeirong. Zone 3 under SSF oversees the garbage collection for the wards 7, 8, 9, 10, 11, 12, 13 and partial areas of ward 24 covering the eastern side of Paona bazar and Babupara. CRED is in charge of Zone 4 for the wards 5 (except the areas of Sanakeithel and Lamphel super market), ward numbers 6, 14, 15, 16, 17, 18, 19, 20, 24 (partial western side of Paona bazar, the western side of NH-39, Nambulane) and ward number 27. The last zone V under WUM oversees garbage collection for the wards 4, 21, 22, 23, 25 (the eastern side of Thangal bazar, southern side of MG Avenue and western area of BT Road) and the eastern side of Chingmeirong ward number 26. Under this new contract system, door-to-door garbage collection is to be taken at least once a week for a monthly charge of Rs. 100.

However, the difficulties are much beyond the mere simple act of contract service and garbage collection. The Workers Union Manipur notes that one of the major challenges of keeping the Khwairamband bazaar area clean is the shortage of pickup vehicles (*The Sangai Express*, 15 February 2009). Thus, even if one were to collect all the garbage, the lack of transportation vehicles impedes the effort itself. Further, the weekly system of garbage collection is not undertaken regularly, and it is also noticed that in some ward areas the responsible organizations have also stopped the garbage collection. T. Lekhendra, Secretary CRED in an interview also revealed that Manipur has no permanent garbage dumping ground; hence, garbage is dumped at the open site and the organizations have to pay some amount of money to the Imphal Municipal Council (henceforth IMC) for dumping wastes. Also, the organizations do not receive any government support or grants or any financial support from the IMC. Bandhs, curfews and the rising fuel prices particularly during economic blockades also hinder the garbage collection. Especially during economic blockades when the cost of fuel shoots up, as the rises lead to increase in costs that make it expensive for CRED to operate when the fees received from households are fixed. Moreover, public response is not helpful/accommodating for CRED. For instance, households tend to group their wastes such as one registered household will group their garbage with other households and ultimately paying for the charge of one household only, to avoid paying the monthly charges. Consequently, the lack of vehicle along with rising operational cost (labour charges, fuel cost others) and the absence of any support from the IMC or the government have forced CRED cut down the garbage collection from twice a week to once a week.

The Imphal urban area therefore lacks the most basic physical infrastructure. This factor by itself is a major detriment to building a resilient urban system. A well-planned and built city system particularly physical infrastructure such as road infrastructure, communications, waterways, etc., ensures that during disasters the physical systems as the foundation of the city must be able to survive and function even under extreme conditions (Godschalk 2003). However, in Imphal the twin combination of increasing population growth and low-quality public infrastructure and services have aggravated the crisis in the urban area.

An Urban System, not Disaster Resilient

A disaster resilient city must therefore have physical infrastructures that are both resilient and adaptable to hazards and disasters. However, in the case of Imphal, not only is the Imphal urban area overpopulated and lacks the most basic physical infrastructure and public services, and the area is also vulnerable to hazards. Imphal is sited in earthquake zone V—the highest level of seismicity—and this overpopulation of the area naturally aggravates the impact of hazards such as earthquake. This vulnerability of the Imphal Valley to disasters was revealed in the 2016 January earthquake, which even though had its epicentre in the town of Noney in the Nungba subdivision of Tamenglong hill district, about 33 kms from the capital, caused extensive damages to the valley. The earthquake affected important government and commercial building such as the Ima Keithel or women's market, Manipur Secretariat, ISBT, BSNL office, bridges, and other private houses, etc. This is because while the damages in the hill districts were relatively limited by its sparse population and the use of traditional wooden housing, in the Imphal Valley increasing population pressure on land, improper building construction and local site effect such as the Ima markets that are located around the riverbanks or the marshy pats in the case of the Central Agricultural University aggravated the impact of the earthquake (Gahalaut and Kundu 2016).

Flooding, particularly flash flood is also major hazard in Imphal. Heavy rain and thunderstorms occur very frequently in Manipur and affect both lives and livelihood. For instance, incessant rainfall in the months of July-August 2015 not only triggered landslides that swept away the village of Joupi in the hill district of Chandel and killed 21 people, cut off the town of Moreh and Jiribam due to landslides and broke down all roads and communication between villages and their districts and towns, washed away bridges, schools, hospitals and other essential infrastructure, submerged hectares of paddy fields, crops and fish farms, and displaced thousands of villagers. The floods also caused extensive damages to the four valley districts of Imphal West, Imphal East, Bishnupur and Thoubal and the two hill districts of Chandel and Churachandpur. However, even as the state was still recovering from the floods, in May 2016 Manipur again experienced heavy rain, thunderstorms and hailstorms that caused large-scale destruction to over 2000 houses, crops, farms and livelihood. The rain further damaged the Imphal-Myanmar Trans-Asian highway that had been affected in the January earthquake and was yet to be repaired.

However, the Imphal Valley being situated below the flood level of Imphal river also means that in case of excessive rains flash floods occur frequently in the Imphal Valley (CDP). However, flash floods have been aggravated by the absence of proper drainage channels, often as a result of unplanned land use patterns and regulation and the non-maintenance and cleaning of drains. As a result, heavy rains often cause flash floods in the centre of Imphal, particularly Nagamapal, Uripok, Waheng Leikai, Keishampat and Keishamthong. Moreover in the absence of an institutionalized system of garbage collection, most citizens continue to throw their waste or garbage on the side of the roads and near the riverbanks. The CDP also notes that this lack of a proper garbage and sewage disposal system has turned the Nambul river into a waste dumping centre (Government of Manipur). Unfortunately during floods and waterlogging, the garbage floats in the water and becomes a potential source of disease and epidemics. This is true again for the areas such as Waheng Leikai, Nagamapal, Ima Keithel and Thangal bazar. Further river embankments even in the heart of the capital at Nambul River around Keishamthong are still not completed.

The dense population growth and unregulated land use patterns have also created heavy traffic congestions in the Imphal urban area. In 2014, the major fire in the heart of the capital at Majorkhul in Thangal bazar where at least 5 people including a 12 year old girl sustained burnt injuries, destroyed over 13 houses and 3 shops causing an estimated loss of goods worth crores of rupees was not controlled immediately as traffic jams on the roads delayed the arrival of the fire service.[3] The heart of Imphal city—particularly the stretch from Kanglapat to Keishampat, Waheng Leikai, Nagamapal and Khoyathong—is most traffic congested. Thus, to mitigate urban risks to hazards there an urgent need to integrate urban planning and disaster risk reduction strategies and policies in the Imphal urban area. However, the present focus of disaster management is upon 'imparting training on basic disaster management to enhance disaster preparedness among the general public under the scheme/programme of capacity building'.[4]

The state has also failed to involve community participation in creating a disaster resilient urban system. Rather during hazards, particularly heavy rains and flash floods most households in Imphal (and in Manipur) have to individually look out for themselves. For instance, though flooding is a perennial problem in the low lying areas of Uripok, Kangchup Road, Tourangbam Leikai, Polem Leikai, Achom Leikai, Naoremthong, in Nagamapal, Keishampat, Keishamthong, the initiatives for de-siltation and dredging of sewers is either a collective responsibility of the Leikais or the individual and not of the IMC. Moreover, the failure to introduce daily garbage and waste collection has also contributed to clogged drains and sewers that fill up rapidly during heavy rains.

Conclusion

A disaster resilient urban Imphal is fundamental, not only because Imphal is the nerve centre of Manipur but also because hazards occur very frequently in the state. According to the Manipur State Disaster Management Plan between 1926 and 2009, Manipur experienced 14

---

[3] Majorkhul Fire Tragedy: Five hurt; 13 houses, 3 shops destroyed. *Hueiyen News Service*, Imphal June 24, 2014.

[4] 'Need to inculcate preventive culture to mitigate problems caused by disaster'. *Source The Sangai Express/DIPR*. Imphal July 29, 2016.

URBAN RISKS TO HAZARDS: A STUDY OF THE IMPHAL URBAN AREA 359

floods and 24 earthquakes (Government of Manipur).[5] However, while these recorded earthquakes have a magnitude of 6.0 Mw and above, the state also frequently experiences minor tremors. Prior to the 4 January 2016 earthquake of 6.7 Mw, a 4.0 Mw minor tremor was felt in the town of Wangjing, in the valley district of Thoubal in the month of October—four months before the January earthquake, and then in September, two minor tremors measuring 4.4 Mw and 4.2 Mw were also recorded. Even before the April 2016 earthquake that had its epicentre in Myanmar and repercussions felt in Manipur, in the morning of that same day Imphal recorded a minor tremor of 4.6 Mw.

However, despite the frequency of hazards, Manipur has yet to institutionalize and integrate disaster risk reduction strategies in government policies and programmes. Disaster risk reduction strategies 'through systematic efforts to analyse and manage the causal factors of disasters, including through reduced exposure to hazards, lessened vulnerability of people and property, wise management of land and the environment and improved preparedness for adverse events' (UNISDR 2009) not only reduce disaster-related vulnerabilities but also build the capability of communities to deal with disasters. For example, a system of information dissemination and sharing can significantly reduce disaster risks by preparing or warning and advising the citizens ahead of impending hazards. In the case of the Imphal Valley, urban risks to hazards can be mitigated by integrating disaster risk reduction strategies such as disaster resilient urban housing, regulated land use patterns, better designed and planned physical infrastructure such as roads, drainage systems, garbage and waste collection, safeguarding public infrastructure from hazards, and capacity building of the community and civic agencies in urban planning. Integrating urban systems with disaster risk reduction strategies is therefore fundamental to building a disaster resilient Imphal. Thus, even though the State Government has designated the Manipur Fire and Emergency Service as the first responders, however as the Majorkhul fire incident revealed in the absence of a disaster resilient urban system, urban risk to hazards remain equally critical.

---

[5] Of which two severe earthquakes measuring 7.25 Mw and 7.2 Mw struck the area of Southern Manipur near Moirang on 1 July 1957, and on 6 August 1988 along the Indo-Myanmar Border region, respectively.

REFERENCES

Gahalaut, V. K., and Bhaskar Kundu. (2016). The 4 January 2016 Manipur Earthquake in the Indo-Burmese wedge, an Intra-slab Event. *Geomatics, Natural Hazards and Risk*. https://doi.org/10.1080/19475705.2016.117 9686.

Godschalk, R. David. (2003). Urban Hazard Mitigation: Creating Resilient Cities. *Natural Hazards Review*. August.

Government of Manipur. City Development Plan: Imphal, Imphal Municipal Council. http://manipur.gov.in/IMC/CDP_Imphal.pdf.

———. Manipur State Disaster Management Plan. Volume 1.

National Institute of Disaster Management, Manipur. Ministry of Home Affairs, Government of India, New Delhi.

Twigg, John. (2015). *Disaster Risk Reduction*. Overseas Development Institute.

United Nations. (2009). *UNISDR Terminology on Disaster Risk Reduction. United Nations International Strategy for Disaster Reduction*. Geneva, Switzerland, May 2009.

Wamsler, Christine. (2006). *Mainstreaming Risk Reduction in Urban Planning and Housing: A Challenge for International Aid Organisations*. http://works.bepress.com/christine_wamsler/8/.

Yambem, Sylvia. (2013). *Wellbeing and Access to Urban Services: Interrogating the Linkage and Disconnect Through the Imphal Experience*. Unpublished thesis, Jawaharlal Nehru University, Delhi.

Community Awareness on Landslide Disaster: 'Experience of Nagaland'

Supongsenla Jamir and N. U. Khan

INTRODUCTION

The Nagaland State is prone to all kind of natural disaster like earthquake, flash floods, landslide and forest fire. Every year the state faces huge losses of property due to natural disaster. The degree of losses due to disaster depends on the type of disaster and its place of occurrence. The geographical location places a major role on the occurrence of natural disaster.

The landslide has become a major constraint in the State of Nagaland. It causes widespread damages in the hilly terrains every year. The landslides disasters are major components of many major natural disasters and are often major contributors to losses of lives, properties and livelihoods of the people. In recent years, the impacts of natural disasters are increasing in terms of frequency, complexity, scope and destructive capacity. The population growth and the expansion of settlements and lifelines over hazardous areas are increasing the impact of natural disasters both in the developed and in the developing world.

S. Jamir (✉) · N. U. Khan
Jamia Millia Islamia, New Delhi, India

© The Author(s) 2018   361
A. Singh et al. (eds.), *Development and Disaster Management*,
https://doi.org/10.1007/978-981-10-8485-0_24

Both districts, Kohima and Mokokchung, have been experiencing a steady growth towards urbanization and attracting more population from rural area and from outside state. Many development activities are taking place in the state but unfortunately without a systematic procedure or considering the instabilities of the soil. Development is very important, but it should not lead to any drastic implications. The implications can be like soil erosion, landslides, loss of flora and fauna. Vulnerability in the state is increased in population and people residing in the prone area not realizing its disastrous impact on their life. Lack of inadequate infrastructure is a major concerning problem in the state because both the government and the privates have constructed buildings without proper planning and expert consultancy. During the research, one of the officials stated that in past the land use planning was never a point for consideration for all kind of development activities in the state by both the individuals and the builders.

According to the geologist, the state is vulnerable to landslide disaster because of its weak soil and unstable land (Anbalagan et al. 2007). A landslide is the downward and outward movement of slope-forming materials composed of rocks, soil, artificial fills or a combination of all these along the surface of separation by falling, sliding, flowing under a fast or slow rate, but under the action of gravitation and where the triggering factor maybe natural or anthropogenic (Varnes 1984). Landslide disaster is very frequent in the Himalayas and Eastern-Western Ghats in India (Avasthy 2006).

Landslide awareness consideration is very important in assessing the quality of life since the environment impact is directly on the well-being and life of the residents.

METHODOLOGY

Research methodology clearly depicts the blueprint of any study and provides an insight into the statement of the problem, aims and objectives of the current study, research design, sample, sampling methods, rating scale, tools adopted, administration and designing of the questionnaire, method of data collection, data analysis and limitations of the study in an orderly manner. The fundamental objective of any research is to uncover results and draw a meaningful conclusion.

The present study adopts a '*descriptive*' research design which includes an in-depth interview schedule and a case study approach designed for the affected communities. It is in alignment with the research objectives, and appropriate study will be drawn based on the results.

Two districts of Nagaland have been chosen as the location of the study because Kohima being the capital of Nagaland is been affected by landslide every year in all the part of the town area. And on the other hand, Mokokchung district has been chosen because in the history of landslide disaster in Nagaland, most causality happened in this Mokokchung town. The universe is narrowed down only to the town area of the two districts.

A total samples size of 140 households that are directly affected by landslide disaster were collected representing both the district, i.e. Mokokchung and Kohima town area. Also, data from officials i.e. District Disaster Management Authority, Geological & Mining Department, Boarder Road Organization & Administrative Training Institute and 3 CBOs have included here.

The sampling design of the current study focuses on the people who have experienced the landslide disaster have been purposively selected and interviewed thereby taking their consent and time. The task was to select the sample area for the research. The two districts, viz. Kohima and Mokokchung town, had been selected purposively on account of their fulfilling the study requirement. Kohima is the capital of Nagaland, and every year most of the community becomes the victim of the landslide disaster. It has become a routine for them especially during the monsoon. Same goes with the Mokokchung town, being the hub of the Ao tribes most of the people settle down in the Mokokchung town for the betterment of the children's education and also because of their job.

Keeping in mind the objectives of the study, many colonies were visited to identify the landslide-prone area as there were no much data records available with the government officials. At the same time, visited few officials in both the districts and taken their interview.

To encourage the respondents to share their experiences, both open- and close-ended questions have been used as a part of the study. Although close-ended questions assist in quantifying data, the open-ended questions generate awareness about their understanding to disaster issues, their source of livelihood, their expectation from the administration and their preventive measures taken to reduce or prevent disaster, etc. The question sets divided into different parts. First set talk about the demographic profile of the respondents, the second set talk about the level of awareness on landslide disaster prevailing in the state and the last set talk about how it impacts on their life as well as what they do to minimize the impact of landslide disaster.

The interview schedule comprised of both open- and close-ended questions. The responses from the close-ended questions obtained were then systematically entered in Statistical Analysis for Social Sciences (SPSS) to quantify the data. The open-ended responses for each respondent, wherever possible, have been quantified with the help of the Microsoft Excel. On the basis of the objectives and the results, relevant tools have been used.

FINDINGS

The respondents from both the districts residing in the prone area are aged 24 years and above. Most of the respondents were aged between 41 and 50 years of age followed by 51 and above. And 95% of the respondents are Christian and tribals and most of them were literate. As the study is done on urban area, 48.6% of the respondents were government employees and the remaining were employed on private firm or doing small business. As many respondents are government employees, 55% of the respondent incomes were more than 15,000 per month. Surprisingly, 50.7% of the respondents are residing in the locality for more than 15 years and 26.4% are residing between 5 and 10 years and 15.7% are residing for 10 to 15 years in the same locality. Only 7.1% are residing for 5 years and below. It can be seen that as compared to elderly people, the younger generation are more flexible to migrate to a safer place than the elderly people compromising with the disaster and continue living in the same locality even after frequent landslide disaster.

Nagaland is situated in a hilly terrain, and it is prone to all kind of natural disaster. Majority of the respondents with 91.4% are aware about the disaster happening in the state. When asked about the trends of landslide, majority of the respondents with 84.3% agreed that the landslide is increasing with time. The reason is mostly encroachment of land in the name of development, intense rainfall, increase of population, migrations, etc., in the past years everyone has one or more experience of landslide disaster.

Awareness on Causes of Disaster

Majority of the respondents with 95% are aware about the causes of the landslide disaster. Only 5% are not aware about the causes of the landslide. Some of the major contributors are mainly because of heavy rainfall, bad drainage system, earthquake as it weakens the soil, encroachment of land, etc. Especially in hilly terrain if earthquake happens, then there will be high chances that it will be followed with landslide because the earthquake shakes the earth and makes the soil weaker. Some respondents even admit that they are also responsible in the contribution because they know it's not a safe place to build a house but they still built it as they did not have any alternatives option to settle down. The places are getting congested day by day, and more people are migrating to the urban area in search of a livelihood. The respondents say that during the monsoon they have sleepless night and they prefer not to go out leaving their children alone. They make sure if there is continuous rainfall they try to be back home if they are outside because landslide is very unpredictable and can happen anytime if there is continuous rainfall.

People living in the landslide-prone area have one or more landslide experiences in the past 20 years. Many of them had scary experience in which they even lost their dear ones during the landslide disaster. Their pain still becomes very fresh whenever they recall those days and talk about it. Still many of them fear for the same kind of incidents because they are in the same home and locality. It is obvious that everyone has experienced the disaster in different ways. Some people lost their dear and loved ones and other lost their whole life earning in the landslide disaster. Still many people live in the same house and place and continue living as a victim of disaster. It seems somewhere people have compromised a lot with their life by continue living in the same locality even after experiencing frequent landslide disaster.

Many people had lost their house and cattle during landslide with 48.6%. Most of the respondents who lose their house and cattle were residing near the drainage area or on a slope. 9.3% has lost their home, cattle and others like their kitchen garden and reduction of land. And 6.4% has loss their dear ones and many more in the disaster. The remaining respondent has lost their farm, portion of their house, their kitchen garden, reduction of their land, etc.

It was found that 75% of the affected families of the respondent did not migrate and continue to stay in the same locality and 25% of the respondents prefer to migrate to a new safer place. Most of the respondents did not migrate to a new place because of many reasons. Some people did not migrate because they didn't have any other option and few of them are very attached to their place and home as they have been living there for decades. Those people who are living in the disaster-prone area have compromised to the situation and risking their life by staying at the same place even after frequent landslide.

Intervention by the Administration

During any disaster, the intervention of the government is very important as it is their responsibility to look after the safety of the people. From the data, it can be seen that 72.9% of the respondents agreed that the administration intervenes them after the disaster. And 27.1% says the administration did not intervene or even visited them after the landslide disaster. It is very important for the administration to act swiftly in times of calamities and also to help the communities in anyways. On the other hand, 62.1% of the respondents did not receive any kind of relief after the disaster. And 30% of the respondents received some kind of relief in terms of financial, in kind or a helping hand after a week or so. 5.7% received relief within a week, and only 2.1% received relief within 2 days. Intervention doesn't mean only visiting the disaster site and vanishing from the scene thereafter but more important thing is to deal with the ongoing problems along with the communities.

One of the officials mentioned that they carry out geological investigation and recommend stakeholders/government to adopt remedial measures and minimize risk as per the report provided. The officials say that poor drainage system is also one of the main contributors to landslide as well as encroachment of land. The officials even blame that there is no proper land use pattern and everyone build or construct the house according to their convenience. They say that they can reduce the risk by creating preparedness; through awareness among the public, campaign, media, local channels, billboards, mock drills, precautionary measures of stone quarrying and deforestation.

In the past few years, the official says the landslide trends have increased with the passage of time. They say it's mainly because of nature of land as well as human error. Human errors are like inappropriate land use planning, poor and unplanned drainage, development without planning, indifferent attitudes of the public, stone quarry, deforestation, loose soil and lack of preventive measures. And some are because of nature like excessive rainfall, weak soil, slope, undulated land and climate change. But they say somewhere it all can be minimized if we take proper preventive measures. The officials agreed that there were no proper safety measures taken in the past due to lack of proper initiative taken by the public as well as the government.

The community people said the administration responded swiftly only the area or place like the main road or the highway because in this condition if they don't act swiftly the public will complain to the higher authority and they will end up in big troubles and whenever it happens in some locality, they are less bothered about it and it is mostly the locality people who clean the mud and all. Many of the respondents don't have much expectation of any kind of help from the administration because in the past many years they didn't receive anything from them except with some false promises. The swift intervention of the government is very necessary in times of any calamities because it is their duty to help and support the people in times of trouble. It is very saddening to see that the administration is less bothered to provide any kind of rehabilitation or counselling to the affected communities after the disaster.

There are countless NGOs working in the state and district level but none of the organization work on disaster issues on both the districts. Majority of the relief or help provider were from church associations and student union. Many respondents expressed that they had received financial help from the church member and also relief in kind like rice, vegetables and utensils; the church associations help all the people who are in need of help through relief in kind and also with prayers. The Nagaland is a Christianity state, and it plays a very important role in the state by helping people and motivating the people to help each other in times of trouble. They don't have any specific criteria or curriculum to help the people but whenever it is necessary, they help the people in needs.

82.1% didn't approach the government for any kind of help or relief. And 17.9% approached the government for assistance after the disaster. Most of the respondents didn't approach the government because they were not aware about the facilities that are available for them to claim. Many of the respondents have accepted their loss as their destiny and didn't approach anyone for assistance. The sad part is even those people who approach the government for help didn't receive much or for few anything. One of the mean reasons for not approaching is lack of awareness of the facilities available for them, and there is no one to guide them.

30.7% has insurance and as many as 69.3% of them didn't have any kind of insurance. Most of the respondents know the importance and value of maintaining the insurance but only a few could maintain it. It is not easy for a single-job holder in a family to insure themselves or the house because they are more concern for their survival and their carrier of their children. Financial problem was also one of the major constraints for not maintaining any insurance or keeping their house insured. Many respondents feel the importance of maintaining insurance but many of their income is very limited and they also did not want to compromise on the education of their children so they prefer to send their children to good school instead of saving some money or maintaining insurance. Many of the people here are dependent on limited salary, and the flow of money is very limited to many families so it becomes a big deal to maintain any kind of insurance.

Efforts to Prevent Landslide

It was pretty clear that most of the residents were aware about the landslide disaster and its impact on their life. Slowly, the people have started taking initiative to prevent its occurrence to some extent and also to minimize its impact on them and stay safe from the landslide disaster. Many of the respondent built wall near the drainage to protect themselves from heavy flow and clean the drainage from time to time voluntarily because they realized that if they don't clean it on time, it will be of their own loss and they cannot rely on the municipal council to clean them. They said the municipal people never come to clean the drainage and whenever there is any disaster, they started pointing fingers on each other and ultimately they will blame the residents for whatever has happened. The respondents said that in anyways they are always at the receiving end. Those respondents who could not effort to build wall have put tarpaulin to cover the land in order to save from the rain because when the soil becomes weak there are more chances of mud flow. Few of the respondents agreed that they have started planting trees in their colony to keep the soil intact. In one locality, all the residents were ordered to plant more than 2 trees from each family by the chairman of the locality and every year from time to time they organize community social work to clean the locality and the drainage since a massive disaster in 2005 and also they impose fine to any residents who throw garbage in the drain especially plastic bags. Many of the respondents are doing their best ability to minimize the landslide disaster or its impact on their life.

Discussion and Conclusions

From the above findings and observation, the present study can summarize that most of the respondents are living in their own house and very less respondents are living as a tenant. The occupants who are living on the prone area are well aware about landslide hazard and its associated risk. From both the districts, people living in their home do not easily migrate to a safer place as compared to those people living in the rented house. Most of the respondents living in their own house do not prefer to migrate because for decades they have been living there at the same time they are attached to their house and locality and for some people it is associated with poverty as well because many of the respondents were compelled to live there as they were not financially sound and have accepted the landslide as part of their life and take it lightly. Many people still believe that it is 'Gods will' and no one can stop it. Literacy also plays an important role in people understanding things.

Majority of the residents are aware about the landslide disaster because they are living everyday with it and the frequency of the landslide disaster is high. As mentioned before, 95% are aware about the landslide disaster, but when asked about the consequences of the landslide disaster, they are less concern about landslide impact on their health. The people are aware only about the basic of the disaster and how it happens. Majority of the respondents are not aware about their rights to approach the government for help or compensation nor do they have much knowledge about Do's and Don'ts during and after disaster.

Another major reason for people migrating to the urban place and residing in the prone area is the result of 'push' and 'pull factor'. Many people in villages have fewer opportunities for their growth, and their employment is very limited so they migrate to the urban area in search of a better livelihood and opportunity. Unlike the rural area, in urban area there are more scopes and opportunities for people like better education and better life with more opportunity to start up a business. Roadside settlement is also another major vulnerable to landslide disaster because many respondents who are living in the roadside and near the drainage are more exposed to landslide disaster and experienced frequent incident of disaster. At the same time, majority of the respondents were not happy with the government on disaster management mechanism.

The finding suggests that the occurrence of landslide disaster is very frequent especially during the monsoon time, and the trend is increasing with time and more massive and destructive. The government needs to act wisely involving the community people. On the other hand, according to the officials, the land owner does not follow the rules of the government while constructing any house. For this, the government needs to come up with strict roles to be implemented for every citizen of the state in order to tackle the disaster.

The most concerning issues in the state need to involve the local communities by providing awareness programs, training by the governments and also involving the NGOs and CBOs to work together with the government on disaster-related issues. Another major concern is that the State Government is not providing any kind of counselling or rehabilitation facilities to the disaster-affected people victims. Counselling is very important to those people who come out of the trauma-like disaster because at this time they are very much devastated and scared as they have experienced the pain and loss at the same time.

A more systematic effort is needed to reach the disaster-affected communities and identify the disaster risk level in the state. Making disaster factors more visible creates incentives for acting on the root causes of disaster rather than the consequences. It is important to reach the grass root level and provide awareness program and training to the mass communities.

REFERENCES

Anbalagan, R., Singh, B., Chakraborty, D., & Kohli, A. (2007). *A Field Manual for Landslide Investigations*. New Delhi: DST.

Avasthy, R. K. (2006). Incidence of Landslides: An Essential Ingredient for Landslides Hazard Zonation in North Eastern India. In R. K. Avasthy, B. Singh, & R. Sivakumar (Eds.), *Landslides: A Perception and Initiatives of DST* (7–14). Kolkata: Indian Society of Engineering Geology.

Benjamin, L. (2005). Disaster Risk Management in Southeast Asia: A Development Approach. *ASEAN Economic Bulletin, 22*(2), 229–239.

UN/ISDR. (2004). *Living with Risk: A Global Review of Disaster Reduction Initiatives*. Genva: *Inter-Agency Sceretariant of the International Strategy for*

Disaster Reduction. Retrived from http://www.unisdr.org/files/657_lwr1. pdf.

Varnes, D. J., & Geology., I. A. of E. (1984). *Landslide Hazard Zonation: A Review of Principles and Practice*. Paris: UNESCO. Retrieved from http:// unesdoc.unesco.org/images/0006/000630/063038EB.pdf.

Risk Perception and Disaster Preparedness: A Case Study of Noney, Tamenglong District

Homolata Borah

Introduction

The State of Manipur, the '*Jeweled Land*', is susceptible to disasters like earthquakes, landslides, floods, torrential rains, fire and accidents, which have over a period of time exposed the vulnerable economic, social, political, administrative profile of the state posing question on the disaster management system. The Northeast region is backward and neglected, grapples national attention burying mysteries of several disasters that have wrecked the socio-economic structures demanding national and international cooperation to manage identified gaps and future stresses. The modern age and the benefit of technologies have not been able to percolate into the region uniformly. The pillars of Sendai Framework for Disaster Risk Reduction (SFDRR 2015–2030) was introduced to tackle disaster risk drivers, aimed to reduce mortality, protect livelihoods by understanding risks, making investments in disaster risk reduction, building better governance to build back better

---

H. Borah (✉)
Centre for the Study of Regional Development, Jawaharlal Nehru University, New Delhi, Delhi, India

© The Author(s) 2018
A. Singh et al. (eds.), *Development and Disaster Management*,
https://doi.org/10.1007/978-981-10-8485-0_25

373

which requires active cordination for creating a sustainable society. The initiation of Disaster Management Act (DM Act), 2005, has also seen a change in the process of synthesis and implementation which has ushered in contextualizing the complexities and loopholes in the earlier existing standard procedure. But still most states remain where they were. The change in the structuring in terms of administrative set-up pertaining to the mandate of the DM Act, 2005, has been existent, but the questions on efficient delivery require intervention by different actors.

Manipur overall has been subjected to several events of earthquake and also other natural hazards like floods, landslides and forest fires. The Manipur State Disaster Management Plan Volume 1 suggests building communities to respond to disaster in a planned manner targeted to reduce losses. The state also has laid mitigation and prevention practices to be followed through Information Education Communication (IEC), training programmes, mock drills and also to build capacity of the stakeholders. Much of the implementation processes have featured slippages in implementation which are necessary to be carried forward to fulfil commitments (Table 1).

The action plan as laid down in the State Disaster Management Plan Volume 2 calls for five different actions inclusive of declaring earthquake as a disaster, institutional mechanism to respond, efficient trigger mechanism, response mechanisms of different departments and provisions of relief to be distributed in the disaster-affected areas.

Besides the recognition and institutionalizing in line roles and responsibilities, it is crucial to assess timely implementation addressing disaster risks.

Risk perception is essentially studied to structuralize efficient communication strategies to cope with the occurrences of disasters. The idea of risk perception is largely controlled by numerous socio-economic and cultural determinants, and its study can be utilized to analyse its effects on the implementation of disaster risk reduction methods. The paper aims to understand the effect on preparedness mechanisms for disaster mitigation and the perception of risk and decision-making during and after the process of natural disaster in Noney village in Tamenglong district, Manipur. Revelations in terms of assessment of the risk of natural disaster and the causes comprising the core idea of risk perception can be unveiled.

Numerous factors like education, social network, interpersonal trust, financial security and employment are studied in association with the perception of risk and its increasing or decreasing impact in terms of acceptance of preparedness and mitigation strategies. From a policy

Table 1 To show the various episodes of earthquake in Manipur (*Source* Manipur, NIDM, National disaster risk reduction portal) http://asc-india.org/seismi/seis-manipur.htm

Major earthquake events of Manipur		
Date	*Earthquake intensity*	*Description*
15 April 1992	Earthquake Mb 6.3	NW of Mawlaik, Chin Division (Indo-Myanmar Border region), 01:32:11.0 UTC, 24.2680°N, 94.9275°E, 130.90 kms depth
18 September 2005	Earthquake Mb = 5.7	A moderate earthquake at Myanmar–Manipur border, 24.653°N, 94.807°E, D = 82 kms, OT = 07:26:00 at 12:56 IST causing isolated minor damage to property in some parts of Manipur. The earthquake was felt at many places in Northeast India and Bangladesh as well as in tall buildings in northern Thailand
4 September 2009	Earthquake	Myanmar–Manipur border, Mw 5.924.381°N, 94.712°E, D = 97.6 kms, OT = 19:51:03 UTC. A moderate earthquake struck the Myanmar–Manipur border, at 01:21IST. It was felt widely in Northeast India and in Bangladesh.
4 January 2016	Earthquake 6.7	Epicentre—Noney, Tamenglong district

perspective, it stands crucial to understand how the community perceives risks and hazards and also what determines its outcomes as people's preparedness in response; recovery is controlled by risk perception along with other important factors.

The paper tries to answer why certain people are opposed to the vulnerabilities of disaster and some are indifferent? And also tries to explore the possible solutions.

The paper attempts to observe risk perception before and after the occurrence of the earthquake in relation to the preparedness strategies as laid down and practised by the government and to find associations

376   H. BORAH

if preparedness and mitigation strategies are available in a high risk perceived situation.

Contextualizing Risk Perception in Disaster

The level of preparedness and mitigation strategies coherently aiming at building community resilience, reducing vulnerabilities, developing adaptive capacity and reducing exposure is determined by a multitude of factors which act as determinants. Risk perception is one of the most crucial factors associated with conception of knowledge, information and behaviour which determines the level of preparedness, response and communication strategies. The process of perception of risk includes stages of collection of information, synthesis and constructing interpretations often influenced by observations and information from external sources. Risks are often perceived by impressions of knowledge, emotions, past experiences, values and attitudes. The mental constructions of algorithms are devised in coalescence with social and cultural milieu attenuated with endogenous agents per se communication media and interactions with the rest of the world like news and other information from peer groups. The risk perception model has been formulated in four different levels: level 1: heuristics of information processing, level 2: cognitive and effective factors, level 3: social and political institutions and level 4: the cultural background (Wachinger and Renn 2010). Each of these levels possesses manifestations which are embedded into the next level represented with interplay of the variables to gain accurate understanding of the process of perceiving risks.

The perceived risk during disaster carries with it the components of assessments of the natural hazard which has direct connotations with the physical environment, accessibility to road and transportation, distance from the actual location of the natural hazard and interpersonal trust, decision-making, etc. The inclusion of social dimension and the role of social scientist in the strategies of disaster risk reduction by exploring the segment of "risk perception" and understanding how such modulations can effectively supplement the study and question of "what affects preparedness" can be an essential part of disaster risk management.

A research study carried forward on the Beijing hailstorm in 12 July 2012 captures examples to elucidate the governmental initiatives in early warning systems and its connection with the human dimension contextualized therein. The findings of the study depicted less effectiveness of the governmental early warning systems when human behavioural attitudes were taken in perspective as they failed to assimilate the technicalities of the messages delivered. The findings also recorded people's difficulties in assessing the extent of the impact of the natural hazard. Disaster risk reduction and its initiatives have mostly drawn attention of technocrats. The emergencies have over the passage of time evolved the ever existing needs of operational interface with social scientist and technocrats to explore the expanded dimensions of social aspects.

'People are prisoners of their own experiences' (XuJianhua 2014) and likewise will create mental maps owing to past experiences to lay impact on preparedness mechanisms.

Noney Earthquake as the Event

The 4 January 2016 tremor shook the Northeast India and some parts of West Bengal in India. The epicentre of the high-intensity earthquake that was marked 6.7 on the Richter scale occurred in Noney in Tamenglong district in the Manipur State.

The earthquake tremors caused cracks causing havoc to both kutcha and pucca houses, and some documented shifting of the house from the base and mud patches falling away from the mud walls. The partially standing fragile structures were anticipated to fall with an aftershock or in case of strong winds or even rain. The earthquake tremors have caused cracks to develop in most of the houses, pucca and kutcha. Few of the walls have fallen down. The standing structure with partial damage is most unsafe and likely to get fully damaged with an aftershock or in case of rains. The sustained structures made the residents vulnerable to mosquito, snake bites, rains and storms.[1]

The records documented from field visit depict the event as 'horrible' and 'unexpected' by the people of the community. The people drowned in fear as they did not know what to do. They first turned to family and friends, NGOs and community for help and tried to reach government agencies and functionaries the last. It was found that the people have not received any training and they feel the need of such measures.

---

[1] Sitrep-7: Manipur Earthquake 14 January 2016.

The damages caused by the earthquake were spread over different districts like Senapati, Imphal West, Imphal East, Tamenglong, Thoubal and Bishnupur, and the worst affected districts remained Senapati, Imphal West and Tamenglong causing 8 deaths, 2100 houses damaged and more than 10,000 people displaced. The State Government and the humanitarian agencies have extended delivery of service to the affected populace. The State Government announced Rs. 500,000 assistance to the families affected and some financial support to those injured.

The 'risk conception' helps to understand and examine preparedness for hazards. Self-protective behaviour by residents of flood-prone urban areas can reduce monetary damage by 80% and reduce the need for public risk management (Grothman and Reusswig 2014). To delve into answering the questions of why certain residents take precautionary measures while others do not (Grothman and Reusswig 2014), the Protection and Motivation Theory (PMT) is developed elaborating private precautionary damage prevention by resident's perception by previous floods experience, risk of future floods and non-protective response. Often at times efforts have been made in the name of 'public good' without knowing how the public perceives disaster. As a result, public did not cooperate and large amount of resources are wasted, without assessing what the public perceives in a disaster (XuJianhua 2014).

RISK PERCEPTIONS AND DISASTER IN NONEY

Significant amount of literature in different disciplines like economics, geography, sociology, psychology and in disaster management has developed the understanding of the determinants affecting the evolution of perception of risks. Risk perception is determined by risk attitudes, risk communication, education, socio-economic background, interpersonal trust and past experiences, and the aspect of risk management stands crucial (Table 2).

The village Noney falls in the subdistrict of Nungba comprising a total of 3854 population with a total number of 635 households with 2140 male population and 1714 female population. The community is Scheduled Caste dominated with 2339 people and 5 persons belonging to the SC category; there are 2953 literates as against 901 illiterate population (census 2011). The percentage of main workers and marginal workers to total workers remain 48.59%.

Table 2 The socio-economic profile of Noney village, Manipur (*Source* Computed from census, 2011)

Socio-economic profile of Noney village, Tamenglong district, Manipur	
India, 2011	
Indicators	*Percentage*
Percentage of male population	56
Percentage of female population	44
Literacy rate	77
Female literacy rate	83
Male literacy rate	69
Sex ratio	801
Child sex ratio	861
Work participation rate	48.59

Through interactive sessions with a social activist of the village Noney, it was revealed that before the occurrence of the earthquake the level of preparedness was almost nil and even thereafter no capacity building exercises were conducted. The experience of the hazard has shaped the perception of the community people, whereby they feel the need of such training programme to avoid situations of panic and anxiety. The people are very strongly knit with each other as during the event they also turned to help each other and also to extend support. The records also reveal that they are aware of their poorly informed situation about disaster, but also agree to the capacity of the community to be resistant enough to bounce back in the post-disaster event with the help of local responders, friends, community people, NGOs and church.

The risk perception of the community is very close to reality based on past experiences and constant interpersonal communication and interactions backed by knowledge and community participation, due to the role of church and NGOs.

FINDINGS

Most of the community members believe to be moderately prepared for any disaster, as the 'bounce back' capacity for one disaster was manageable but frequent events will paralyse due to the losses and damages.

It was also found that lack of awareness about the high vulnerability and susceptibility of the region was not perceived as a risk factor until the occurrence of the event on 4 January 2016. The event has educated the people of the level of intensity and the damages that can potentially reoccur in the region.

The judgments, evaluation and experience of the community people have moulded 'risk perception' which appears to adhere to the belief of high risk of the village to the occurrence of a disaster event and hence calls for the urgent need of preparatory measures, interventions and institutional set-up by the government to offer concrete and comprehensive solutions.

COMPARING RISK PERCEPTION AND PREPAREDNESS AND VULNERABILITIES

Given the state of high susceptibility and vulnerability of the region falling in a high seismic zone, it is interesting to also look back into the history that narrates to us the stories of disaster and several occasions of disaster events. The risk perception of the villagers is not in tandem with the institutional arrangements. The perception of risk is much higher than what the disaster management set-up has offered. There is need for strong interface between the perceived risks as felt and the elements of professional and expertise efforts aimed towards disaster risk reduction to save a thousand lives by smart investments in building community resilience.

WHY DO WE NEED TO BE CONCERNED (ROLE OF A SOCIAL SCIENTIST)?

The role of the social scientist forms an important wing in the whole construct of the disaster management to cover under its umbrella the social aspect and the determinants that questions who is the most vulnerable and why approaches concerning the most vulnerable have to be sensitive; it is also very important to bring about the crucial and neglected factor of 'human touch' which is lacking in most solution models. But understanding societies in different spaces and times, estimates can be interfaced to realize understanding. This understanding has to be created and utilized further for enhancement and delivery of efficient services by inclusivity of social scientists.

The department of relief and disaster management of the Manipur State has documented two volumes of Manipur State Disaster Management Plan. The plans standardize the institutional management of disaster events. But creating spaces in planning for answers pertaining to perception of risks to enhance community's willingness to participate in preparation mechanism and taking responsive actions calls for new initiatives and a change in the traditional ways of dealing with disasters. The socio-economic, behavioural and educational limitations can be addressed through new dimensions of plans and involvement which will touch upon several slippery areas including risk perception.

BIBLIOGRAPHY

Bewer, N., Weisnstein, N., & Cuite, L. (2004). Risk Perception and Their Relation to Risk Behavior. *The Society of Behavioral Medicine.* 125–130.

Government of Manipur. (2013). *Manipur State Disaster Management Plan.* Available at: http://manipur.gov.in/wp-content/uploads/2013/02/manipur-sdmp-061113-vol1.pdf.

https://link.springer.com/content/pdf/10.1007%2Fs11069-005-8604-6.pdf.

http://www.teriin.org/index.php?option=com_featurearticle&task=details&sid=960&Itemid=157. Accessed on 15 November 2016.

National Academic Press. Workshop Summary (1997). *Risk Communication and Vaccination.*Available at: https://www.nap.edu/read/5861/chapter/3.

NIDM. (2012). *Manipur State.* Available at: http://nidm.gov.in/PDF/DP/MANIPUR.PDF.

Wachinger, G., & Renn, O. (2010). Risk Perception and Natural Hazards. CapHaz-Net WP3 Report, DIALOGIK Non-Profit Institute for Communication and Cooperative Research, Stuttgart. Available at: http://caphaz-net.org/outcomes-results/CapHaz-Net_WP3_Risk-Perception.pdf.

Slovic, P. et al. (2005). Affect, risk and decision making. *American Psychological Association, 24*(4). S35–S40.

Studer, A. J. (2000). Vulnerability of infrastructure. *Zurich, 18.* April. 1–7.

UNISDR. *Sendai Framework for Disaster Risk Reduction, 2015–2030.* Available at: http://www.unisdr.org/we/inform/publications/43291.

Usuzawa, M. et al. (2014). Awareness of disaster reduction frameworks and risk perception of natural disaster: A questionnaire survey among Philippine and Indonesian health care personnel and public health students. *Tohoku Journal Experimental Medicine, 233,* 43–48.

Institutions of Faith in Resilience Building During Disasters: Church and Communities in Lamka Vis-à-Vis Churachandpur

Priyanka Jha and G. V. C. Naidu

INTRODUCTION

Institutions of faith provide support where the state fails. Church in Churachandpur is an example of sustaining faith in times of crisis and destruction. Its role has only expanded as a security cover to local communities because the district continues to be the country's most impoverished[1] districts which is surviving since 2007 on the Backward Regions Grants Fund. Lamka vis-à-vis Churachandpur is ancient historic areas which have got one name[2] of Churachandpur around the Second World War. It is alleged by some local people that merging Lamka into Churachandpur was part of a strategy of the Meitei tribe to control the

[1] Ministry of Panchayati Raj Report (2006).
[2] Notes written by Col. Maxwell, Col. Woods and Col. Shakespear. 29 July 1937, File No: G.S. 2753 of 1940 Dilip K. Lahiri & Binal J. Dev: *Manipur: Culture and Politics*, 1987, pp. 111–112.

P. Jha (✉) · G. V. C. Naidu
Jawaharlal Nehru University, New Delhi, India

© The Author(s) 2018
A. Singh et al. (eds.), *Development and Disaster Management*,
https://doi.org/10.1007/978-981-10-8485-0_26

smaller tribes such as Kuki, Paitei and Zo. To worsen the situation, this district faces from some of the worst kinds of disasters like earthquakes, landslides and floods as it nests in the Zone V of Seismic India. Recurrent disasters have somehow made people resilient to these upheavals and help them sustain these difficult conditions. So with the minimum support and help coming in from the institutions of the state, the community over a period of time have started drawing help from community-based organizations such as the church in coping with disasters.

This paper attempts to look at how the poverty stricken, unemployed broken families overcome the trauma of disasters. This has been reworked and reshaped by bringing in institutions that are of non-state kind and takes the discourse of Disaster Risk Reduction (DRR) outside the realm of state and NGOs, pitching the role of religion and community-based organization in a big way.

COMMUNITIES BEYOND HYOGO

There is a marginal shift in the discourse on disaster risk reduction due to changes that have been brought about from the Hyogo Framework for Action 2005–2015: Building the Resilience of Nations and Communities to disasters[3] to the Sendai Framework,[4] which makes community a key player in the situations of disaster and attempts at building community resilience. The need to draw resources from within the communities implies direct involvement within institutions. So, the question that this paper seeks to raise is whether the church (since most people in the region under discussion are strong followers of Christian faith) can be perceived as a community institution in the building of community resilience for providing skills for mitigation and adaptation.

Church is a dominant reality in Churachandpur which is trusted as an institution of faith. This has led to the marginalization of other community institutions which had worked as a shadow state. However, in some societies, there has been an erosion of the role of the state and its institutions, and our field area of Lamka in Churachandpur district is a classic example of this phenomenon of erosion of state's legitimacy. Some of the reasons that explain the erosion of trust and legitimacy in the perception

[3] http://www.preventionweb.net/files/8720_summaryHFP20052015.pdf.
[4] http://www.preventionweb.net/files/44983_sendaiframeworkchart.pdf.

of the masses towards the government are as follows: First, the state has failed to deliver the key/basic infrastructural services such as electricity, roads and drinking water. Out of these, road infrastructure seems to be poorest and hence is a great concern in mitigating the effects of disasters. It can be seen that the condition of roads are deplorable is precisely because the state has failed to recognize the salience of road infrastructure. The national highway 39, the road that connects Imphal the capital of Manipur with Churachandpur, is the road that reveals the truth of the backwardness of this district, No question that, as people in the area have repeatedly underscored, the condition of this highway was really bad. This national highway connects to Khuga Dam, which took us about four hours, it became obvious that if in the normal times, the road is in such a bad condition, one can imagine the response time would escalate in times of a disaster if the road was badly damaged since this is the only connecting road.

Ethnic and community disharmony weakens resilience in this region. Identities which have made this north-eastern state a site of insurgency and disharmony. The literature on the problem of insurgency and ethnic clash is manifold, but the ground is still germane to know how this impacts the relation between the citizen and the state during disasters. The problem of insurgency caused by ethnic identities has had a huge impact on the paradigm of development in Manipur as for many years this state has suffered from this problem.

As a result, the position of the state as welfare and service provider has weakened with the insurgent groups emerging as key actors with the control of the resources making them pivotal in decision-making forums. The infrastructural capacities of the state have been deeply affected by these tensions which make conditions extremely unstable. The state is inhabited by different tribes like Paitei, Simte, Zou, Ghangte, Thanka and Vaiphei who rally around different kinds of identity demanding separate administrative units and in some cases separate statehood. Now, this complex situation between the electorally elected representatives and people chosen by insurgents is creating a situation whereby the state seems to be losing out what it is meant to do. As a result, the need to look at the role of other non-state actors, who have replaced the state in providing the services that the state should have offered, need greater focus.

ROLE OF CHURCH

Christianity entered Manipur through the routes of colonial Christianity as missionaries played a key role in establishing it. The onset started in the early nineteenth century through the works of Americans, and Welsh missionaries consolidated by the early twentieth century. Christianity made inroads in the tribal communities whereby Nagas, Kukis, Paitei and other tribal groups took to Christianity. According to the 2011 census, Christianity is the second largest religion in Manipur. The Protestants outnumber the Catholics. Some of the important denominations are as follows: the Manipur Baptist Convention (MBC), Kuki Baptist Convention (KBC), Kuki Christian Church, Reformed Presbyterian Church, The Presbyterian Church in India, Church of Christ and the Seventh-day Adventist Church.

Over the period of time, the Church has succeeded in establishing itself as a key player in both normal and abnormal times. There are many reasons that could be attributed to the success of church's functioning in situations of disaster. Some of them are as follows: First, Church emerges as a true community organization as it is centrally located in the working and functioning of the community. There are other community-based organizations that are working but the kind of legitimacy that church has amongst the people, it appears to easily exert its influence and control. Church has enormous control in the lives of the people and is seen as institution of social capital and trust. There are many associations that work on the guidance of the church.

Second, since religion is a very important source of the self and also in bringing collective conscience, one needs to understand and locates its role in creating consciousness towards safety and community resilience. This can be understood from the fact that the attendance of Sunday masses is very high. This is also the time that the community members get to meet each other. The Sunday masses are very attended depending on one's denomination. The Churches witness huge turnout, PCI (R) has 400–500 people attending while Salvation Army has 700–1000 people coming in while EBCC has around 1000 people attending since it is one of the biggest churches in this region. Sunday masses play a very crucial role in sending out messages it wants to the community. The values and morals that need to be sent out are done during this process. The church, as a result, plays the role of educator and litigator as part of disaster risk reduction and as a facilitator in post-disaster relief operations. The Sendai Framework emphasizes the role of community-based organizations in the processes of adaptation and mitigation. In a true sense, the Sunday masses can be used as the platform where information regarding disaster mitigation and adaptation can be provided. Church is a crucial institution as it transforms itself from being religious entity to that of a social cohesive being as it provides an array of services which are of material, physical and spiritual kind. This could be explained as it is an institution that succeeds in generating trust in a context where the other state institutions have failed. Trust and social capital become the defining reasons for the legitimating of its role and place in the eyes of the community.

CHURCH AS AN EPITOME OF TRUST

One has to understand that Church has succeeded in establishing itself as a legitimate institution in the region because it has the faith and trust of the people as the realization that the state and the government have failed miserably in providing and up keeping the responsibilities which were promised when voted to power. Marred by non-service, corruption, failure to provide in time bound manner empowered institutions of community with Church being one of them to take up these responsibilities especially during the times of disasters. This could understand better by locating the reasons as to how church succeeded in establishing itself as an institution of trust and faith. The Churches in Lamka, Churachandpur, are Vengnuam EBCC, New Lamka Presbyterian Church, Lamka Presbyterian Church, Church of Christ, Hebron Church these have followers depending on one's denomination.

First, Churches of various denominations are well equipped with resources and infrastructure to house assemblies and have a network that could cater to all especially the most marginalized like women, children and old people in pain, distress and abandonment. Theses Churches run many orphanages which help the destitute children in their education and skill development. Many of the women are trained and skilled in different vocations empowering them to have their own enterprises. Wete, a brand that sells bags, shawls, wallets and other items, was the initiative of EBC, Lamka CCpur which provided employment opportunity as well as skill development in weaving using indigenous patterns and style which has brought attention to Paitei weave.

Second, Churches have regular assemblies like the Sunday masses as mentioned earlier and use this platform to encourage people to bond with each other as instil the feeling of commonness in a context like theirs which is marred with ethnic tensions. The state has completely failed to bond in a manner which a church did. Third, the Church has certain solutions to provide and which is well accepted even if it is religious or utopian such as get together and pray for deliverance and almighty will send help. This spiritual service can work wonderfully in instilling will in the lives of people when nothing seems to work out. One witnesses in Manipur that the government have been unable to instil the kind of political will and vigour that the church has been working towards. And this explains that instead of the state and state institutions, it is the church that fathoms the trust and cooperation of the masses.

CHURCH AS AN INSTITUTION OF THE COMMUNITY

Church is central in instilling the sense of commonness and belonging in the lives of people. Symbolically it becomes a public space where people interact with each other. The webs of significance are drawn by church in the community. Also as a unifier of community, the individual loss and sorrow is transformed as the distress of the community instilling the ethics of what we owe to others especially in the times of disaster. Church formalizes the communitarian ethics which instils sensitivity in the discourse and situation of a disaster making it humanistic. The top-down statist and technical approach to disaster is inserted into one which is now emanating from within organically which saves itself from being remote to the needs of the community. The community in alliance with church works with indigenous knowledge making use of locally available resources.

The church had been instrumental in establishing number of community-centric institutions like Youth Paite Association, Young Mizo Association, Hmar Youth Association, Chin Youth Association and Simte Youth Organization which have been helped by the church and would provide counselling and support to the young upholding the Christian ethics and principles of love, peace and brotherhood. These associations play a crucial role in student counselling and supporting the educational needs. Zomi Human Rights Foundation (ZHRF) another organization has been working closely with EBC and Central Lamka Baptist Church in running programmes for Youth by organizing seminars, talks and conventions motivating them. One fact that stems out is that the role and responsibilities that the state should be undertaking are now replaced by such youth and philanthropic organizations.

The communitarian ethics surfaces very strongly during the times of disasters as noticed in the context of earthquake of January 2016. The institution of Church is trusted by communities because they are treated as a reliable institution when the state fails to deliver services and securities of various kinds. In the same manner, it can be compounded that it was the communities with the central location of the church emerged as the first respondents instead of the state. The communal bonds and norms existing through the age-old customs and traditions inherent in the society and formalized by the Church allowed the community to come forward in times of needs and difficulty notwithstanding joy. The communal harmony and service delivery in colloquial term are referred to as 'Tlawmngihna' which in English refers to Altruism on which many of the youth organizations are grounded upon, one such organization is Young Paite Association.

CHURCH AS AN EPITOME OF SERVICE

The Church thus becomes a core and nodal agency in terms of reaching out to the masses in Manipur. It has been recognized that before the entry or action by the state institutions in a situation of disaster, it is the church that reaches people as it is located at the heart of the community and society. This reaching out could be in the form of being the physical structure or building or in the form of the organization that provides the relief aid and material. In many disaster-hit areas, it has been recognized that church is the sole brick and concrete building which acts as the emergency centre in case of flood and/or earthquake. As far as its role of relief provider is concerned, it was found out that almost all churches in Churachandpur have their relief arm which provides the people with ration like wheat and rice.[5] In the course of field visit, it was found that each church has relief fund at the local, district or division level, and in

---

[5] This information was provided by Rev. Dr. Luaichinthang who is the Pastor of Bungmual EBC and worked very closely with the people during the Earthquake of January 2015.

some cases also at the level of headquarters. This implies that the coordination and organization of churches are well established.

In the course of field visit, it was also found out that the church succeeded in generating some amount of money which was allocated towards providing relief to the poorest. Indeed, it is noted globally that in situations of disasters a large number of religion-based, in particular church, organizations worked very rigorously on the field in relief activities. One such organization that has been working in 127 countries is the Salvation Army, which received attention during the course of rehabilitation and reconstruction of Haiti earthquake. Important information that was shared by the pastor of the EBC situated in Rayborn College, Lamka was that the church from time to time sent money to disaster-struck areas outside Manipur in India, as the larger spirit that guides the church is brotherhood and sisterhood. The Church was also involved in the acts of reconstruction and redevelopment like road repair, roadside drainage and digging of wells. It was informed that activities of similar kinds have been undertaken by the church with the active support and participation of the community since 2008. As the state had abandoned many of similar tasks, the church took the onus of finishing these activities on itself and has been working on it. Whether floods or earthquakes, Church is one institution that the community relies on.

One of the key ideas that emanated from the interview with Mr. Khenpi Tombing, the pastor of EBC, is with respect to the concerns about mental health that get missed out during the times of disasters. During the earthquake disaster of January 2015, many people came to the church seeking emotional and spiritual assistance and more than anything else what they seeked was emotional comfort. 'To see the loved ones dead, could be the most difficult emotion to pass'.[6] It is interesting to note that apart from the material, physical and infrastructural support that the Church has been catering to, it has enamoured the task of supporting the masses emotionally which makes it a true institution of community resilience. The church is very forthcoming in disseminating information regarding disaster risk reduction. In future, they are very keen to participate in awareness creation and training. They have attempted to undertake capacity building as at their own levels and are working towards making the community resilient. In the interview,

[6] Interview with Reverend Tombing undertaken on 6 June 2016 in Lamka, Churachandpur, Manipur.

it was indicated their keenness of working with the government in making disaster risk reduction a reality. Making people aware of what is expected thus training is certainly imperative.

Conclusion

Manipur gives us an example as a model of best practice which can be followed since churches have played a very important role in community resilience. As the missing link between the state and the community, the role of church needs to be discussed in greater detail. The key understanding that this paper has arrived at is that the church as a community-based organization has the capacity of drawing the mass base and support. The positive ground that it gathers for its success can be drawn from the fact that, as an institution of trust and social capital, it can effortlessly direct the communities to navigate the paths of disaster risk reduction that the state has not been able to do so.

However, one cannot deny that there are some problems in the logic that they give to people with regard to the cause and origin of disaster, as it was found out that in the course of interviews with the pastor that the reason that was given to the people with regard to earthquake was God's Fury. For many, it might appear as illogical and superstitious or pushing the people to non-reason, but one needs to recognize that people are religious and take the idea of 'god' very seriously. So, one should not keep religion out in the discourse of disaster but bring it and its institutions on board. If risk reduction can be worded in a language that people understand, it works. Eventually, these policies and ideas are for the people, and if people are embedded in the idioms of religion, then these are key and central to our discourse.

The Sendai Framework uses the category of 'Resilience' which means the capacity to stand against the disaster, stand against fear. It also means the capacity to withhold has the courage to deal with. Religion and its institutions train and teach the same ideas. The point is to bring the very foundation into the discourse of DRR. It makes one wonder through the course of the fieldwork how is it that these people who are constantly reeling under so many disasters, both man-made and natural, ethnic clashes, floods and earthquakes derive their strength and resilience from. The question is how is it that they do not break down and how is it that they 'smile in distress'. It seems that when the roots are strong, one should not fear the wind. But they cannot be left to deal and fight with the wind, instead the institutions that carry them should be taken seriously and brought in the framework of DRR.

REFERENCES

Dev, J. Bimal, and Lahiri, Dilip. (1987). *Manipur: Culture and Politics.* Delhi: Mittal Publications.

Information gathered here is on the basis of first-hand experience gathered in the course of field work undertaken in April 2016 in Lamka, Churachandpur, Manipur and telephonic conversation with the Pastors of different denominations. This article is based on ethnographic insights, detail, observations and interviews. The methodology adopted was qualitative in nature.

Ministry of Panchayati Raj Report. (2006).

http://www.preventionweb.net/files/8720_summaryHFP20052015.pdf.

http://www.preventionweb.net/files/44983_sendaiframeworkchart.pdf.

Relating Resource Extraction Pattern with Disasters: Political Ecology of Barbhag Block, Nalbari

Snehashish Mitra

INTRODUCTION

India's North-Eastern Region is a frontier for the thrust eastward, both in terms of political imagination and logistical implications. Comprising of eight states, the region shares border with Nepal, Bhutan, Bangladesh, China and Myanmar. Over the decades since India's independence, the North-Eastern Region has been deemed as underdeveloped, backward and strife torn with ethno-militant violence. The region is one of the six biodiversity hot spots of the world, given its possession of abundant natural resources. The resources have provided sustenance to both indigenous and migrant communities, while serving as objects of extraction

---

The paper is a partial outcome of the project 'The Indian Underbelly: Marginalisation, Migration and State Intervention in the Periphery' supported by Stockholm University. I am indebted to Sanjay Barbora for his supervision and the members of the Gramya Vikash Mancha, Nalbari, for their constant support.

S. Mitra (✉)
School of Social Sciences, National Institute of Advanced Studies, Bengaluru, India

© The Author(s) 2018   393
A. Singh et al. (eds.), *Development and Disaster Management*,
https://doi.org/10.1007/978-981-10-8485-0_27

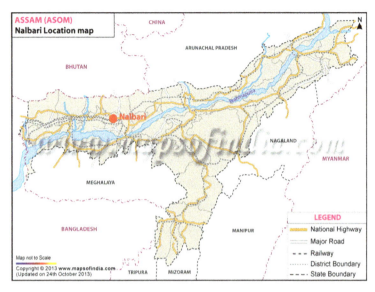

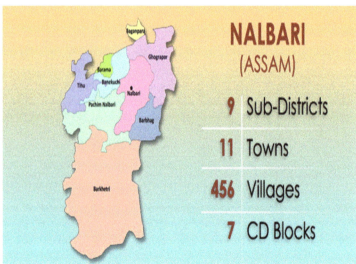

Fig. 1 Locating Barbhag Block, Nalbari district (Top: *Source* http://www.mapsofindia.com/india/where-is-nalbari.html. Accessed December 8, 2016) (Bottom: *Source* http://www.districtnalbari.com/Index.aspx. Accessed December 9, 2016)

for British and Indian States. Tea, timber and crude oil were the main resources which went into colonial accumulation, while post-independence focus has gradually moved to additional resources like water (for hydroelectricity), uranium, coal, rubber plantation and jatropha plantation. Such activities have serious implications on the use and extraction of resources of the region, which involves alteration of environmental landscapes and brings in the concept of disaster by causing environmental calamity such as landslide, floods and earthquakes. Northeast India is a high seismic zone area, and therein it is essential to factor in the nature of resource exploitation that holds a direct impact on the cause and probability of disasters affecting especially the marginalized and vulnerable communities of the region (Fig. 1).

Nalbari district is located in western Assam. The district falls between the hills of Bhutan and the Brahmaputra River. Nalbari has numerous water channels criss-crossing through its landscape as water from the hills moves downstream to join Brahmaputra River in form of streams and rivers. This has led to creation of large number water bodies in Nalbari which have enabled agriculture and sustained the '*machmorias*' or the fishing community in the district. Agriculture is the mainstay of major section of the populace. Proximity to Guwahati city (around 70 km from Nalbari town) and early spread of education had created a working class involved in the governmental and non-governmental service sectors.

The Barbhag block in Nalbari has been a significant area in terms of history. Etymology of Barbhag suggests the prosperity of the region, '*Bar*' means big, while '*bhag*' means share—as during the Ahom[1] rule the region used to pay a high tax due to the prosperous yield from agriculture and fishery. The prosperity of the region was also responsible for the stiff resistance of the British taxation policy. During the British period, farmers of the area regularly revolted against the tax impositions. According to a report published in Hindoo Patriot on 5 February 1894 (Guha 2014), British Government found it difficult to collect land revenue in *tehsils*[2] of Patidarang, Nalbari, Barama, Bajali and Upar Barbhag (present Barbhag). Farmers of Barbhag region organized Raiz Mel (people's convention) against the British Government, which indicates the

---

[1]Ahom kingdom ruled large parts of Brahmaputra Valley for around 600 years, 1228–1826.

[2]A colloquial term for an administrative unit.

396   S. MITRA

strong control of the farmers over local politics in those days. In continuance with the rebellious attitude during the colonial period, Nalbari was a hub of United Liberation Front of Assam (ULFA)[3] since the 1970s. Under the influence of ULFA, Nalbari turned into a liberated zone by the late nineties

Analysis of the current situation in Barbhag will remain inadequate unless the environmental history of the region is taken into account. Works of David Arnold (1993) and Jean Dreze (1988) have shown how the different grids of power surrounding management of environmental resources have been crucial in influencing the relation between ecology, politics and survival in colonial India. Ranabir Samaddar (1998) in his work on the *Junglemahals* in West Bengal and Jharkhand has investigated the politics of community formation and how environmental happenings would churn the different kinds of social reaction through unrest, revolt and migration. Samaddar emphasizes on the post-colonial senses to understand the new kind of biopolitics and biopower:

> It will help (a) bring back the issue of the colonial dynamics, which continues of course with changes; (b) point out in this context how a new science of governance tries to make sense of the phenomenon, (c) and, understand how the migrant, through the act of crossing borders and boundaries (borders of hunger, starvation, death, and life; of places and countries; of stations in life and occupations; finally borders of positions in the discrete map of division of labour) copes with a system that teams up with nature's calamities to turn millions into perishable lives.

While management of disaster is a major task of governance, especially in Northeast India, which frequently experiences with earthquakes and floods, it is also necessary to interrogate the mechanisms of governance, which creates human-induced disaster. Therein, the nature of economics, i.e. neoliberal economic policy, needs to be unravelled as it would give a clearer picture of how resources such as land are viewed by planners and how does the local communities in tandem with grassroot organizations adapt by either migration or reinventing and reconfiguring patterns of resource extraction for sustenance.

---

[3] ULFA was formed to carry out an armed struggle against India to form a socialist Assam.

Barhbag's Trajectory

The network of water bodies in Barbhag block comprises of the rivers named Baralia and Pagladia, which flow across the eastern and western border of the Barbhag block. There are also several lakes in the area locally known as '*beels*'. The hydrological cycle of the rivers has been majorly influencing the livelihood pattern of Barbhag block, hinging around the flood trajectory in the area. In 1950, when Assam was struck with earthquake, the nature of water bodies including that of Brahmaputra River changed drastically as the river bed rose due to massive siltation. The first devastating floods occurred after the great earthquake in the area between rivers Pagladia and Baralia, in Barbhag, in the then Kamrup district. From 1951 onwards, floods continued to wreak havoc without any let-up. By the end of the seventies, it had taken over entire Assam.[4] As a response towards the flood problem, the governing institutions of both the State and Central Government have taken several steps, but those are yet to offer respite to the people of Assam living with the rivers and water bodies. The flood issue in Barbhag was managed by constructing embankments on both sides of the Baralia and Pagladia River despite the opposition of the local people. The local people however construed such step as a misjudged one as they thought rather than containing the water, it should be allowed to flow towards the Brahmaputra river. The concern, however, went unheeded. The state planners considered oppositions of local communities as irrational much like in other similar instances, for example the resistance against imposition of control over commons in Africa was viewed as irrationalism by an ecologically destructive and ignorant native population (Grove 1990).

For the first few years, things remained unchanged; however, with time, the embankments suffered infringement due to lack of maintenance. Again in 1990, the Baralia River was channelled southwards towards Hajo which made the river seasonal from perennial in the stretch in Barbukia village and left the area under the whims of Pagladia River. During the monsoon, when Brahmaputra would exceed its carrying capacity, it would push back sand-filled water through Pagladia and flood the villages under Barbhag blocks, and this would lead to two significant outcomes:

---

[4] http://www.telegraphindia.com/1020918/asp/northeast/story_1207577.asp. Accessed May 31, 2016.

398 S. MITRA

1. The embankments hindered the free flowing of the flood water and resulted in inundation of farmlands within Barbhag.
2. The sand-filled water of Pagladia eventually filled up the water bodies, which severely compromised the livelihood of *machmorias*. As the sands increased the height of water beds, it further aggravated the flood problems by delaying the water passage during monsoon.

These environmental changes triggered several livelihood challenges by putting a curtain of uncertainty over the yields from the agricultural fields and the water bodies. According to a study by Swarupjyoti Baishya,[5] the Khoponikuchi village was inundated for 20 days in 2012 and incurred a loss of Rs. 1,00,000 in terms of agricultural lands and cattles. Due to the flood, 30 school days and 40 'man days' were lost. Baishya cites sediment discharge, destruction of wetlands, expanding human settlements and partly managed embankments as the main factors behind the persistence of flood inflicted misery on the local communities.

The current Nalbari district administration has come up with a 'Flood Response Plan'[6] elaborating on the different mechanisms to tackle flood hazard. Focus of the plan is on technical expertise to prevent flood and alert the local people in case of an imminent flood. Land use pattern is supposed to be restricted around the water bodies, which are envisaged to be 'water retention basins'. Nalbari is covered under the 'Flood Early Warning System' (FLEWS) project where the district would be alerted for an upcoming flood situation by North-Eastern Space Applications Centre, Shillong, with 6–7 hours of lead time. The plan rightly points out that the flood situation in Nalbari depends jointly on the rainfall in Bhutan and water level of Brahmaputra in Upper Assam; flow of water from both northern and eastern side locates Nalbari in a vulnerable position. The document accepts that the embankments maintained poorly and are responsible for aggravating Nalbari's flood issue. Apart from the citing the formal and mechanized responses to the flood issue, the plan document brings in the importance of local sociopolitical organization like NGOs, women's organization and students'

---

[5] http://www.ijirset.com/upload/september/12_Vulnerability.pdf. Accessed May 31, 2016.

[6] http://nalbari.nic.in/files/Flood%20_Response_Plan.pdf. Accessed June 1, 2016.

organization with regard to volunteering and reaching out to affected peoples during the floods. Due regards to local opinion would perhaps undo the past mistakes of implementing projects out of urban elitist consideration resting their dependence on the complex network of technology, infrastructure and logistics.

FROM PROSPERITY TO PERNICIOUSNESS

Barbukia is one of the villages in Barbhag block. The village comprises of 4 hamlets, one inhabited by Hindu general castes, another by Hindu scheduled castes while the other two by Axomiya Muslims.[7] Barbukia is bounded by two rivers, Pagladia in the west and Baralia in the east. Both the rivers originate in Bhutan, flow through Baksa and Nalbari, join in Kamrup district and eventually join the Brahmaputra. Until the 1950s, villages like Barbukia were prosperous in nature due to flourishing agriculture and fish yields from the water bodies.

Birendra is a farmer from the Barbukia village, he said, 'Farming is like a lottery out here, if it floods early we can't even harvest the crops. Even if it rains less out here, but Baksa receives abundant rainfall; the water bodies will be filled up and flood the area. On the other hand if Brahmaputra exceeds its carrying capacity it will push back water through Pagladia river'.

Boralibeel is the largest of the water bodies in the area. While earlier it was a common property resource, it was transformed into a government property following disputes between the villages of Barximulia and Arangamou. Now, the Boralibeel ceases to exist, except the monsoon season. The Soil Conservation Department had recently proposed an investment of Rs. 40 lakhs to build fisheries in the Boralibeel area. This would involve raising embankments to protect the fishery water from flood waters. Birendra apprehends that such activities might further delay the water outflow during monsoon and render the returns from agriculture in front of further uncertainty which would force him to migrate to support his family.

---

[7] Axomiya literally means the citizens of Assam; however, I use the term Axomiya Muslim to differentiate them from the Bengali Muslims, which is a commonly used categorization in Assam. This differentiation on the basis of indigeneity has created social unrest in Assam mainly directed towards the Bengali Muslims who are often targeted as illegal Bangladeshi citizens.

400   S. MITRA

Measures to control the flood in the Nalbari district were envisaged through the 'Pagladiya Dam Project' (PDP), to be set in Thalkuchi near the Indo-Bhutan border, in Baksa district. The project has been in the pipeline due to the resistance of the local communities predicated on the apprehension of losing their livelihoods and kinship network through loss of agricultural lands and displacement. The estimated cost of the project has escalated from Rs. 12.60 crore in 1968–1971 (based on report of the Central Water Commission) to Rs. 1136 crore in 2004. While the official report claims that PDP would protect 40,000 hectares of land from flood, the social cost would involve submergence of 38 settled villages, affecting about 12,000 families. Unclear resettlement policy of the government has further antagonized the people from the project. While the people in and around the Barbhag floodplains were in support of the project, organized and consolidated opposition has thwarted any proceedings for the PDP.

Government interventions like such have been analysed deeply in the 'hazards approach', an important building block of the discipline 'political ecology'. Focusing both on 'natural' and 'technological' problems faced by human communities, hazards research took as its goal the rational management and amelioration of risk (Robbins 2011: 33). Gilbert White challenged the conventional way of thinking about and dealing with floods. Writing his thesis in the 1940s, White concluded that the traditional way of dealing with flood hazards—building more engineered structure, like that in Barbhag—is expensive, irrational and does little to solve the human problem. Rather he called for better land use and change in people's behaviour to mitigate future impacts of natural flood events (White 1945). Farmers like Birendra in Barbhag and communities in Thalkuchi echo the observations of White, in a different place and time. However, such floodplain investment, which defies the logic of local people, can be better understood by delving into the political economy behind it, role of capital in agricultural development and the control of legislative processes through normative ideologies, vested interests and campaign finance. Empathizing with local communities, Monirul Hussain (2008) notes that centralization of the decision-making process has blocked its access to the voice of the subalterns living at the margins of society and geography.

Locating the '*Machmorias*'

The changes in the local landscape in Barbhag have affected the fish-catching scheduled caste community (*machmorias)* to a large extent, as the water bodies were severely compromised due to the continuous siltation. Now, a considerable number of the *machmorias* have indulged in fish trade only rather than catching fishes from water bodies. The supplies of fish in the local markets now come from the fisheries mainly controlled by the Bengali Muslim community. A member of the *machmoria community*, Dulal Namasudra states that 'earlier only we would be involved in the fish trade as it was looked down by others, but now the Brahmins, Muslims are also in the business'. Such competitions from fellow communities have reduced the scope of the *machmorias* in their ancestral trade. The involvement of women in the fish trade has reduced among the *machmorias* with the impartment of education, which has somehow conveyed to the community that it is not respectable for a woman to indulge in such activities. This trend is in continuance with findings from different studies which suggest that *ceteris paribus*, labour force participation of women, declines in education and increases again only at higher levels of education (Kingdon and Unni 1997; Das and Desai 2003). It might be floated as an anecdote that the reduced participation the women in the workforce had reduced the income of a *machmoria* household, pushing them to the margins of livelihood certainty. Economic, gender, environmental factors have compounded into reduced dividends from the local environmental resources for the *machmorias* and the communities in the Barbhag, hence forcing migration from the area to different places within Assam and India.

Migration and Adaptation

The flood-prone Barbhag block has depleted its reliance on local livelihood opportunities. The state of perniciousness has created outmigration of labour from the region. The pattern of outmigration is different among the hamlets owing to their community structure, land holdings and their location with regard to the water bodies and embankments. General trend is that migrants from one hamlet generally work in the same city or industry. Sandal industry, fish packaging industry, security services are some of the common sites of employment for the migrants. As the city of Guwahati is nearby, there are both shuttling and seasonal migrants. In this context, migration is not only a response but also an act of resilience where the actor is not only escaping the site of disaster but also exploring market access where labour and skills can be traded against remuneration. Thus, one comes across Barhmin suppa in Barbhag block, a hamlet earlier inhabited by Brahmins has no Brahmin residents at the moment. While the nation-state boundary offers multiple sites for outmigration, uncertainty looms over one's security beyond the North-Eastern Region. Racial treatment of north-easterners in the mainland in India is widely common, and the mass exodus of north-eastern labourers from Southern India at the backdrop of the Bodo-Bengali Muslim riots in Western Assam, 2012, would indicate the precariousness of migration predicated on themes of security, citizenship and neoliberal economy.[8] Such incidents show how calamities, both natural and political, back at the native place can condition the lives of the migrants faraway through the grid of labour mobility, disaster management, resource politics and policies.

The battle of survival in the Barbhag block has generated several responses from the local communities, Apart from migration, one of the significant responses has been a change in the nature of resource extraction. Taking a cue from the activities from the Bengali Muslims in the neighbouring villages of Kamrup district, the people have Barbhag block has also started to set up fisheries which involve erecting mud walls of considerable altitude to prevent flood water from seeping in. While the investment in the fishery might earn profit individually, given the topography and environmental nuances of the area, it holds the potential to negatively affect agriculture by further delaying the water passages and eventually leading to further migration from the Barbhag block by creating a socio-ecological trap. Such tendencies of extraction from natural resources are central to the theme of marginalization theory in political ecology. Marginalization is a process whereby the politically and socially marginal people are pushed into ecologically marginal spaces and economically marginal social positions, resulting in their increasing demands on the marginal productivity of the ecosystems—resulting in degraded landscape returning less and less to an increasingly impoverishes and desperate community.

Legislations like Brahmaputra Board Act, 1980, have been unsuccessful in controlling the issue of flood.[9] Rather failure of the Board has shifted some of the responsibilities such as that of Dihang and Subansiri rivers in favour of the Central Water Commission. The incompetency of the governing institutions over the years worsened

---

[8] Refer to http://www.thehindu.com/opinion/op-ed/home-is-hardly-the-best/article3796017.ece. Accessed June 2, 2016.

[9] For the detailed functioning of the Brahmaputra Board Act refer to http://www.assamtribune.com/scripts/detailsnew.asp?id=apr3010/state05. Accessed June 1, 2016.

the situation and accelerated the level of outmigration. It is then some local youth came together to form a Non-Governmental Organization (NGO) Gramya Vikash Mancha (GVM) which is now functioning in different sites of Nalbari, Kamrup and Baksa districts.[10] Funded by donors like Tata Trust, GVM has fostered the practice of 'Diversion Based Irrigation' and 'System of Rice Intensification' in the areas they function. GVM has also embarked on the endeavour of restoring livelihood in several villages, some of them being in the Borigog-Barbhag block. The water channel running through the blocks originates in the Bhutan hills and is known as Ghogra jan. The Ghogra jan had been silted for over 25 years due breach in embankment, and as a result, agricultural lands depending on water supply from the jan laid barren with no cultivation. GVM stepped in 2008 by arranging funds from donor agency Action Aid and involving the farmers through daily wage of Rs. 100 for cleaning up the jan along with 30,000 bighas[11] of paddy field. The activity involved no mechanical inputs rather the indigenous knowledge about the local topography of the farmers and fisher folks. Since then, there has been successful harvest of paddy in the area which has also plugged the stream of outmigrants from the region to a considerable extent.

GVM has also resurrected the indigenous network of irrigation channels in the villages of Bhutan foothills known as *dongs*, which not only provide water for irrigation in the villages but also control the water flowing down to Nalbari district. Such initiatives can be considered as acts of resilience wherein the local communities collectively manifest the environmental resurrection, enabling them to ensure their livelihood, food security and insulation from the flood hazards to an extent. Activities of GVM in the region show that what Ranabir Samaddar (2015) opines about governance—'Clearly economic reason is not enough to govern society, which needs ecological reason to be governed – because if dangers are to remain with us and if marginal existence is going to be the characteristic of our life then we need a new ecological understanding of our precarious existence' (Fig. 2).

---

[10] http://infochangeindia.org/agriculture/stories-of-change/nalbari-farmers-resume-cultivation-after-25-years.html. Accessed June 1, 2016.

[11] 1 bigha equals to 14,400 square feet in Assam.

Fig. 2 'Diversion Based Irrigation' canals constructed by Gramya Vikash Mancha (GVM) with the labour participation of the villagers in Nagrijuli Block, Baksa district (*Source* Photograph taken by the author during fieldwork)

CONCLUSION

Earthquakes and floods have been the two recurring natural hazards of Northeast India. South Asian nations as a whole face the brunt of environmental calamities, which compromise life, livelihood and security. The intensity of the calamities is often compounded by faulty planning of disaster management and the ongoing course of neoliberal development heavily hinging on accumulation of resources like land, water, forests and extracting energy from the same. The neoliberal dreams often meet a *cul-de-sac* in Northeast India through resistance and resilience often orchestrated through popular movements. In Barbhag, we see that people who otherwise exhibit the multiple fault lines of Assam's sociopolitical milieu, have come together for a common environmental cause to reclaim their space of resource extraction which inextricably influences their survival and location in terms of mobility.

Barbhag, along with Nalbari and numerous districts of Assam, has been in the quagmire of insurgency, conflict, resource crisis and subsequent migration of labour. While issue of insurgency and alternate ideas of nation state and sovereignty are 'solved' through army operations and a constant regime of militarization and vigilance, environmental issues are unable to extract the equal level of attention from the state. The onus, thus, *prima facie* rests solely with the local communities to deal with hazard through alternate patterns of resource extraction/sharing and migration. Migration from sites of natural disaster is a common occurence, as shown by a study titled 'Ecosystems for Life—A Bangladesh-India initiative: Ecology, Politics and Survival in India's Northeast and Deltaic Bengal'[12], how the cyclone 'Alia' in the Sunderban belt of West Bengal in 2009, turned the agricultural land saline rendering them unsuitable for cultivation and therefore trigerring outmigration of males for search of a viable livelihood elsewhere. This trend of migration helped the declining economy of the region to rejuvenate and solve the problem of seasonal unemployment. Such migration becomes a reality in disaster-stricken region for the sake of sustenance, as observed in both Barbhag and Sunderban. Vulnerability and precariousness often lead to trafficking, another mode of mobility.

Barbhag block of Nalbari has been criss-crossed with both water bodies and histories of conflicts. Brazen memories of encounters are still deeply inscribed in the psyche of the people and their narratives. The nature of governmentality in the region has restricted the voice of dissent largely. The infrastructural experiments have failed to contain the flood problem, while grass root initiative by organization like GVM has shown that it is not beyond the means to put an end to the state of environmental despair. However, the reach of an NGO cannot serve as an alternative to the state duty and mechanisms. There are large swathes of agricultural lands, which are yet to be freed from weeds and silts. The inevitability of dependence on the formal economy is also turning futile as a large sum

---

[12] Refer to http://www.mcrg.ac.in/IUCN/IUCN_Concept.asp. Accessed June 2, 2016.

406   S. MITRA

of money is often required to enter the arena of the salaried class according to the local narratives, a practice not uncommon in the post-colonial societies of South-east Asia, as observed by Tania Murray Li during her work in the Sulawesi highlands plains of Indonesia (Murray Li 2014: 40). Therein, it is necessary to understand the requirement to maintain and sustain the local environmental resources for harnessing ingredients for survival and accumulation as well. Local reliance would also serve the means to deal with intermittent flood hazard through involvement of the local communities of Barbhag through stakeholding mechanisms, without depending on the centrality of the centre to solve the problems of the periphery.

REFERENCES

Arnold, D. (1993). Social Crisis and Epidemic Disease in the Famines of Nineteenth-Century India. *Social History of Medicine, 6*(3), 385–404. https://doi.org/10.1093/shm/6.3.385.

Baishya, S. (2013, September). Vulnerability Assessment and Management of Flood Hazard in Baralia River (Bhairatolajan): A Case Study of Khopanikuchi Village of Hajo Revenue Circle, Kamrup District Assam. *International Journal of Innovative Research in Science, Engineering and Technology, 2*(9), 4257–4270.

Das, M., & Desai, S. (2003). *Why Are Educated Women Less Likely to be Employed in India? Testing Competing Hypothesis.* Washingtom, DC: Social Protection, World Bank.

Drèze, J. (1988). *Famine Prevention in India.* Helsinki, Finland: World Institute for Development Economics Research of the United Nations University. Also to be found as WIDER Working Paper 45, May 1988. http://www.wider. unu.edu/publications/working-papers/previous/en_GB/wp-45/. Accessed June 6, 2016.

Grove, R. H. (1990). Colonial Conservation, Ecological Hegemony and Popular Resistance: Towards a Global Synthesis. In J. M. Mackenzie (Ed.), *Imperialism and the Natural World* (pp 15–50). Manchester, UK: Manchester University Press.

Guha, A. (2014). *Planter Raj to Swaraj: Freedom Struggle and Electoral Politics in Assam.* New Delhi: Tulika Books.

Hussain, M. (2008). *Interrogating Development: State, Displacement and Popular Resistance in North East India.* New Delhi, India: Sage.

Kingdon, G., & Unni, J. (1997). *How Much Does Education Affect Women's Labour Market Outcomes in India?* Ahmedabad: Gujarat Institute of Development Research.

Murray Li, T. (2014). *Land's End: Capitalist Relations on an Indigenous Frontier*. London: Duke University Press.

Robbins, P. (2011). *Political Ecology: A Critical Introduction*. West Sussex: Wiley Blackwell.

Samaddar, R. (1998; reprint, 2013). *Memory, Power, Idenity: Junglemahals: 1880–1950*. Hyderabad: Orient Longman.

Samaddar, R. (2015, December). The Precarious Migrant: Issues of Ecology, Politics and Survival. *Refugee Watch, 46*, 51–72.

White, G. F. (1945). Human adjustment to floods. Research Paper 29. Department of Geography, University of Chicago.

Communities and Disasters in Nagaland: Landslides and the Cost of Development

Imkongmeren

Nagaland is situated in the Northeast of India comprises of hilly terrain. Landslides occur frequently in the hilly mountains of Nagaland due to development process as well as other factors. Major cause of landslides in the state is triggered by rainfall infiltration. Buildings are constructed with poor engineering input without any regulatory framework. Meanwhile, government's remedial policies were seen as insufficient to tackle economic losses and human casualties. This article aims to guide community leaders and state administration to address landslide hazards through inclusive policies and implementing strategies to help reduce risk to life and properties. It also evaluates the cost of development and socio-economic impacts on the community due to landslides disaster.

Cost of Landslide to Socio-Economic Development

Most part of the Naga Hills is prone to landslides due to geodynamically sensitive geologically young mountains formation. Landslides are triggered mostly by heavy rainfall during pre-monsoon and post-monsoon,

---

Imkongmeren (✉)
Centre for the Study of Law and Governance,
Jawaharlal Nehru University, New Delhi, India

© The Author(s) 2018   409
A. Singh et al. (eds.), *Development and Disaster Management*,
https://doi.org/10.1007/978-981-10-8485-0_28

it is also intensified by irrational and unregulated rapid development process that has been placed in all parts of the state. There have been much developmental related causes affecting the vegetation cover, unregulated mining, unplanned constructions of roads and buildings and inappropriate management of flood water. There has been extensive removal of forest cover over the years either by fires or by timber harvesting. Heavy rainfall causes frequent cloudbursts and flash floods.

The communities are faced with challenges such as lack of technical know-how about factors contributing the calamity; lack of funding and negligence of the government; poor communication and infrastructural facility constraints the capacity of government and local communities to effectively mitigate landslide risks in Nagaland.

Landslides in the state take a heavy toll on life and properties every year causing excessive economic damages that worth lakhs and crores of rupees. The monetary costs associated with landslides result from damages to building structures, loss of land and disruption to communication routes. Rapid growth of urbanization in mountain tops has increased the number and frequency of the natural disaster in the recent decades. Developmental activities disturb large volumes of earth materials in construction of buildings, transportation routes, reservoirs and communication systems. Development on and near steep mountain slopes have been a primary factor escalating lope failures causing frequent unprecedented man-made natural disaster an alarming threat to the livelihood of the communities.

In Nagaland, both rural and urban communities suffer from acute water shortage problem. There has been an increase in unplanned construction of underground water reservoirs without considering scientific pre-treatment measures, or no precautions were taken considering the weak geological formation of the land slope (Public Health Engineering Department Nagaland 2008). Massive deforestation has added to the fragile geological and adverse environmental degradation. Large areas of natural forest in the Assam–Nagaland border and in the interior districts were destroyed by extensive timber harvesting and jungle clearing for development purposes. Nagaland Economics & Statistics Department (Annual Administrative Report of 2014–2013) has cautioned over the adverse effects suffered by the deforestation that,

COMMUNITIES AND DISASTERS IN NAGALAND ... 411

Unfortunately, over the years, degradation of forest and natural resources has been issues of concern primarily caused by unsustainable development practices, increase in population, migration, urbanization and increased used of forest products for economic activities,

Construction of roads requires scientific planning and methodical construction, but roads are constructed manipulating the construction codes by contractors without obtaining the basic data on geological formation, topography and drainage pattern. Hilly geologic formations and heavy rainfall (average annual rainfall being 2500 mm) are more susceptible to landslides than others. Most of the new urban towns and old villages are situated in steep slopes which are most susceptible to landslides. Steep slopes landslides tend to move fast rapidly and thus more dangerous than other landslides.

MAJOR LANDSLIDE AND ITS IMPACTS

In August 2001, Dimapur area experienced a cloudburst which lasted almost for one hour. This gave rise to so many landslides in that area, particularly the Paglapahar region which experienced the heavy downpour. In a stretch of just 4 kilometres on NH-39, seven major slides occurred which brought traffic to a standstill. In this incident, one commercial vehicle was crushed where 3 people were killed and some injured. In August 2003, the whole New Market colony in Kohima Town was affected by landslides. Property worth lakhs were destroyed by this slide.

The most tragic landslide that affected Nagaland in the recent past was 26 May 2005 landslide occurred in Mokokchung Town where 14 people were buried alive, many more injured and damage to property was extensive. The national highway-29 near Kiruphema and Phesama has been affected by massive landslides every year blocking the highway during the monsoon season, the NH-29 the main lifeline connecting two states (Manipur and Nagaland) has been constantly deteriorated due to landslides.

A number of displaced people were increased though the State Government could not announce any relief immediately aftermath of the incident, the local community members from different organizations always come forward to give emergency relief and helped during such emergency operation. Displaced families were given temporary shelters by relatives and other villagers while students were provided temporary accommodation in various college campuses. The Phesama Village Council had been desperately making appeals to the State Government agencies to provide shelters for the displaced villagers and monitor the situation.

Landslides cause immense strategic setbacks to the implementation of developmental programs in the state. Landslides are responsible for greater economic losses than is generally perceived. General estimation of the total incurred cost associated with landslide damages could be unfeasible since adequate data cannot be collected from all remote villages. However, the landslide disaster and the cost of development suffered by the local communities in Nagaland have been colossal and unrelenting. Though actual damage, destruction and casualties are directly caused by huge landslides, the enormity of the economic losses upon the community is being overlooked by the government. Most of the roads leading to remote villages in Nagaland remained cut-off during rainy seasons; landslides do occur in many parts of the state even in winter seasons. Rain-triggered landslides have been affecting thousands of people along village roads and highways. "India flood and landslides 2016", has reported that even in the month of April, Mon district in Nagaland has been affected by floods and landslides where over 100 houses were severely damaged forcing residents to evacuate (http://floodlist.com/asia/india-floods-landslides-april-2016).

Since statehood in the 1963s, successive State Governments have been confronted with the landslide disasters. According to the *Morung Express News* (Dimapur 17 July 2011), the State Government has been allegedly giving only temporary quick fix solution with no permanent solution rather blaming it on bad weather and geography or bad road constructions. In 2006, the landslide hazard zonation map (LHZM) indicated the Kohima Municipal Council dumping areas under very high-hazard zone. Prof. G. T. Thong, Dept. of Geography, Nagaland University, disclosed that,

> It is the makeup of the rocks and also human activity and interference with nature, which is to be blamed for recurrent landslides in and around Kohima town. Kohima is primarily made up of two kinds of rocks namely Barail Sandstone and Disang Shale. A characteristic of the Disang Shale is that when water enters the rock, after a period of time it breaks and on

weathering forms fine dark clay. This clay on being saturated lose their shearing strength. Another factor has been the surrounding paddy fields. These fields trap water for 2-3 months and this leads 'pore water pressure' to escape slowly and seep through the land gradually making it unstable. (*Morung Express* 17 July 2011)

CHALLENGES TO COMMUNITY-BASED LANDSLIDE MANAGEMENT

In the landslide susceptible hillside areas, local people apply their own landslide risk reduction measures like—retaining wall construc-tions and making wider drainage system. This mitigation technique is commonly practice focusing on immediate visible soil erosion and slope failure threat rather than invisible cause like under surface water movement. Malcolm G. Anderson and Elizabeth Holcombe (Karnawati et al. 2005, p. 590) observed that building retaining walls is very expensive measures but not effective in cases of inadequate drainage in weak slopes. 'It is rather proper management of surface water infiltration and grey water from households which constitutes the real threat to trigger landslides'.

Government agencies are expected to take initiative to mainstream drainage and retaining wall constructions by providing adequate fund-ing, technical support and get involve enhancing the maintenance of the community-based landslide protection measures. Capacity building of the local authorities like the Village Council, the Village Development Board and communitised village society would be vital to make the risk reduction policy a sustainable one. Although some officials and relief workers may be on the disaster strike spot, it is obvious that they may not afford to help everyone immediately, and thus the community should be made aware and prepared to cope with the emergency. State agencies should therefore realize the need to ensure the involvement of communities, particularly in the context of Nagaland where the many tribal villages have an inherent and inbuilt system of traditional ties and bonding. Government can take advantage of this strong social capital and utilize resources locally available to ensure cost-effective and quicker response by involving the active participation of tribal organizations, institutions, churches and voluntary organizations.

Social capital in the context of the Naga community implies strong social bonding and community spirit, social accountability and loyalty

414 IMKONGMEREN

within the fellow members and the absence of caste and social discrimination. Social capital that has stemmed out of traditional institutions and practices is prominently reflected in all other social organizations like Naga Hohos, and youth and women bodies and religious institutions. Tribe and clan traditions and loyalties play an important part in the life of Nagas (Nagaland State Human Development Report 2004).

STREAMLINING LANDSLIDE DISASTER MANAGEMENT

Central and State Governments have not been doing enough to mitigate the problems faced by the communities due to landslide hazards. In his keynote address during the India Disaster Management Congress in New Delhi, R. K. Bhandari (2006, p. 3) suggested that state should create a dedicated and committed network of multidisciplinary expert teams to carry out cost-effective landslide investigation, slope remediation and preventive measures. Government is obligated to protect communities from man-made disasters lifeline and strategic roads from landslide disasters by proactive remediation.

Proper regulation is required for construction and development activity, concern departments such as PWD, Road and Bridges, Housing, Irrigation and Municipal Development Authorities to strictly monitor simultaneous implementation of protective regulations. Research projects scientifically investigated, and monitored, studied and analysed, could lead to forecasting and prediction of landslides and landslide disasters that would have otherwise occurred could be averted. It is imperative for government to commission institutions for landslide studies.

Landslide Hazard Zonation and Geohazard studies of townships (phase-I, 14 August 2011) report noted that six districts—Phek, Zunheboto, Mokokchung, Wokha, Kohima and Dimapur are prone to landslides and under threat by geohazards like earthquakes and water contamination. Similarly, the Department of Geology and Mining had carried out a project on 'Landslide Hazard Zonation and Disaster Management and Mitigation plan using Remote Sensing (RS) and Geographic Information System (GIS)', a case study along the highway NH-29 during 2009–2011. Detailed landslide inventory map by collecting all existing information and current data on landslide was generated including preliminary investigation of major landslides along the NH-29 of Dimapur–Kohima section was generated for understanding and suggestion of remedial plans.

NAGALAND STATE DISASTER MANAGEMENT

The NSDMA has been coordinating other state departments allocating appropriate roles related to their departmental powers, functions and facilities. While designating the senior officials at the state secretariat as team leaders, the district and subdivisional level officers are assigned to coordinate with supporting agencies such as municipal cooperation and private organizations, community volunteers. Some of the 'Landslide Emergency Support Mechanisms' are:

a. Provide all of the types of communications devices needed during emergency response operations.
b. Conduct damage assessment, casualties.
c. Provides Transportation, Search & Rescue, administering first aid and evacuation.
d. Debris Clearance & Equipment support, restoration and repair water supply system and relief camps.
e. Mobilizing support from various organizations and facilitating donations.

Nagaland State Disaster Response Force (NSDRF) constituted by comprising of 5 (Five) companies drawn from the existing 5 battalions IRBs was launched on the 10th of October 2011; each battalion has 5 Specialized Search & Rescue Teams and Dog Squad. Emergency Operation Centres (EOC) has been set up in all district HQs to coordinate the flow of information with respect to activities associated with relief operations which look after rehabilitation, preparation, prevention and mitigation and co-ordinates with different organization. NSDMA has taken initiatives to provide training on disaster response to the different units of the IRB Nagaland Police. However, it is felt that there is a need to create a full-fledged professionally trained full time 'State Disaster Response Force' since NSDMA officials and a handful of trained personnel are not sufficient enough to afford a professional approach towards addressing landslide problems.

INVOLVING THE COMMUNITY

The local governments such as village and town councils can play vital roles when empowered with adequate funding and facilities and appropriate administrative structures to manage disaster response. The legislation empowering the traditional administration system of the Nagas through the Nagaland Village and Area Council Act 1978, the constitution of the Village Development Boards and evolution and enactment of the communitisation of public services as agencies for development have strengthened the traditional governance constituency (Nagaland Village and Area Councils Act 1990). Since then the community as the responsible owner and government as facilitator have become integral to the new aspect of governance and its service delivery role under the Nagaland local governance.

Strong social community bonding among the Naga communities is an added advantage to get the communities involved in controlling landslide disasters. Therefore, policies and strategies should be formulated, revised, modified and applied considering the socio-economic context of the indigenous Naga culture and traditions. Some of the thrust areas for community involvement are: Church bells are most effective as all villages have a church. The Naga Log Drums could be revived as backup to support church bells for emergencies in the event of destruction of bell towers. Drainage systems should be properly planned and taken care of. Community and private landowners should offer to utilize their land for emergency rehabilitation measures.

CONCLUSION

Nagaland is a landslide-prone state which threatens the livelihood of the communities hindering specially the road communication facilities. Many localities, roads and other infrastructures are seen as vulnerable to landslides disaster and create a serious threat to life safety. Landowners and construction companies must be aware of the implications in developing land identified as landslide risk areas. They must conform to policies and norms regulated by authorities and government to reduce risk from landslide hazards to ensure protection of life and properties of communities. Every responsible citizen living in landslide-prone zones and disaster vulnerable areas should remain alert and be prepared at all time and be

COMMUNITIES AND DISASTERS IN NAGALAND ... 417

willing to extend any possible help whenever such disaster strikes in their community. Such prompt coordination and synergy among various agencies at various levels in dealing with any disaster can minimize loss of life and properties. Meantime, the state authorities should integrate institutional arrangements, facilitate accurate art of forecasting, early warning systems and protect the threatened communities.

REFERENCES

Bhandari, R. K. Keynote address on "India Disaster Management Congress", New Delhi 29–30 November 2006, Session A2, Organized by National Institute of Disaster Management, p. 3.

Eastern Mirror, 25 August. "Phesama Landslide Continues to Haunt: 45 Families Displaced". http://www.easternmirrornagaland.com.

Government of Nagaland. *Annual Administrative Report of 2014–2013*. Department of Economics & Statistics Kohima.

http://floodlist.com/asia/india-floods-landslides-April-2016-arunachal-pradesh-assam-nagaland.

http://indianexpress.com/article/india/india-others/more-than-100-people-hit-by-landslides-in-a-naga-village/.

http://nsdma.gov.in/Docs/Newsletter/Newsletter-2013.pdf.

http://nsdma.gov.in/history_of_disasters_in_nagaland.html.

http://www.theshillongtimes.com/2011/08/14/landslide-hazard-zonation-mapping-for-nagaland/#VAHXsk30Q7t36cpU.99.

http://www.uniindia.com/images/share_icon.jpg.

https://www.nagaland.gov.in/portal/portal/StatePortal/AboutNagaland/StateProfile.

Karnawati, D., Ibriam, I., Anderson, M. G., Holcombe, E. A., Mummery, G. T., Renaud, J. P., et al. (2005). "An Initial Approach to Identifying Slope Stability Controls in Southern Java and to Providing Community-Based Landslide Warning Information." In Malcolm G. Anderson, Elizabeth Holcombe, & James R. Blake (eds.), "Reducing Landslide Risk in Communities: Evidence from the Eastern Caribbean". *Journal of Applied Geography*, pp. 590–599.

Morung Express, 17 July 2011. http://morungexpress.com/troubleshooting-landslide-disasters-in-nagaland/.

Nagaland State Disaster Management Authority (NSDMA), Home Department Government of Nagalan Civil Secretariat Kohima-797001, Nagaland.

Nagaland State Human Development Report 2004, GOI—UNDP Project, Published by Department of Planning & Coordination, Government of Nagaland, p. 71.

Nagaland Village and Area Councils (Second Amendment) Act, 1990 (Act No. 7 of 1990), This Act Amends the Nagaland Village and Area Councils Act, 1978 by changing the title of the Act into Nagaland Village Council Act, 1978 and by deleting sections 23–42 abolishing the Area Councils.

NSDMA Newsletter, Volume-I, Issue II, July 2011–March 2012, p. 28. http://nsdma.gov.in/Docs/Newsletter/Newsletter-2011.pdf.

Oregon Department of Land Conservation & Development, "Planning For Natural Hazards: Landslide," Capitol Street Ne, Suite 150 Salem, July 2000. https://www.oregon.gov/LCD/HAZ/docs/landslides/05_landslide.pdf.

Public Health Engineering Department Nagaland, Communitisation of Water Supply and Sanitation Model Rules and Guidelines 2008.

The Government of Nagaland notified the Nagaland Disaster Management Rules, 2007, on 30 October 2012, in exercise of the powers conferred under Section 78(1) of The Disaster Management Act, 2005 (Central Act of No. 53 of 2005). NSDMA 2007, see NSDMA manual 2007. http://nsdma.gov.in/Docs/Acts_Plans/NSDMR%202007.pdf.

The Village Development Board VDB Model Rules 1980 (revised). In 1990, Directorate of Rural Development Nagaland, a Manual of Rural Development Department Nagaland, Kohima 2009, pp. 5–12.

GLOSSARY

AAMDANG	wet paddy cultivation during monsoon time.
ALI-A:YE-LIGANG	sowing festival of Mising community.
BAO	deep-water cultivation in flood inundation area.
CHANG/KARE	Rise platform constructed by bamboo and wood.
DERMI OYING	a kind of tree used for vegetable.
DINGORA	a kind of instrument for catching fish.
DOBUR UIE	fertility festival of Mising community.
DORBUM	a kind of grass.
JEING RIBY	cane rope.
KHOLIA	box-type instrument for catching fish.
KHUTI	temporary settlement for cattle and buffalo grazing.
KIRATA	a generic term in Sanskrit literature for people who lived in the mountains, particularly in the Himalayas and Northeast India, and who are postulated to have been a Mongoloid in origin.
KUMSUNG OKUM	a granary.
LAI-AAM	rain-fed paddy cultivation, generally cultivated between January and June.
MERAM	fire place of Mising people.
MONGOLOID	constituting or characterizing of a race of humankind native to Asia and classified according to physical features.

© The Editor(s) (if applicable) and The Author(s) 2018
A. Singh et al. (eds.), *Development and Disaster Management*,
https://doi.org/10.1007/978-981-10-8485-0

OKUM	a house, home.
OTUNG	a hollow bamboo container.
PIMUK	a kind of tall grass.
PIRO	a kind of grass.
PORANG	a kind of bamboo fish traps with flat ends.
LAI-AAM	rain-fed agriculture practice.
SAPORI	River Island use for temporary settlement.
SILTATION	the leaping movement of sand or soil particles as they are transported in a fluid medium over an uneven surface (Vocabulary.com).
TASE	thatching grass.
TIBETO-BURMAN	a language family that includes Tibetan, Burmese and related languages of southern and eastern Asia.
TULI	a pot-shaped cane container.
UBOTI	a cylinder-type instrument for catching fish.
UIE	a supernatural being.

INDEX

A

AISHE, 8
Ali-Aye-Ligang, 305, 306, 316
APHRODITE, 46, 55, 56
APHROTEMP, 46–48
'Aradhal', 63
Arangamou, 399
ARCGIS, 174
Axial Ridge, 26

B

Babupara, 108, 355
Barak in Assam, 34
Baralia River, 397
Barbhag block, 22, 394, 395, 397, 399, 401–403, 405
Barximulia, 399
Bhuban and Bokabil Formations, 162
Bhuj M-7.9 (2001), 31
Brahmaputra, 19, 28, 34, 35, 41, 59–61, 63, 67, 68, 70, 74
Brahmaputra Board Act, 1980, 402

C

Cachar Hills, 95
Central Industrial Security Force (CISF), 16, 135
Centre for Research on Environmental Development (CRED), 355, 356
Centre for Research on the Epidemiology of Disasters (CRED), 12
Chakpi River, 71, 72, 234, 253, 254
Chang Okum, 268, 269
Cherrapunji (Sohra), 33
Chingmeirong, 355
Chin Youth Association, 388
Churachandpur Mao Fault (CMF), 153
Church of Christ, 387
cloudbursts, 29, 410
Construction Federation of India (CFI), 143
Coordinated Regional Downscaling Experiment (CORDEX), 46, 48, 52, 55
crustal deformation studies, 157, 159

© The Editor(s) (if applicable) and The Author(s) 2018 421
A. Singh et al. (eds.), *Development and Disaster Management*,
https://doi.org/10.1007/978-981-10-8485-0

422 INDEX

CSIR-NEIST (N-E Institute of Science & Technology), 31
cyclone 'Aila', 405
cyclonic disturbances, 29, 35

D
Delhi Centric Development, 4
Dhakuakhana subdivision, 304
Disaster Management Act (DMA) 2005, 15, 36, 77, 107, 108, 111, 130, 335, 374

E
Early Warning Systems (EWS), 30, 113, 115, 131, 188, 231, 347, 376, 377, 417
Earthquake Early Warning (EEW), 92
earthquakes, 3, 4, 9, 20, 25, 28, 30, 32, 33, 96, 107, 115, 116, 120, 129, 130, 132, 149, 150, 153, 157, 161, 164, 169, 221–223, 230, 241, 242, 286, 320, 337, 346, 352, 359, 373, 384, 391, 395, 396, 404, 414
Emergency Events Database (EM-DAT), 12
Emergency Operation Centre (EOC), 108, 109, 415
environmental determinism, 302, 316
Epicenter, 150, 164
ERDAS, 174
excavator machines, 138

F
flash floods, 21, 29, 37, 59, 68, 71, 97, 129, 222, 242, 254, 301, 321, 322, 343, 352, 357, 358, 410
Flood Early Warning System (FLEWS), 34, 35, 37, 80, 398

Flood Response Plan, 398
floods, 262, 263, 265, 266, 268, 269, 271, 274–277, 301, 302, 304, 309–315, 322

G
Gangetic floodplains, 115
Geological Survey of India, 177
Gilgit–Baltistan area, 9
Global Agenda Council on Humanitarian Response, 143
Global Climate Observing System (GCOS), 12
Global Ocean Observing System (GOOS), 12
global seismic monitoring network, 157
Global Terrestrial Observing System (GTOS), 12
GLOBK software, 155
Gogamukh, 64
Gramya Vikash Mancha (GVM), 403, 404
Great Shillong Earthquake, 30
greenhouse gases (GHG), 43
GSHAP data, 96

H
Himalayan Tsunami, 59
Hindustan Construction Company (HCC), 143
Hmar Youth Association, 388
Honolulu Workshop, 14, 25
Hurricane Katrina, 142
hurricanes, 3, 142

I
ICIMOD Terrestrial Observation Panel for Climate (TOPC), 12

INDEX 423

Imphal river, 101, 102, 253, 254, 352, 357
Imphal River, 70, 71
Imphal Valley, 96, 153, 163, 242, 253, 349, 352, 353, 356, 357, 359
India Meteorological Department, 77, 83, 177
Indian Plate, 26, 28, 29, 153, 162
IndoBurmese Arc (IBA), 150
Indo-Myanmar border, 26, 28, 99, 100, 242, 288, 359, 375
Input Thematic layers, 174
Iril River, 71, 253, 254

J

Janji-Arkep, 306
jatropha plantation, 22, 395
Jiadhal River, 19, 63, 64, 74

K

Kabui Khullen, 163, 164, 166
'Kampong', and 'Bajaus' houses, 61
Kangla or Kanglapat, 352
Karakoram pass, 9
Kare Okum, 68
Kashmir M-7.9 (2005), 31
Kedarnath, 59
Keishampat, 246, 352, 357, 358
Keishamthong, 357, 358
khankho, 19, 283–285, 293
Kiruphema and Phesama, 411
Koppen's classification of climate, 42
Kosi Nepal, 59
Kuki Baptist Convention (KBC), 386
Kuki-Chin, 95
Kuki Christian Church, 386
Kuki Inpi, Manipur (KIM), 286, 288
Kuki Khanglai Lawmpi (KKL), 286
Kukis, 19, 281, 283, 294, 386
Kuki Students Organisation (KSO), 286, 289

L

Laiphulia, 20, 304, 306, 307, 313
Lamka, 22, 117, 202, 208, 210, 211, 332, 334–337, 339–341, 343–345, 347
Landslide Hazards Zo nation (LHZ), 17
landslides, 3, 4, 8, 9, 20, 30, 32, 33, 37, 67, 68, 91, 93, 101, 116, 129, 130, 132, 139, 140, 149, 161, 162, 170, 171, 173, 181, 185, 216, 230, 231, 233, 241–243, 251, 253, 257, 286, 288, 320–322, 332, 336, 346, 357, 361, 362, 373, 374, 384, 409–414, 416
Latur M-6.2 (1993), 31
lom and som, 283–285, 293
Lushai Hills, 95, 293

M

machmoria community, 401
Majuli Island, 8, 36
Manipur Baptist Convention (MBC), 386
Maranging, 163, 164
marginal farmers, 6, 257
Matmora, 20, 304, 306–308, 310, 313–316
Meiteis, 95
Mimbir-Ya:me, 305, 316
Mising community, 19, 20, 61, 62, 68, 69, 71, 75, 303, 305, 306, 308, 310, 312
Moreh, 72, 150, 288, 322, 353, 357
morung, 283
Munich Re, 12

N

Naga Hills, 95, 409
Nagamapal, 357, 358

424 INDEX

Nagas, 95, 283, 294, 386, 414, 416
Namsing, 71
National Bureau of Soil Survey and Landuse Planning geological maps, 177
National Flood Commission report, 301
National School Safety Programme (NSSP), 108
National Disaster Relief Force (NDRF), 11, 16, 39, 135
North-Eastern Space Applications Centre (NESAC), 39
NNW–SSE trending, 162
nokpante, 283
Noney Area, 164
Noney Earthquake, 377
Nubra River, 9

P
Pagladiya Dam Project (PDP), 400
Paites, 20, 334, 336, 347
Paona bazar, 246, 352, 355
plate tectonics theory, 149
Po:rag, 305, 306, 316
pre-monsoonal precipitation (MAM), 53, 61
The Presbyterian Church in India, 386
Protection and Motivation Theory (PMT), 378

R
rainfall, 3, 33, 35, 44, 45, 53, 59, 68–72, 74, 76–78, 93, 101, 105, 115, 126, 140, 164, 170, 173, 233, 242, 254, 287, 312, 321, 322, 336, 352, 357, 364, 365, 367, 398, 399, 409–411
Reformed Presbyterian Church, 386
regional climate model (RegCM3), 45

Riverine erosion, 34, 35
'river loving' people, 19, 61

S
Saikul, 86–88, 247, 288–290
satellite bell error, 153
Scientific and Technological Advice (SBSTA), 12
Scrippe Orbital and Positioning Analysis Centre (SOPAC), 155
seismic vulnerability, 14, 26, 29
Sendai Framework, 21, 75, 93, 235, 347, 373, 384, 387, 391
Seven Security Force (SSF), 355
Seventh-day Adventist Church, 386
Siachen Glacier, 9
Siang River, 60
Simte Youth Organization, 388
Skardu, 9
Slope modification, 181
soil maps, 177
State Disaster Response Force (SDRF), 108
Subansiri, 63, 64, 216, 402
Sumatra–Andaman earthquake, 153
Sunda Plate, 162
Swiss Re, 12

T
Tawang Valley Case of A.P., 131
Thangal bazar, 246, 355, 357, 358
Thangmeiband Assembly Constituency Development Forum (TACDEF), 354
Thoubal River, 71, 74, 78, 253, 254
tomngaina, 19, 281, 283–285, 293
traditional knowledge systems, 309
Transdisciplinary, 11, 12, 18, 22
tsunami affecting Andaman M-9.3 (2004), 31
tsunamis, 3, 30, 33

INDEX **425**

U

United Liberation Front of Assam (ULFA), 396
United Nation's Framework Convention on Climate Change (UNFCCC), 12
Unmanned Aerial Vehicles (UAV), 10
Upper Brahmaputra Valley, 28
Uripok, 252, 253, 255, 357, 358
United State Geological Survey (USGS), 157, 161–164

V

Vulnerability Atlas of Manipur, 68, 322

W

Waheng Leikai, 357, 358
Women Vendors' market, 85, 86

Workers Union Manipur (WUM), 355
World Climate Research Programme (WCRP), 12, 46
World Economic Forum, 143

Y

Young Mizo Association, 388
Youth Paite Association (YPA), 388

Z

zawlbuk, 283, 285, 293
Zo, 22, 384
Zomi Human Rights Foundation (ZHRF), 388

CPSIA information can be obtained
at www.ICGtesting.com
Printed in the USA
LVHW02*1911210618
581517LV00014B/363/P